# Components and Services for IoT Platforms

Georgios Keramidas • Nikolaos Voros
Michael Hübner
Editors

# Components and Services for IoT Platforms

## Paving the Way for IoT Standards

Springer

*Editors*
Georgios Keramidas
Technological Educational Institute
of Western Greece
Embedded System Design and Application
Laboratory (ESDA)
Antirio, Greece

Nikolaos Voros
Technological Educational Institute
of Western Greece
Embedded System Design and Application
Laboratory (ESDA)
Antirio, Greece

Michael Hübner
Chair for Embedded Systems
for Information Technology (ESIT)
Ruhr Universität Bochum
Bochum, Germany

ISBN 978-3-319-42302-9 ISBN 978-3-319-42304-3 (eBook)
DOI 10.1007/978-3-319-42304-3

Library of Congress Control Number: 2016950523



Printed on acid-free paper

This Springer imprint is published by Springer Nature
The registered company is Springer International Publishing AG Switzerland

# Preface

The Internet of Things (IoT) is upon us and touches almost every industry! Current forecasts predict that there will be more than 50 billion devices connected to the Internet by 2020 and it looks like this is just the beginning. What is highly interesting with this technology trend is that it is more defined by what it enables rather than the technology itself. This is because IoT enables all kinds of smart objects (e.g., smart sensors and actuators) to communicate and interact with each other across different networks and domains. As such, a plethora of new services and applications are possible and must be created revealing the potential to create substantial new markets and to stimulate existing ones.

However, it is also clear that on the way to a fully connected cyber-physical world, numerous challenges have to be addressed. Do we have the technology and experience to make IoT reality? What kind of technology is actually missing? Do we need new hardware technology, new tools and programming languages, new network protocols, new services, and new data management techniques? To sum up, do we really need new technology or is IoT just a matter of ensuring interoperability among existing technologies or do we maybe need a paradigm change in technology and design methods? As with every new technology wave, we are witnessing a debate regarding industry standards striving to find a position in the IoT marketplace.

The intention of this book is to shed some light on all those questions. Certainly, one book cannot cover all topics related to the trends in IoT. Since IoT is in fact in its infancy, we see this book as an ongoing effort to offer the current trends and open issues in the area and also to sketch the future of IoT; we are strongly convinced that the IoT industry will not stop until your tablet, smartphone, or smartwatch is downscaled enough so they fit on your eyes or your wrist.

The book is organized in five main sessions covering the whole spectrum of the IoT technologies. The first four parts of the book are:

- Platforms and Design Methodologies for IoT Hardware
- Simulation, Modeling, and Programming Frameworks for IoT
- Opportunities, Challenges, and Limits in WSN Deployment for IoT
- Efficient Data Management and Decision Making for IoT

The last part of the book is devoted to the description of four IoT use cases and includes also an additional chapter offering our view on various IoT projects funded by the European Commission under the Horizon 2020 work program.

The editors would like to thank all the authors of the individual chapters of the book and also the program committee of the International Conference on Field-Programmable Logic and Applications (FPL), 2015. The conception of writing this book was born in the course of the First International Workshop on Components and Services for IoT Platforms (WCS-IoT 2015), held in conjunction with FPL 2015 (http://esda-lab.cied.teiwest.gr/wcs-iot/).

We would especially like to thank the members of the organizing committee of the WCS-IoT workshop: their contribution and participation in this effort as well as their experience have led to a very successful program for the workshop. In this book, we included all papers presented in the workshop.

Finally, we would like to state that we strongly believe that this effort should be followed up by additional workshops and conferences in the future focusing on the existing and evolving IoT standards: every industry can benefit from IoT deployments but must adhere to a myriad of existing technology and regulatory constraints plus a new set of policy considerations. The latter will be especially true in the new IoT vertical markets.

Antirio, Greece — Georgios Keramidas
Antirio, Greece — Nikolaos Voros
Bochum, Germany — Michael Hübner

# Contents

# Part I
# Platforms and Design Methodologies for IoT Hardware

# Chapter 1
# Power-Shaping Configurable Microprocessors for IoT Devices

**Fabio Campi**

## 1.1 Introduction

Despite being a technology little less than 50 years old, our capability to design and manufacture integrated circuits (ICs) has deeply affected our society and the way we live. The concept of computer has somehow pre-dated the advent of IC manufacturing technology, but as a consumer product the "computer" only makes sense as a product of the CMOS industry. If we define a "microprocessor" as a computer device that is miniaturized to be part of an integrated circuit, we can safely claim that the concept of integrated circuit and that of microprocessor have advanced hand in hand in the last 50 years, influencing each other's development, as they both constantly expanded to new markets and applications. In the last 50 years, integrated circuits have entered our houses first as desktop computers, then as mobile phones, and then again as portable WWW, communication, and infotainment terminals: we can consider the advent and diffusion of the "Internet-of-Things" as the latest pivotal change enabled on our society by the interaction between IC technology and computer design.

It may be interesting to evaluate in this book, on top of all the aspects already introduced in previous chapters, what is the innovation that the "Internet of Things" may trigger on microprocessors. And, to evaluate what impact the advent of the "Internet of things" may have on the way we design microprocessors (plural) today on CMOS IC technology. We can identify three phases in the history of commercial ICs: the first phase took place in the 1970 with the advent of very large scale integration (VLSI), when a mass market for ICs started to appear. The second phase lasted through the 1980s and the 1990s: the diffusion of the personal computer, symbolized by the iconic role of companies such as Apple, Intel,

F. Campi (✉)
Simon Fraser University, Metro Vancouver, Burnaby, BC, Canada
e-mail: fcampi@sfu.ca



G. Keramidas et al. (eds.), *Components and Services for IoT Platforms*,
DOI 10.1007/978-3-319-42304-3_1

IBM, and Microsoft. This was the time when microprocessors entered households, and became a commodity we started to take for granted. The third phase of the evolution of digital technologies is the system-on-chip (SoC) era, triggered in the early 1990s by the diffusion of cellular phones and the advent of the World Wide Web. From the circuit design perspective, *low power design*, that was before a specialized niche of research emerged in this period as a mainstream requirement. Also, severe real time requirements in digital communications, embedded system control, multimedia processing, triggered a wave of innovation in digital design: traditional microprocessor architectures were not able to sustain computational demands and power constraints introduced by emerging application fields. The new market dictated new attention to alternative computer architectures, and more effective solutions for high-performance, low power computing. This led to the advent of Systems-on-Chip (SoC), defined as ICs embedding in a single silicon die all functionalities required by a given application environment: a processor is not considered an electronic "product" in its own terms any more, but a building block (although a very crucial one) in the design of a larger system. Microprocessors have found their way out of the desktop, and creeped into our cars, our phones, our communication infrastructures, and our industrial machinery.

Today, we stand at the dawn of a fourth era of IC design, that of pervasive computing and the Internet of Things (IoT) [1]. Observers agree that the largest market for integrated circuits will be in the near future the production of IoT devices: small, massively distributed ICs including sensor circuitry and related signal processing logic embedded in the physical space surrounding us. As they occupy the lowest hierarchical level in the "connected" communication chain, we will define such nodes as IoT "LEAVES." Of course, the IoT connectivity chain also requires advanced and performant synchronization and processing nodes ("HUBS") that will collect, synchronize, and combine information collected by the ubiquitous sensing leaves (Fig. 1.1). From an economic perspective, it is evident

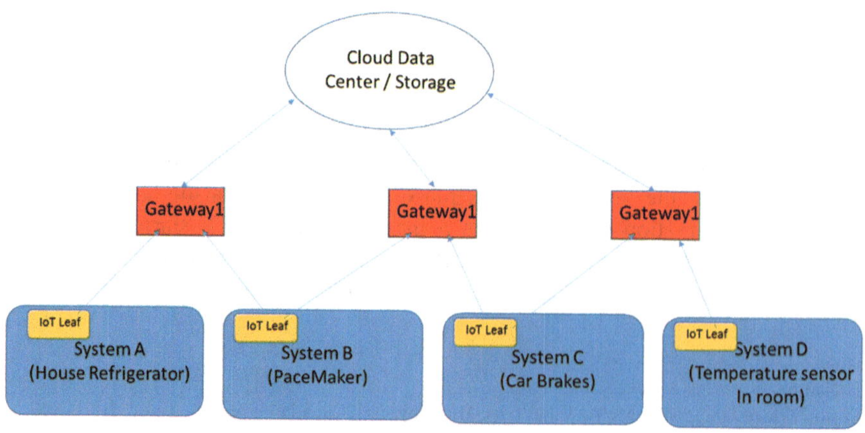


**Fig. 1.1** Schematic description of hardware infrastructure levels in the IoT

how the ratio between ubiquitous small processors/sensors (leaves) and centralized communication/synchronization nodes (hubs) is extremely low. In fact, the decentralization of information processing is one of the most significant aspects of the IoT itself.

Pervasive computing will change our design style and constraints: we will need tiny, flexible devices where electrical and micro-mechanical sensors/actuators and digital signal processing share small silicon areas and limited power resources. Key to the design success will be the synergy between analog sensing circuitry and digital logic. And since such pervasive, embedded devices tend to be both prone to malfunctioning AND difficult to substitute in their embedded locations, reliability of circuits in an increasingly mixed signal context will become of paramount importance.

## 1.2 Microprocessor Design Constraints for the Internet of Things

"IoT" is a "wide" concept, encompassing a large set of applications and protocol layers. In its more general meaning, we can define IoT as a network of physical objects (things) embedded with electronics, software, sensors, and network connectivity. These objects collect, process, and subsequently exchange a varied amount of information from our physical world.

IoT is a protocol stack that enables a "connected" collection and processing of information. In this chapter, we focus at the "lowest" level of the stack that is the *hardware support utilized for the computation*. In particular, we have defined in Sect. 1.1 two levels of computation in the "IoT":

1. "Centralized" gateway *nodes*, where computationally powerful ICs, most likely plugged to a reliable power supply are used to connect, merge, synchronize, and route the flow of information.
2. Local *leaves*, where small embedded microprocessors are tightly coupled to a sensing system, managing the sensing activity and performing localized processing on the sensor's outputs (e.g., monitoring, classification, pattern detection, and encryption).

The devices introduced at point (1) will need the most computational efficiency, in order to handle appropriately multiple flows of information and related computation. Hence, from the architectural perspective, point (1) is where we can expect the more innovation. Most of those devices do not need to be embedded on a physical system, and will most likely be provided with a reliable connection with power supply sources. IoT node architectures are likely to borrow many aspects from today's network routers, and will probably feature large amount of processing core parallelism, and a high degree of internal reconfigurability. In this context, it is likely that high-end FPGA devices, possibly coupled with high-performance embedded microprocessor cores will play a significant role. This chapter will focus on the devices introduced in (2), describing the "leaves" of the hardware infrastructure: what is the hardware that will be needed by the distributed, ubiquitous devices that represent the "IoT leaves," and what features and potentialities would be desirable in such hardware?

While the spectrum of functionalities that can compose the last level of the IoT is by definition very wide, we can determine few guidelines that can be applicable to all:

1. IoT leaves need to *be small devices* that can be embedded in physical systems without affecting the system features, design, and functionality.
2. IoT leaves need *to support connectivity to other leaves and the hub nodes in the network*. Depending on the host system features, this connectivity could be wired (e.g., in the case of a car or household appliances) such as I2C, CAN, USB, or wireless (e.g., in the case of temperature or pressure sensors distributed in a room) such as Zigbee, Bluetooth, radio frequency identification, and wireless sensor network.
3. IoT leaves are first of all a vehicle for collecting/transferring information from/to the "cloud" to/from a physical system. As a consequence, they are likely to *share the same IC, or be in very close proximity with an analog sensing system* (or an actuating system if the flow of information is from the cloud to the device).
4. IoT leaves may be embedded in conditions where it is not possible to provide reliable power supply connection. This is not the case for home appliances or automotive devices, but may be the case for sensors implanted in a person's body or massively distributed in a room or in a public space. In such cases, IC must *consume as little energy as possible*, being connected either to a small battery or to an energy harvesting system.
5. In all cases, the number of distributed leaves and their pervasiveness will grow exponentially with the diffusion of the IoT. Given their large number and sometimes difficult location, it would be unpractical and costly to reach the leaf IC to fix it or repair it in case of malfunctioning or upgrades. For this reason, possibly the most desirable feature of a leaf IC for the IoT would be *programmability* coupled with *long term reliability*.

### *1.2.1 Microprocessors for the Internet of Things*

Another feature that is recurrent in IoT leaves is that of carrying one or more embedded microprocessor core(s). This is not a requirement per-se, but an indirect effect of the guidelines outlined above: the primary scope of an IoT leaf is that of sensing, gathering, storing, processing, and transmitting information to/from a physical system. This action must be enabled in a small chip, with limited IO connection, possibly limited bandwidth and utilizing a very limited power budget. As a consequence, *IoT LEAVES MUST PERFORM A FIRST STAGE OF*

*PROCESSING LOCALLY*. This local processing may include input monitoring, classification, pattern detection, and encryption of collected data. Applying a first level of signal processing on the same IC that is collecting information, the bandwidth required by the communication will be reduced by several orders of magnitude. Also, depending on different parameters, the type of sensing/actuation may be modified depending on external commands or even on the data being collected: for example, a surveillance system, once triggered by an anomalous input, may be required to significantly augment the detail of its monitoring activity. A temperature sensor in a refrigerator/heating system may be activated by a critical condition, and the next sample interval may need to be shorter.

A processor system is also the best way to ensure that the leaf functionality can be fixed or adjusted on-location, with remote software upgrades, without requiring physical access to the device. As a consequence, *we can consider the presence of one (or more) embedded microprocessor(s) as a baseline feature that will be common to any IoT leaf architecture*.

Another advantage of featuring an on-board microprocessor is that of supporting a real time operating system (RTOS). While requiring a hardware cost overhead, RTOS ensure the most efficient use of the hardware resources and greatly ease the user interface to the device and the utilization of the IC, its connectivity and its eventual upgrades. In fact, RTOS for the IoT is a very relevant research field that deserves particular attention. Every IoT device, with only few exceptions, will run in the immediate future some degree of local RTOS. This chapter will focus on the hardware design aspects of the IoT leaves, and not on the OS details. Nevertheless, the presence of RTOS on all levels of the IoT stack is implied and acknowledged in any step of the hardware architecture definition.

In conclusion, integrated circuits that compose the lowest, and numerically most significant level of the IoT hardware infrastructure need to be small, highly reliable,

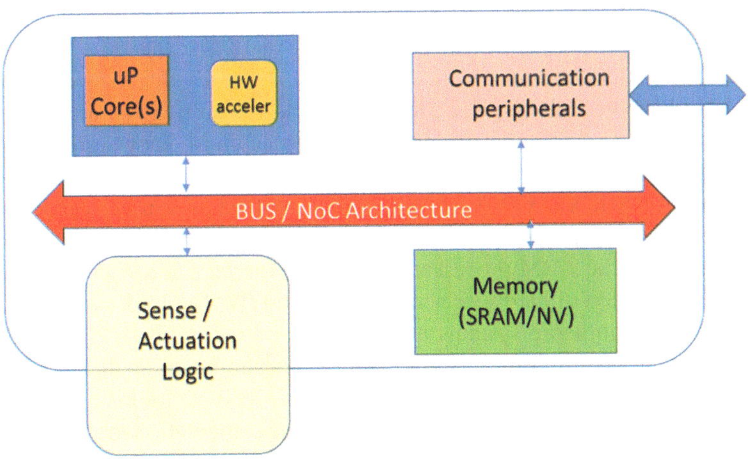


**Fig. 1.2** Leaf node template architecture

with small energy requirements. As summarized in Fig. 1.2, they will be composed by analog sensing/actuating logic (or a direct interface to the same), a digital signal processing unit centered around one or more microprocessors and the related bus/memory architecture, one or more connectivity protocols (wired or wireless). Some of such devices may be coupled to a small battery, possibly supported by an energy harvesting system. We summarize in the following the constraints to be met in their hardware design in a strictly hierarchical order of priority:

1. *Real time processing capabilities*, in order to accurately monitor and report the status of the physical system
2. *High Reliability*, in order to avoid costly and unpractical on-the-field intervention and substitutions
3. *High Programmability*, at two levels:
   a. In terms of run-time reconfiguration: it must be possible to perform different data processing task of different computational cost in different moments in time.
   b. In terms of friendliness to upgrade: it must be possible to refine/upgrade the service provided by the leaf by means of remote software upgrades
4. *Low cost*, as leaf IC may be produced and pervasively distributed in massive numbers
5. In some cases, *Limited Energy consumption* related to the lack of a reliable power supply connection due to the nature of the physical system where the leaf is embedded
6. In some cases, *low area and low Pad count* due to the nature of the physical system where the leaf is embedded.

## 1.3 The Role of Power Consumption in IoT Devices

The constraints outlined above represent an interesting shift with respect to the constraints that have been driving the processor market. Traditionally, microprocessor used to be highly performant, general purpose machines. They were plugged to power supply, and needed to perform at the highest possible speed very generic application benchmarks. With the diffusion of embedded systems, new computation paradigms have been proposed to mitigate the cost of microprocessor computation: the emergence of the ARM processor architecture targeted at low power computation, the introduction of hardware acceleration and HW/SW co-design in SoC, and the diffusion of multicore systems are examples of such trends. Processor-based systems have evolved, in order to negotiate the constraints of application-specific, real time embedded systems. But in itself, a microprocessor still is an intrinsically power inefficient, highly redundant general purpose computation machine.

Given the set of constraints outlined at the end of Sect. 1.2, hardware designers targeting leaf devices for the IoT find themselves balancing a difficult trade-off:

constraint 3 (programmability) imposes the utilization of processors. On the other hand, constraint 1 (real time response) requires relevant computational capabilities, imposing the design of a computationally strong architecture, possibly supported by hardware acceleration. In turn, constraints 4, 5, 6 (area, energy, and design cost) underline several limitations on what a designer can do to ensure high performance. Traditionally, performance can be obtained either architecturally, or by aggressive use of technology: in the first case, the processor design can exploit DSP/VLIW/Superscalar/MultiCore features, to ensure high computation parallelism. In the second case, the processor can use advanced technology nodes to ensure small geometries, and micro-architectural or circuit features such as heavy pipelining and high transistor driving strengths. In all strategies outlined above, the immediate drawback is large area and high power consumption.

### *1.3.1 Localized Effects of Power Consumption*

In the budgeting of an IoT leaf IC design, power consumption may be a significant issue. Many IoT leaves may be distributed in critical environments, so that they may have no power supply or easy battery replacement: this could be the case of carry-on bio-medical devices, or sensors distributed in a room or a public space with no direct power supply. If the IC must be supplied by non-replaceable batteries and/or an energy harvesting system, its requirements in term of consumption will be severe.

Given the small size of most IoT leaves, power consumption may also indirectly affect IC area, and impose undesired complexity in the package and the board supporting the chip. The average consumption of the IC will dictate the number of VDD/GND supply pads it needs to support. This in turn, may increase chip area: many small microprocessor-based ICs are pad-limited, which means that the floorplan area is determined by the number of pads, rather than the amount of logic it contains. In addition, more pads will require more pins in the package and increased complexity of the PCB board, thus increasing costs and making the system more difficult to embed in space-critical locations.

### *1.3.2 Globalized Effects of Power Consumption*

Energy consumption in IoT hardware is a relevant constraint, beyond the context of the single product design: projections forecast the production and diffusion of billions of IC devices in our living space [2]. In 2013, the energy consumed by ICT was already more than 10 % of the total consumption in many developed countries, and is on track to increase to 20 % [3]. Studies predict that by 2030, if current trends continue, electricity consumption caused by Internet-related devices will increase up to 30×, and energy prices will grow substantially [4]. We are indeed approaching a paradox, where the IoT will be essential to enable smart energy utilization to minimize greenhouse gas emissions [1, 4], but in turn will be in itself cause for severe increase in energy consumption and consequent GHS emissions. IoT leaves are numerically largely dominant with respect to other ICs that contribute to the IoT infrastructure. Hence, a careful attention to low power design is essential. On the other hand, this cannot be obtained compromising performance, due to real time constraints.

### *1.3.3 Reliability of Integrated Circuits*

A relatively recent issue in IC design is RELIABILITY: as technology nodes scale towards smaller channel lengths, physical effects that used to be negligible in the MOS transistor start coming into play [5, 6]. On top of being a constraint in itself, reliability severely impacts design and manufacturing costs: on the one hand, established design flows need to be patched and adapted in order to take such sub-micrometer effects into account. Concurrently, the lowest reliability of latest CMOS technology nodes affects yield, decreasing profit margins, and consequently increasing manufacturing costs per die.

According to the classification in [6], CMOS reliability issues can be categorized into *spatial and temporal unreliability effects. Spatial unreliability effects* (random dopant fluctuations, line edge roughness, gradient effects, etc.) are immediately visible right after production. They depend on the circuit layout and process conditions. Impacting the geometry and structure of the circuit, such effects can lead to yield loss (malfunctioning circuits or circuits with degraded performance). While impacting manufacturing costs by means of low yields, these effects can be detected by post-fab testing, so that they do not imply the need to reach the ICs while embedded in their physical system for fix or substitution.

*Temporal unreliability effects* are time varying, and change depending on operating conditions such as supply voltage, temperature, computation workload, and activity of neighboring circuitry. These are classified in wear-out/aging effects and transient effects [6]. Examples of the first category include negative bias temperature instability (NBTI), hot-carrier injection, and electro-migration (EM). Examples of the second include electro-magnetic interference (EMI), static voltage drop (a.k.a. IR-Drop, IRD), dynamic voltage drop (DVD), simultaneous switching noise (SSN), thermal runaway (TR), and temperature-related timing degradation.

In the context of IC design for the Internet of Things, temporal effects are way more critical, as they create unplanned, unpredictable and often undetectable malfunctioning (EMI, DVD, and SSN), permanent ruptures (EM, TR), or permanent malfunctioning (NBTI). Such occurrences are very cumbersome to diagnose or fix, requiring direct access to embedded devices, and can potentially affect the sense/actuation patterns of the IoT. Also, temporal effects are strongly related to parameters that in the IoT are almost impossible to control at design time, such as computation activity, ambient temperature, and thermal conductivity of the physical system.

The large majority of the temporal unreliability effects outlined above are strictly related to the power consumption in the IC (NBTI, TR, EM, and IRD), and more in particular to the time-varying peaks in the current distribution across the IC power grid (SSN, DVD, and EMI). As a consequence, especially for low cost, embedded devices that need to deliver significant computational power, energy consumption is not only a constraint in itself: with the latest nodes in technology scaling, *energy consumption has become a severe reliability issue*: high power dissipation, and worse still high supply current gradients may create malfunctioning and even permanent ruptures. Designers challenging the next generation of IoT devices simply cannot confront architecture design without carefully considering supply currents, and their impact on reliability [7]. Moreover, IoT processors are almost invariably coupled to critical analog sensing circuitry: strong current gradients on the digital supply lines may create different kind of disturbances (coupling, substrate currents, and radiated and conducted electro-magnetic interferences) that can jeopardize the correct functioning of the analog and A2D circuitry, leading to further unreliability.

For the reasons mentioned above power consumption, and in particular the shape of the supply currents over time need to become a fundamental design parameter, negotiated at every level of the design: at circuit level, at architectural level, but even, as it will become apparent in the next sections, at the level of the operating system and the application software.

## 1.4 Low Power Design Opportunities for IoT Processors

In VLSI literature, many approaches to low power have been proposed. The most dramatic reductions can be obtained by means of specific low power circuits (e.g., high-Vt transistors, near-threshold computation) and/or micro-architecture design (e.g., synthesis strategies for low speed). Most of such approaches are not applicable to IoT leaves, because they require modification of libraries and design flows, and/or affect peak performance. As described at length in Sect. 1.2, for IoT leaf ICs described in this chapter, low power is a constraint that is very stringent, but cannot jeopardize real time performance or design cost. IoT processors typically cannot sustain the costs dictated by the application of specific low power circuits: in order for the IoT paradigm to be economically sustainable, small cost is indispensable. Standard VLSI design flows must be applicable, and it is unlikely that project budgets could afford design of customized cell libraries enabling aggressive power optimizations. Moreover, most strategies for low power design trade peak performance to decrease consumption: again, this is not acceptable for IoT devices.

Other well-known strategies for low power design are power gating and clock gating. *Power gating* is the practice of shutting off the supply network of a digital block when unused for a relatively long time. Turning off and on the supply voltage of a digital block implies very strong current transitories to “charge-up” the significant parasitic capacitance of the Vdd supply network. As described in

Sect. 1.3.3, on-chip current gradients are dangerous, especially for mixed signal devices such as IoT processors. In high-end SoC design, specific current limiting regulators are embedded in the power supply to avoid current peaks, but this increases design costs, and implies very long "charge-up" times that may be unacceptable for IoT processors. On the contrary, clock gating can be applied with good results for IoT processors, although for most applications, power saving enabled by clock gating is useful but not sufficient. Clock gating is fully embedded in modern HDL-based design flows and as such can be added on top of other strategies discussed in the following. Clock gating will applied in the test cases that will be introduced, and every gain that will be discussed should be considered on top of the clock gating effect.

### *1.4.1 Dynamic Voltage/Frequency Scaling*

Among existing approaches to low power design, supply voltage scaling is the most effective technique to reduce energy consumption with a limited impact on the design flow. In particular, dynamic voltage and frequency scaling (DVFS) is based on the principle of adapting dynamically clock and voltage supply of a system to the chosen computation workload, so that at every moment in time a given part of the circuit will be delivered only the exact amount of power necessary to meet its timing constraints. The same processing core may be run at high voltage (e.g., 1.2 V) when it needs to perform at a high clock speed to handle severe computational requirements, and then be switched to a significantly lower voltage (e.g., 600 mV) when lower performance is acceptable, leading to energy consumption that can be up to 20 %/30 % of the consumption for the same computation at higher speeds. From the VLSI design flow perspective, the overheads related to an extensive application of DVFS are

1. *Architecture Design*: The definition of independent clock/voltage entities requires explicit synchronization in order to support the crossing of clock domain boundaries. Strategies for fully asynchronous communication are well established in VLSI and FPGA design communities, and commonly implemented in most communication IPs available for reuse [8, 9, 28]. A widely utilized example of architecture strategy for asynchronous domain synchronization is described in [9].
2. *Physical Design*: defining separate power supply grids induce small overheads on the IC design. After synthesis, during the floorplan phase, different supply regions (a.k.a. "Voltage Islands") need to be defined. Also, Level Shifter cells must be inserted in between domains to ensure a smooth electrical transition between different voltage levels. It should be underlined that such methodology is usual in any industrial design flow, its only drawback being a slight area overhead.

3. *Timing Characterization*: in the standard VLSI flow, timing constraints are verified only in two "operating conditions": Worst/Slow, for setup violations, and Best/Fast, for hold violations. DVFS requires the digital design to be operative in a significant range of Vdd conditions. Timing closure then must be repeated for each condition. Most importantly, timing libraries will have to be characterized in all desired voltage operating points.

DVFS matches very well with the specifications of IoT processor design: first of all, it only requires minimum adjustments to the design flow, due to the necessity of defining multiple power domains. More importantly, it does not impact peak performance: whenever an IoT leaf node faces tight real time constraints and is in need of high computational power, frequency and voltage can be dynamically adapted to deliver the highest performance. On the other hand, when high-performance is not required, the node can be switched to a lower consumption mode where clock and voltage are small enough to enable energy saving. In fact, referring to Fig. 1.2, it could be relatively straightforward to determine different DFVS domains, so that every sub-component in the IC architecture can explore its own ideal DVFS working point at each moment in time.

### 1.4.2 GALS Design Style and Its Advantages for IoT Processors

As described above, extensive application of DVFS requires architectural adjustments, so that different regions of the same IC can be run with independent supply and clocking. This design style is defined GALS (*Globally Asynchronous Locally Synchronous*) and is largely applied in today's SoC design [11, 12, 26, 27]. In the specific context of IoT leaves, the GALS pattern has appealing advantages that largely justify the limited design overhead: as the separate regions of the IC feature independent, fully asynchronous clocks:

1. Current peaks induced by the clock edge are actually distributed in time, and compensate each other providing a much smoother frequency behavior. A smoother current demand plot is a precious asset to ensure the reliability of the IC, limiting phenomena described in Sect. 1.3.3 such as DVD, EM, EMI, and SSO. Figure 1.3 shows voltage and current waveforms differences obtained applying asynchronous clocking patterns to a GALS SoC.
2. Every region (processor core(s), hardware acceleration, communication peripherals) can run at the clock frequency imposed by real time constraints, thus avoiding any need to over-design a component to meet the clock requirements of another.
3. If, as very common in IoT processors, real time requirements change suddenly due to external events, the performance of each block can be easily scaled by means of DFVS. Reliving the design from dependence between the

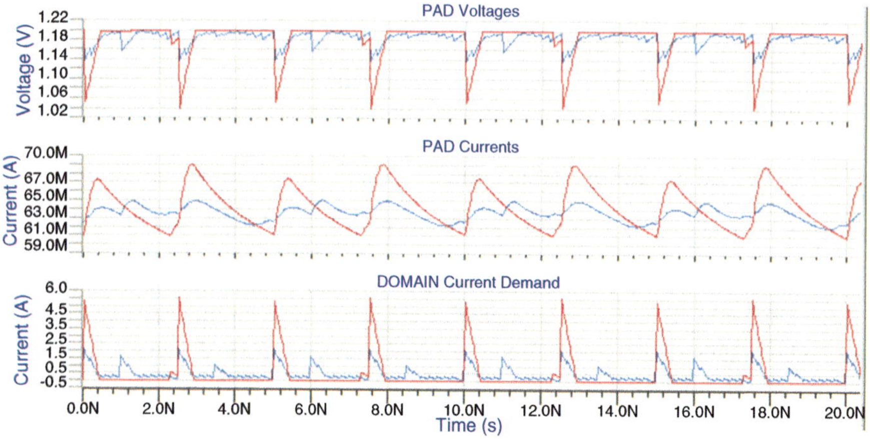


**Fig. 1.3** Power supply current waveforms in a 45 nm GALS design divided in 14 independent DFVS islands. *Red* waveforms are currents due to a synchronous clock; *Blue* waveforms are currents due to fully asynchronous clocks

sub-components clocks, GALS enable a full range of frequency scaling, greatly increasing the potential for power mitigation [13].

The silicon area overhead required by the adoption of the GALS pattern depends on the design features, but is in the range of 5–10 %, due to the synchronization logic. On the other hand, GALS enables smaller clock trees, as the skew balancing needs to be applied on a much smaller number of Flops. Also, timing closure in GALS is much easier, and involves a smaller amount of buffering as critical paths are localized in one of the clock domains and not distributed over the full chip area.

In conclusion, applying GALS to IoT processors allows the designer to build "*Power-Shaping Configurable Micro-Processors*," capable to select according to the current application the ideal working point for each design component, thus meeting real time constraints with the smallest possible consumption. Using GALS and DVFS, one *can shift the power vs performance trade-off from the design space to the application space*. The hardware designer is not responsible for selecting between high speed or low power circuits: the choice is left to the application developer, that with appropriate software instructions may dynamically increase clock frequency and supply voltage of a given processor core, hardware accelerator, or communication peripheral to accommodate sudden demands in computation efficiency, only to tune it down when the transitory has exhausted or is not a priority any more. In processor systems featuring a solid RTOS, power as a design parameter would be simply one of the different factors involved in task scheduling.

The application of GALS and DVFS is not innovative, and is largely established in high-end SoC design [14–16], but the advent of the IoT and its very specific constraints (real time demands, variable computation and connectivity load, reliability, low area and low pad count, and low power consumption) makes IoT processors an ideal opportunity for utilizing this design concept in an innovative context. In fact, IoT processors are an appealing test case to apply DVFS at a finer granularity than what normally applied in large state-of-the-art SoCs. Moreover, recent evolutions in the CMOS technology manufacturing process offer new and exciting opportunities for applying DFVS to small and low cost products, offering further substantial performance peaks AND low power potential at a relatively small design cost. The following sections will focus on these innovations, and the opportunity they offer for IoT processor design.

### *1.4.3 Body (Substrate) Biasing*

Body biasing is an appealing technique that is encountering significant success in modern IC design. By means of appropriate bias voltage applied to the silicon substrate, it is possible to alter the threshold voltage of the MOS transistors, thus modifying the trade-off between consumption and timing performance. In the specific case of IoT leave nodes, this strategy is especially appealing because, similarly to DVFS, body biasing can be applied run-time, by the application developer or, better still, with the support of the RTOS, tuning the biasing voltage of different regions of the architecture depending on the required computation workload (Fig. 1.4).

In order to enable independent biasing of separate regions in the IC, a slight modification of the standard CMOS process is required, defined *triple-well process* (Fig. 1.1): separate and localized P-type and N-type substrate regions are designed around the logic. With this manufacturing option, it is possible to set bias voltages independently for nMOS and pMOS belonging to a given region of the IC. This allows to tune at every moment during the chip lifetime the behavior of each

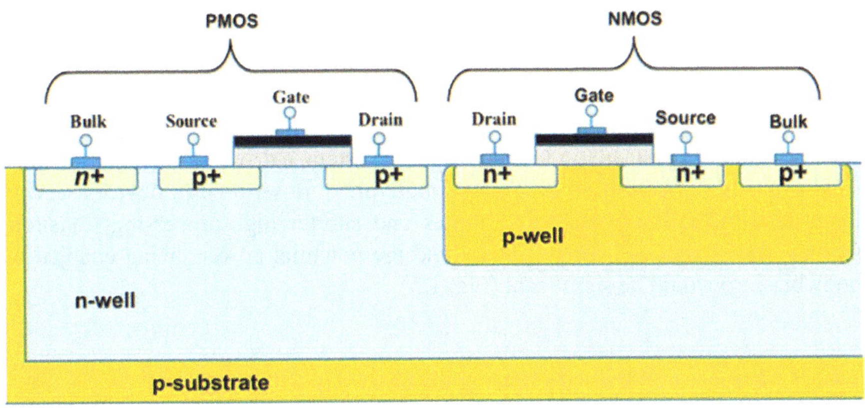


**Fig. 1.4** Triple-well process cross-section

sub-portion of the chip architecture depending on the local workload. The triple-well option is a very straightforward modification of the standard CMOS, and does not imply significant additional design or manufacturing costs.

Biasing the substrate regions (i.e., wells) requires an additional power grid network. But such network is only used for biasing, and not for power supply, and carries negligible currents. As such, it is not particularly critical to design, and does not induce specific design congestion problems or reliability concerns. When applying body biasing, the P and N wells are biased symmetrically with respect to the reference power supply levels (VDD for N-type substrate and GND for P-type substrate). The power/ground supply lines for well biasing are usually named VDDS (pMOS devices, N-type substrate) and GNDS (nMOS devices, P-type substrate). In particular:

- Forward body biasing (FBB) of the substrate (VDDS < VDD, GNDS > GND) induces lower MOS threshold voltage, enabling faster timing behavior of standard cells, while introducing higher leakage consumption.
- Reverse body biasing (RBB) of the substrate (VDDS > VDD, GNDS < GND) induces higher MOS threshold voltage, enabling smaller leakage consumption, while imposing higher delays in standard cell activity.

Body biasing can be used as a powerful tool to fine tune speed/leakage ratio depending on the application workload being deployed at every moment in time in the device lifetime. Differently from DVFS, that significantly affects dynamic power, body biasing in itself mostly impacts leakage, while dynamic power remains essentially unchanged. As a consequence, BB is only convenient in design environments where the leakage component is dominant.

As a rule, IoT nodes are likely to have very leakage-intensive computation patterns. On the one hand, it is not reasonable to assume that most IoT IC would adopt in the short term very advanced technology nodes: the majority of IoT leaves are likely to converge on established, lower cost technologies where leakage is not dominant as compared to dynamic power. On the other hand, it is in the nature of IoT processors to be powered-on for their whole life cycle. Most of the time, such devices would be "monitoring" a physical system: it reasonable to expect that IoT leaf nodes will be mostly of the time "dormant," collecting information at low rates, only to be "awaken" in case of sudden attention from the "cloud" or in case of specific events (e.g., an alarm in a surveillance system, a driver action in a car, and a sudden change of condition in a temperature/pressure sensor). In devices serving long stand-by or low-performance times and shorter high-processing transitory periods, the impact of leakage power (and the potential power saving enabled by body biasing) would be significant (Fig. 1.5).

#### 1.4.3.1 Impact of Substrate Biasing on Latch-Up

We will define in the following the standard CMOS process technology as "bulk" CMOS. Unfortunately, widespread application of body biasing is limited in bulk

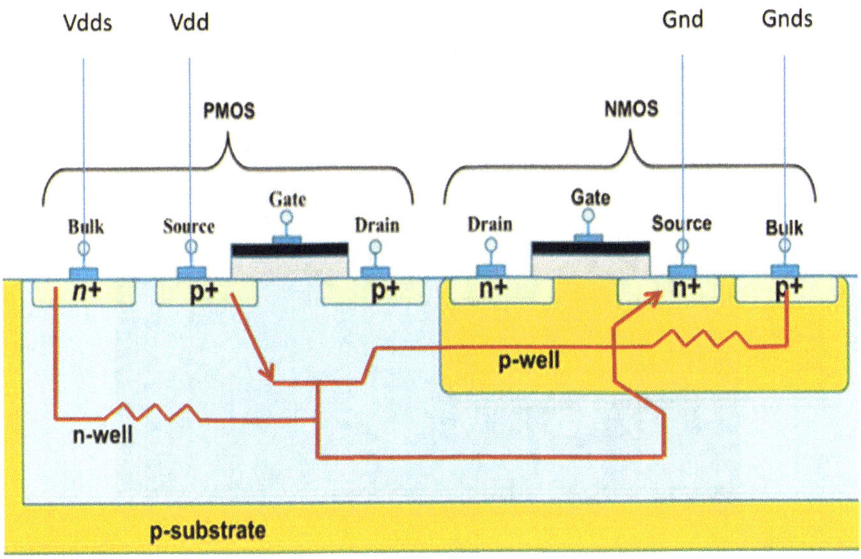


**Fig. 1.5** Schematic description of parasitic BJTs causing latch-up in a triple-well MOS process

CMOS by physical issues. The first reason for concern is the presence of P-Well/N-Well diodes, which must not be forward biased. Especially in technology nodes below 100 nm, a second significant reliability concern is caused by latch-up phenomena [18], [29]: latch-up is a disruptive failure mechanism induced by the presence of a parasitic stack of BJT transistors that is "built-in" in the CMOS structure. Latch-up is triggered when BJTs are turned on due to voltage mismatches between substrate and the diffusion regions of the MOS. Once BJTs are turned on, they create a parasitic conductive channel between the VDD and GND in the chip, and they may cause domino effect on the neighboring CMOS stages, leading to very high currents and ultimately the chip destruction (Fig. 1.6).

While in standard MOS functioning P- and N-substrate are connected, respectively, to VDD and GND, the presence of independent VDDS and GNDS voltage severely increases the risk of latch-up. In order to avoid such occurrence, in bulk CMOS processes the range of biasing voltages must be limited to ±400 mV, limiting the potential for leakage mitigation.

#### 1.4.3.2 Case Study 1: DVFS and Body Biasing on Standard "Bulk" Processes

This section describes the application of DVFS-BB in the implementation of a simplified processor core, described in [17]. The 20 k-gate design is implemented using a 40 nm triple-well CMOS technology with nominal supply voltage of 1.1 V. The core was designed at a 250 MHz operating frequency. The processor

**Fig. 1.6** Example of the routing of VDDS line in a design supporting FBB/RBB biasing (from [17])

architecture, size, and frequency are compliant with "typical" requirements of a processor core embedded in an IoT sensor node. In order to verify the cost of the body biasing the design was placed and routed both with and without the additional power distribution network circuitry that supports body biasing. The substrate P/G grids are similar to the primary supply grids, only with narrower widths, since they do not distribute large current loads (VDDS currents are at least two orders of magnitude smaller than those flowing in the primary power distribution network).

After implementation, the two versions of the design show very little differences: the design implementing body bias utilizes a few more buffering cells to compensate for a smaller routing space due to the congestion induced by the additional grid. Also, the floorplan area had to be made a little larger in order to accommodate the additional wiring. The peak frequency of the two designs is identical, and the area overhead of the solution implementing BB is 1 %.

In order to assess the performance vs. leakage trade-offs related to the design implementing BB, power and timing analysis was performed on the design on a set of different supply and biasing operating points:

- Primary power supply values (VDD): 0.9, 1.1, and 1.3 V;
- For every operating voltage, the following body biasing configurations (VDDS/GNDS) were applied: 0, ±0.4, and ±0.2 V.

Experimental results shown in Fig. 1.7 confirm that leakage reduction is strongly dependent on the RBB/FBB configuration: starting from the nominal case as reference, utilizing RBB to minimize consumption, the design shows a reduction of overall power dissipation up to 84 %. As expected, we observe a concurrent timing

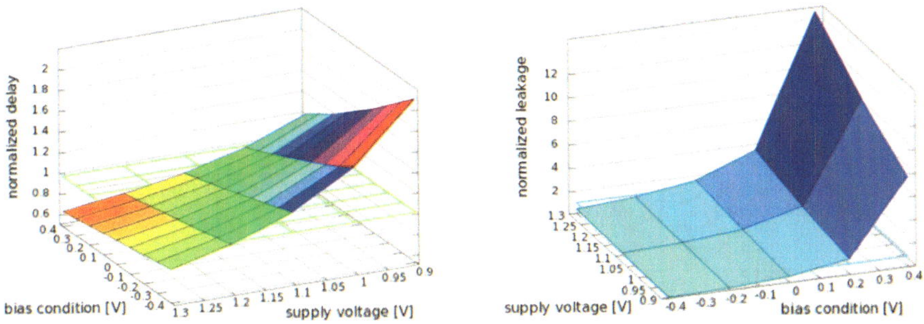


**Fig. 1.7** Propagation delay and leakage power variation across different VDD/VDDS operation points (from [17])

degradation at the lower leakage points, where propagation delays can be almost doubled with respect to the reference: but, DVFS-RBB is a dynamic strategy, and in case the IoT node would be in need of a temporary performance boost, a different operation point could be chosen for a transitory time.

In particular, FBB could be applied to run the design even faster than in the zero-biasing reference condition, enabling localized "overclocking" of the design. This should be applied with care: Fig. 1.7 shows strong asymmetry between timing and leakage behavior. Leakage dissipation has a larger variation range with respect to propagation delay: FBB induces a greater and non-constant increase in leakage, and the ratio between timing and leakage is strongly unbalanced. For this reason, application of FBB must be very limited in time (i.e., to satisfy sudden but temporary computation demands). Continued application of FBB may create significant dissipation that may in turn create reliability issues, in particular thermal runaway [19]. Thermal phenomena on-chip depend on many external factors, that for IoT leaves are difficult to control at design time (e.g., thermal conductance of the physical system embedding the IC and ambient temperature). In case widespread use of FBB is planned, the IC should be equipped with an evenly distributed set of on-chip temperature sensors, and appropriate thermal management software, preferably embedded in the operating system to guarantee timely and immediate intervention in case of unplanned dissipation peaks.

## 1.5 CMOS FD-SOI Technology and Related Opportunities for IoT Processor Design

As described in Sect. 1.4.3.1, the most significant limitation to the application of body biasing is due to the fact that substrate bias increases the risk of latch-up phenomena. If latch-up could be avoided, then VDDS/GNDS could be biased on a very wide range of values ensuring very high flexibility in dynamically tuning consumption.

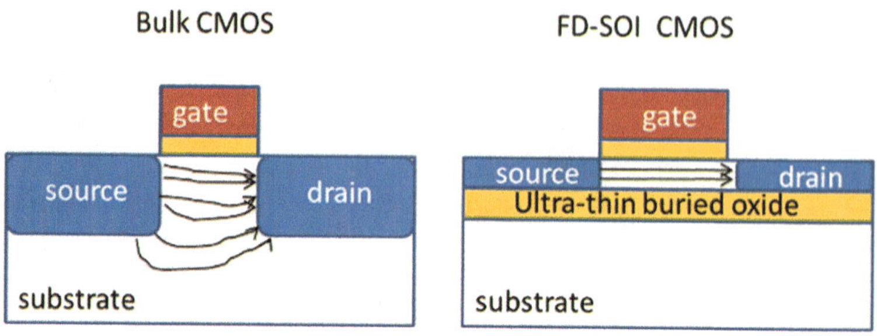


**Fig. 1.8** Silicon-On-Insulator (SOI) technology process

Silicon-on-Insulator (SOI) [21, 30] is a modification of CMOS manufacturing that implies the addition of a thin oxide (i.e., insulator) layer buried below the conducting channel (Fig. 1.8). The layer is normally referred to as BOX (Buried Oxide). In particular, if the conductive channel above the buried oxide is so thin that is fully covered by the depletion region, the technology is defined "Fully-Depleted Silicon-on-Insulator" (FD-SOI). In this case, the channel does not need to be doped making the manufacturing steps cheaper and simpler. While introducing very minimal manufacturing change, SOI has several advantages as opposed to bulk CMOS. First of all, the buried oxide strongly limits currents between source/drain and substrate, dramatically reducing leakage currents. Even more significantly in the context of this study, the buried insulator effectively makes latch-up impossible, enabling a higher range of biasing values. As discussed in Sect. 1.4.3.2, the application of FBB/RBB enables to dynamically trade performance with leakage: FBB improves switching speed at the expense of higher leakage, while RBB is used to decrease leakage when low transistor speed does not affect real time constraints. Being inherently latch-up free, FD-SOI provides much greater control over body bias than is possible in bulk CMOS. The sole limiting factor comes from P-Well/N-Well diodes, which must not be forward biased.

The reminder of this section will focus on a specific FD-SOI manufacturing process, STMicroelectronics FD-SOI-UTBB (Ultra-Thin-Buried-Oxide) 28 nm CMOS. This technology represents a significant test case for the application of FD-SOI strategies to IoT processors, as it has been the first SOI process applied commercially [20]. ST FDSOI-UTBB 28 nm CMOS is currently utilized in various commercial products and also available to small companies and research organizations in form of Multi-Project Wafer (MPW) access. It also represents a relevant example of the potentialities that other SOI options may offer to integrated circuits for the IoT in the near future: in fact, the SOI option is available also for other relevant technologies such as finFET [22, 23]. The considerations that follow, with some distinction due to the specific details of each node, would be applicable for several other contexts, and are in general relevant for a large range of present and future IoT-related designs. In ST FDSOI-UTBB 28 nm CMOS, standard cells are designed with a dual approach [21, 24]: in case a strong reverse bias is required (RBB) cells are designed with a conventional approach, featuring a P-substrate under NMOS and N-substrate under PMOS devices. This allows full range on RBB (VDDS < VDD, GNDS > 0), while FBB is limited to 300 mV, because going beyond this range would create parasitic diodes between the n- and p-wells. In case strong forward biasing is required (FBB), cells are designed featuring a flip-well configuration [21, 31], where N-substrate is used under nMOS transistors and P-substrate under pMOS transistors. This allows full range on FBB, while RBB is limited to 300 mW again due to P-Well/N-Well junctions. The two approaches are mutually exclusive and cannot be applied to the same region of the design, but different regions in the same IC, related to different architectural blocks in the same system could adopt either approach.

### *1.5.1 Application of Forward Body Biasing in FD-SOI*

If we look back at the specifications for IoT leaves that were outlined in Sect. 1.3, the need for sudden, significant performance for short transitories was underlined as the main constraint. Power mitigation is an important feature, but in the IoT context it is only significant if it does not compromise management of real time events. As a consequence, the choice of supporting RBB for enabling very low leakage appears unsuitable, as it would implicitly require to renounce to the temporary performance boost enabled by FBB. While RBB may still be accessible for low speed IPs, the majority of the IoT leaf IC blocks, and in particular the processing core(s) would have to be designed in order to exploit FBB.

Interestingly, the choice of using strong FBB to support high-performance transitory peaks does not necessarily imply high consumption during normal computation: in fact, *the potential speed up made available by consistent exploitation of FBB is such that, applying FBB, it is possible to run a processor (or any other digital IP such as a DMA, an hardware accelerator, a bus or NoC architecture) at the frequency imposed by real time constraints while featuring a lower supply voltage than what would be required without FBB*. In turn, running the application at a lower voltage has a very beneficial effect not only on leakage consumption, but also on dynamic power.

In conclusion, SOI technologies offer a new, innovative strategy for power mitigation that is particularly interesting for IoT devices. This strategy is different and much more effective than the standard RBB strategy outlined in Sect. 1.5.2: on top of applying reverse biasing up to −300 mV for quasi-stand-by conditions, FDSOI flipped-well designs may be forward biased up to 3 V. FBB enables to supply the circuit at much higher speed, with a dual advantage:

1. In case of high peaks of computational requirements, running FBB with high voltage supply would offer speed-ups in the range of 30 % with respect to the non-biased conditions

2. In all other cases, using FBB with lower voltages would allow to reach the same speeds as the non-biased conditions with a significantly lower Vdd, thus enabling lower leakage as well as lower dynamic power consumption.

### *1.5.2 Case Study 2: FBB on FD-SOI Technology*

As a reference example, let's consider the processing system composed by the 32-bit RISC core described in [32] and an FPU unit derived from the one described in [25]. FPU processing is relevant for IoT nodes, as many sensor data classification/processing algorithms are based on FP numbers, and may offer low accuracy or unsatisfactory performances when ported to integer numbers. At the same time, FPU computation may be costly from a power perspective. Both designs were synthesized, and placed and routed on ST FDSOI cmos28nm flipped-well technology in worst case conditions (ss_0.95V_125C). The design supports two fully asynchronous DVFS domains with independent power biasing. The RISC core occupies 35Kgates, and was designed at the target frequency of 230 MHz WITHOUT FBB. The FPU core occupies 45 Kgates, and was designed at the target frequency of 125 MHz without FBB. The Kgates overhead due by the domain synchronization is around 5 %, while the support of independent voltage and substrate islands accounts for another 2 % of the overall area. This overhead is largely justified, as without synchronization the RISC core will need to run at a severely reduced speed given the different performance offered by the two designs. In order to limit design costs related to timing closure and library characterization in different operating points, the number of working points in the design has been limited to

- FBB: 0 (No Bias), 1.1 V (Forward Biasing)
- VDD: 0.6, 0.65, 0.7, 0.75, 0.8, 0.85, 0.9, 0.95, 1

Although in a different project different working points may be necessary/relevant, in the specific case under analysis these conditions cover a broad range of working voltages, covering all possible states that would be relevant for the processor functioning.

Figure 1.9 shows frequency vs total power consumption behavior of the two cores, with FBB = 0 (blue) and FBB = 1.1 V (red). For each frequency, the reported power is relative to the Vdd supply value that meets the frequency constraint with the smallest power consumption. In both cores, the red plot offers smaller power numbers than the blue one: this is because using FBB, we are able to run our circuit at the same frequency but with smaller Vdd, hence we can obtain smaller total consumption. We resort to utilizing FBB = 0 only at lower frequencies, where the leakage component induced by FBB becomes dominant on the dynamic component. The application of FBB on processor core and arithmetical coprocessor requires an overhead of 7 % on circuit area, related to asynchronous cross-domain communication and additional routing overhead related to the biasing power

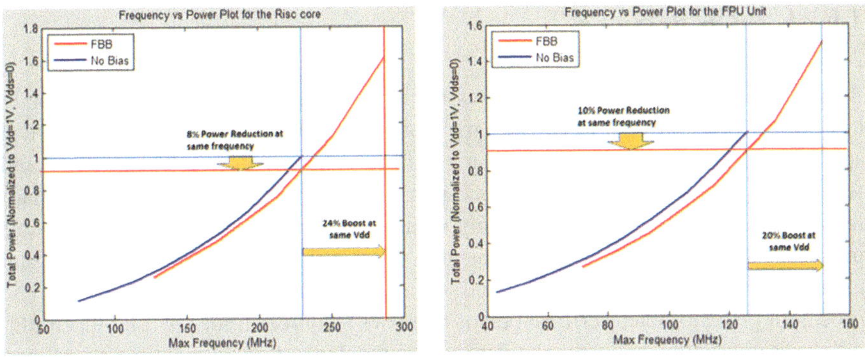


**Fig. 1.9** Power vs frequency behavior of RISC core and FPU unit

network. But this design option can provide two levels of advantage: if applied at fixed VDD voltage, FBB can provide a 20 % boost in performance at the price of higher leakage (as also reported in [20]): in the IoT context, this could be very useful to manage temporary computation peaks due to asynchronous events in the environment. If applied at a fixed frequency, FBB allows to run the design at a smaller VDD, offering an average reduction of 10 % in power consumption when operated the range of the "native" range of frequencies where timing closure was obtained without FBB. Again, in the IoT context this can be very useful to limit power dissipation and related reliability issues, and mitigate energy consumption.

Finally, at the lowest frequencies RBB could be applied up to 300 mV to at the price of performance, obtaining leakage reduction in the same range as the case of "bulk" CMOS already reported in Sect. 1.4.3.2, featuring a 3–5× reduction in leakage at 0.6 V at 10 MHz. Once again, this option is particularly significant as leakage is the largely dominant when the IoT processor is in a dormient or quasi-static state, monitoring the environment in non-critical conditions.

On top of the gains enabled by DVFS-BB thanks to FD-SOI flipped well technology, the application of a fine-grained GALS allows the RISC core to run 80 % faster than it would with a single clock, thus enabling much more effective run-time servicing when the FPU is not heavily exploited. Moreover, having independent tuning of FPU and Core allows to adapt the working point of each to the external events.

Finally, it should be noted that although degrees of DVFS can be applied with relevant advantages also to embedded memory cuts, memories do not follow the same scaling pattern of stdcell logic. In particular, in the technology node of this example, memory cuts should not be scaled below Vdd = 800 mV to preserve the memory state. Measurements reported above only considered digital logic and did not include memory cuts, for whom detailed characterization libraries were not available. In the design, memory cuts had independent voltage, but shared same clock as the RISC core.

## 1.6 Conclusions

A significant implication of the IoT is the insurgence of a new IC market: small, low power, pervasive devices that are distributed in any aspect of the physical space we live in. These IC must be programmable, small, highly reliable, and feature low power consumption. Yet, such aspects must be subordinated to hard real time constraints that can suddenly change depending on asynchronous events presented by the environment. Typically, such ICs are composed by a digital microprocessor system tightly coupled with analog sensing/actuating circuitry. The processor presents a very critical performance vs resources trade-off, and this trade-off varies continuously during the device lifetime depending on real time events and related service constraints. In this chapter, an appealing solution to this time-varying trade-off has been suggested based on "*Power Shaped Configurable Processors*": applying aggressively fine-grained voltage and substrate scaling, the application developer (or the RTOS) can chose at every moment in time for every component in the processor system the minimum consumption that meets RT constraints. Recent developments in the CMOS manufacturing process also offer valuable support to such methodology. While voltage/frequency scaling (DVFS) and body biasing (BB) are established methodologies in high-end SoC designs, IoT processors offer an exciting innovative context for its application. This chapter investigated the application of DVFS/BB on different cores in the same microprocessor system (core(s), co-processors, HW acceleration, and peripherals) an interesting future work would be evaluating the applicability of the same strategy at a finer grain (e.g., on different function units in a superscalar core). One of the main challenges of the proposed methodology is that it requires a holistic approach to power saving at all levels of the design chain: architecture designers must be able to manage manufacturing and circuit issues, and application/OS designers must be able to consider power consumption as a new metric in application mapping and scheduling.

## References

1. R. Bolla, R. Bruschi, F. Davoli, F. Cucchietti, Energy efficiency in the future internet: a survey of existing approaches and trends in energy aware fixed network infrastructures. IEEE Commun. Surv. Tutor. **13**, 223–244 (2011)
2. A. Zanella, N. Bui, A. Castellani, L. Vangelista, M. Zorzi, Internet of things for smart cities. IEEE Internet Things J. **1**(1), 22–32 (2014)
3. A. Manuzzato, F. Campi; V. Liberali, D. Pandini, Design methodology for low-power embedded microprocessors. 23rd International workshop on power and timing modeling, optimization and simulation (PATMOS), 2013, pp. 259–264. doi:10.1109/PATMOS.2013.6662187
4. E. Rodriguez-Diaz, J.C. Vasquez, J.M. Guerrero, Intelligent DC homes in future sustainable energy systems: when efficiency and intelligence work together. IEEE Consum. Electron. Mag. **5**(1), 74–80 (2016). doi:10.1109/MCE.2015.2484699

5. G. Gielen, P. De Wit, E. Maricau et al., Emerging yield and reliability challenges in nanometer CMOS technologies. Proceedings of the IEEE symposium on design and test in Europe (2008)
6. E. Maricau, G. Gielen, *Analog IC Reliability in Nanometer CMOS* (Springer, New York, 2013)
7. D. Draper et al., Signal and power integrity for SoCs. 2010 IEEE international solid-state circuits conference digest of technical papers (ISSCC), pp. 520–520. doi:10.1109/ISSCC.2010.5433856
8. Synopsys Design Ware FIFO, www.synopsys.com/products/designware/docs/doc/dwf/datasheets/dw_fifo_s2_sf.pdf
9. C. Cummings, Simulation and synthesis techniques for asynchronous FIFO design. Synopsys users group conference (SNUG), San Jose, CA, 2002
10. A. Wang, A.P. Chandrakasan, S.V. Kosonocky, Optimal supply and threshold scaling for subthreshold CMOS circuits. Proc. ISVLSI, Apr 2002, pp. 5–9
11. M. Krstic, E. Grass, F.K. Gurkaynak, P. Vivet, Globally asynchronous, locally synchronous circuits: overview and outlook. IEEE Des. Test Comput. **24**(5), 430–441 (2007). doi:10.1109/MDT.2007.164
12. D. Rossi, F. Campi, S. Spolzino, S. Pucillo, R. Guerrieri, A heterogeneous digital signal processor for dynamically reconfigurable computing. IEEE J. Solid State Circuits **45**(8), 1615–1626 (2010). doi:10.1109/JSSC.2010.2048149
13. J.W. Tschanz, S.G. Narendra, Y. Ye, B.A. Bloechel, S. Borkar, V. De, Dynamic sleep transistor and body bias for active leakage power control of microprocessors. IEEE J. Solid State Circuits **38**, 1838–1845 (2003)
14. D.N. Truong, W.H. Cheng, T. Mohsenin, Z. Yu, A.T. Jacobson, G. Landge, M.J. Meeuwsen, C. Watnik, A.T. Tran, Z. Xiao, E.W. Work, J.W. Webb, P.V. Mejia, B.M. Baas, A 167-processor computational platform in 65 nm CMOS. IEEE J. Solid State Circuits **44**(4), 1130–1144 (2009). doi:10.1109/JSSC.2009.2013772
15. R. Airoldi, F. Garzia, J. Nurmi, Improving reconfigurable hardware energy efficiency and robustness via DVFS-scaled homogeneous MP-SoC. 2011 IEEE international symposium on parallel and distributed processing workshops and PhD forum (IPDPSW), 2011, pp. 286–289. doi:10.1109/IPDPS.2011.160
16. S.M. Martin, K. Flautner, T. Mudge, D. Blaauw, Combined dynamic voltage scaling and adaptive body biasing for lower power microprocessors under dynamic workloads. Proc. ICCAD, Nov 2002, pp. 721–725
17. A. Manuzzato, F. Campi, D. Rossi, V. Liberali, D. Pandiniin, Exploiting body biasing for leakage reduction: a case study. Proc. ISVLSI, Aug 2013
18. N.S. Kim, T. Austin, D. Baauw, T. Mudge, K. Flautner, J.S. Hu, M.J. Irwin, M. Kandemir, V. Narayanan, Leakage current: Moore's law meets static power. IEEE Trans. Comput. **36**(12), 68–75 (2003). doi:10.1109/MC.2003.1250885
19. A. Vassighi, M. Sachdev, *Thermal and Power Management of Integrated Circuits* (Springer, Berlin, 2006). ISBN 978-0-387-25762-4
20. D. Jacquet et al., A 3 GHz Dual Core Processor ARM CortexTM-A9 in 28 nm UTBB FD-SOI CMOS with ultra-wide voltage range and energy efficiency optimization. IEEE J. Solid State Circuits **49**(4), 812–826 (2014)
21. F. Arnaud, N. Planes, O. Weber, V. Barral, S. Haendler, P. Flatresse, F. Nyer, Switching energy efficiency optimization for advanced CPU thanks to UTBB technology. IEEE int. electron devices meeting (IEDM), Dig., 2012
22. R.C. Johnson, *FinFETs + FD-SOI Proposition: May Save Power*, http://www.eetimes.com/document.asp?doc_id=1327035. Accessed Feb 2016
23. T. Matsukawa et al., Lowest variability SOI FinFETs having multiple Vt by back-biasing. 2014 Symposium on VLSI technology (VLSI-technology): digest of technical papers. doi:10.1109/VLSIT.2014.6894393
24. FD-SOI technology innovations extend Moore's law. A GlobalFoundries white paper, September 2015, http://globalfoundries.com/docs/default-source/brochures-and-white-papers/globalfoundries_fd-soi_wp_sep2015.pdf. Accessed Feb 2016

25. C. Brunelli, F. Campi, J. Kylliainen, J. Nurmi, A reconfigurable FPU as IP component for SoCs. Proceedings of 2004 international symposium on system-on-chip, 2004, pp. 103–106. doi:10.1109/ISSOC.2004.1411160
26. AMBA AXI and ACE protocol specification. ARM, http://infocenter.arm.com
27. M. Coppola, M. Grammatikakis, R. Locatelli, *Design of Cost-Efficient Interconnect Processing Units: Spidergon STNoC* (CRC Press, Boca Raton, FL, 2008)
28. A.J. Martin, M. Nystrom, Asynchronous techniques for system-on-chip design. Proc. IEEE **94**(6), 1089–1120 (2006). doi:10.1109/JPROC.2006.875789
29. Keshavarzi et al., Technology scaling behavior of optimum reverse body bias for standby leakage power reduction in CMOS IC's. ISLPED, 1999, p. 252
30. N. Planes et al., 28 nm FD-SOI technology platform for high-speed low-voltage digital applications. Proc. symp. VLSI technology (VLSIT), 2012
31. Best-in-class standard-cell libraries for high-performance, low-power and high-density SoC design in 28nm FD-SOI technology. STMicroelectronics white paper, 2015, http://www.st.com/web/en/resource/sales_and_marketing/presentation/technology_presentation/Standard_Cell_White_Paper_20150928.pdf. Accessed Feb 2016
32. F. Campi, R. Canegallo, R. Guerrieri, IP-reusable 32-bit VLIW RISC core. Proceedings of the 27th European solid-state circuits conference (ESSCIRC 2001), 2001, pp. 445–448

# Chapter 2
# Formal Design Flows for Embedded IoT Hardware

**Michael Dossis**

## 2.1 Introduction

Embedded systems are usually relatively small/medium computing platforms that are self-sufficient. Such systems consist of all the software and hardware components which are "embedded" inside the system so that complete applications can be implemented and run without the aid of other external components or resources. Usually, embedded systems are found in portable computing products such as PDAs, mobile, and smart phones as well as GPS receivers. Nevertheless, larger systems such as microwave ovens and vehicle electronics contain embedded systems. Nevertheless, here embedded systems are considered that can communicate with each other by means of a wired or wireless communication protocol, such as Zigbee, IEEE.802.11 standard, or any of its derivatives. Therefore, special attention is paid here to embedded Internet-of-Things (IoT) hardware, its design methodology, and implementation requirements.

An embedded platform with such communication features can be thought of as a system that contains one or more general-purpose microprocessor or microprocessor core, a number of standard peripherals, along with a number of customized, special function co-processors, accelerators or special function engines on the same electronic board or integrated inside the same system-on-chip (SoC). Already a number of IoT manufacturers include hardware encryption and cryptography blocks in order to increase the security capability of their devices. Normally, specific blocks such as encryption/decryption and video processing engines can be attached on the embedded system's bus and offer hardware accelerated functionality to the IoT module. An embedded and portable system that includes all of the above hardware/software modules can be thought of as a complete, standalone IoT node.

M. Dossis (✉)
TEI of Western Macedonia, Kastoria Campus, Fourka Area, Kastoria, 52100 Greece
e-mail: dossis@kastoria.teikoz.gr



G. Keramidas et al. (eds.), *Components and Services for IoT Platforms*,
DOI 10.1007/978-3-319-42304-3_2

Currently, such embedded systems are implemented using advanced field-programmable gate arrays (FPGAs) or other types of programmable logic devices (PLDs). Alternatively an embedded IoT node can be implemented with ASIC + SoC logic plus a microcontroller core, but this is financially viable only for large sale volumes. For smaller volumes, or at least for designing the prototype, FPGA logic is the best solution in terms of price/functionality yield. Nowadays, FPGAs offer a very large integrated area, circuit performance, and low power capability. FPGA implementations can be seamlessly and rapidly prototyped, and the FPGA can be easily reconfigured when design updates or bug fixes are released.

Advances on chip integration technology have made possible such a complexity of embedded and custom integrated circuits (ICs) that often their spec-to-product time exceeds even their product lifetime in the market. This is particularly true for commercial embedded electronics with product lifetimes compressed to less than a quarter of a year sometimes. Of course this is expected to be true for current and future IoT devices as well. This, in combination with the high design cost and development effort of such products, they often even miss their market windows, generating in turn, competitive disadvantages for the producing industries. The current engineering practice for the development of such systems includes to a large extent methodologies which are semi-manual, add hoc, segmented, not-fully integrated, empirical, incompatible from one level of the design flow to the next, and with a lot of design iterations caused by the discovery of functional and timing bugs, as well as specification to implementation mismatches late in the development flow.

This chapter reviews previous and existing work of HLS methodologies for embedded systems. It also discusses the usability and benefits using the prototype hardware compilation system which was developed by the author. Section 2.2 discusses related work and bibliography. Section 2.3 presents HLS problems related to the low energy consumption which is particularly interesting for embedded system design. The C-Cubed hardware compilation design flow is explained in Sect. 2.4. Section 2.5 explains the formal nature of the prototype compiler's formal logic inference rules. In Sect. 2.6 the mechanism of the formal high-level synthesis transformations of the back-end compiler is presented. Section 2.7 outlines the structure and logic of the PARCS optimizing scheduler which is part of the back-end compiler rules. Section 2.8 explains the options for generating target FSM + datapath micro-architectures and the communication of the accelerators with their computing environment. Section 2.9 outlines the execution environment for the generated hardware modules as accelerators. Section 2.10 discusses experiments with synthesizing hardware within the C-Cubed framework such as formal synthesis and verification, and Sect. 2.11 draws useful conclusions and proposes future work.

## 2.2 Background and Existing Work

### 2.2.1 *High-Level and Logic Synthesis*

All of the issues reported in the introduction and elsewhere have motivated industry and academia to invest in automated, integrated, and formal design automation methodologies and tools for the design and development of embedded ICs and systems. These methods are based on the idea of transforming a textual software-program-code-like description into a netlist of logic gates. Nowadays, a higher level of code abstraction of hardware description formats, such as VHDL and Verilog, C, SystemC, and ADA, is pursued as input to automated high-level E-CAD tools. Methodologies such as high-level synthesis (HLS) and electronic system level (ESL) design employ established techniques, borrowed from the computer language program compilers and established E-CAD tools as well as new algorithms such as advanced operation scheduling, loop unrolling, and code motion heuristics.

Currently, the design of digital systems involves programming of the circuit's functionality at the register-transfer level (RTL) level in hardware description languages such as VHDL and Verilog. However, for custom designs that are larger than a hundred thousand logic gates, the use of RTL code for specification and design usually results into years of design flow iterations and verification simulations. Combined with the short lifetime of electronic products in the market, this constitutes a great problem for the industry. Therefore, there has been a pressing need for more abstracted input formats such as C, C++, and ADA. However, the programming style of the (hardware/software) specification code has an unavoidable impact on the quality of the synthesized system. This is deteriorated by models with hierarchical blocks, subprogram calls, as well as nested control constructs (e.g., if-then-else and while loops). The complexity of the transformations that are required for the synthesis tasks (compilation, algorithmic transformations, scheduling, allocation and binding) of such high-level code models increases at an exponential rate, for a linear increase in the design size.

During HLS, the input code (such as ANSI-C or ADA) is first transformed into a control/dataflow graph (CDFG) by a front-end compilation stage. Then various optimizing synthesis transformations are applied on the CDFG to generate the final implementation. The most important HLS tasks of this process are scheduling, allocation and binding. Scheduling makes an as-much-as-possible optimal order of the operations in a number of control steps or states, parallelizing as many operations as possible, so as to achieve shorter execution times of the generated implementation. Allocation and binding assign operations onto functional units, and variables and data structures onto registers, wires, or memory positions, available from an implementation library.

A number of commercial HLS tools impose their own extensions or restrictions on the programming language code that they accept as input, as well as various shortcuts and heuristics on the HLS tasks that they execute. Such tools are the CatapultC by Mentor Graphics, the Cynthesizer by Forte Design Systems, the

Impulse CoDeveloper by Impulse Accelerated Technologies, the Synfony HLS by Synopsys, the C-to-silicon by Cadence, the C to Verilog Compiler by C-to-Verilog, the AutoPilot by AutoESL, the PICO by Synfora, and the CyberWorkBench by NEC System Technologies Ltd. The analysis of these tools is not the purpose of this work; however, they are tuned towards linear, dataflow dominated (e.g., stream-based) applications, such as pipelined digital signal processing (DSP) and image filtering.

In order to guarantee that the produced circuit implementations are correct-by-construction (with regard to the input specification functionality) it is mandated that the HLS tool's transformation tasks (e.g., within the scheduler) are based on formal techniques. In this way, repetitive execution of the design flow verification process will be avoided and important time will be saved within the development project. Correct-by-construction also means that by definition of the formal process, the functionality of the implementation matches the functionality of the behavioral specification model (the source code). In this way, the design will need to be verified only at the behavioral level, without spending very long and time-consuming simulations of the generated RTL, or even worse of the netlists generated by a subsequent commercial RTL synthesizer.

Behavioral verification (at the source code level) is orders of magnitude faster than RTL or even more than gate-netlist simulations. Releasing an embedded IoT product with bugs can be very expensive, when considering the cost of field upgrades, recalls, and repairs. Another thing, less measurable, but very important as well, is the damage done to the industry's reputation and the consequent loss of customer trust. However, many embedded products are indeed released without all the testing that is necessary and/or desirable. Therefore, the quality of the specification code and the formal techniques employed during transformations ("compilations") in order to deliver the hardware and software components of the system are receiving increasing focus in IoT chip application development.

### 2.2.2 HLS Scheduling

The HLS scheduling task belongs into two major categories: time-constrained scheduling and resource-constrained scheduling. Time-constrained scheduling aims to result into the lowest area or number of functional units, when the task is constrained by the max number of control steps (time constraint). Resource-constrained scheduling aims to produce the fastest schedule (the minimum number of control states) when the maximum number of hardware resources or hardware area is constrained (resource constraint). Integer linear programming (ILP) solutions have been proposed, however, their run time grows exponentially with the increase of design size, which makes them impractical. Heuristic methods have also been proposed to handle large designs and to provide sub-optimal but practical implementations. There are two heuristic scheduling approaches: constructive solutions and iterative refinement. As-soon-as-possible (ASAP) and the as-late-as-possible (ALAP) scheduling both belong to the constructive approaches.

Operations that belong to the critical path of the design are not given any special priority over other operations in both ASAP and ALAP algorithms. Thus, excessive delay may result on the critical path, resulting into bad quality of the produced circuit implementation. LIST scheduling utilizes a global priority function to select the next operation to be scheduled. This global priority function can be either the mobility of the operation [1], or its urgency [2]. Force-directed scheduling [3] calculates the range of control steps for each operation between the operation's ASAP and ALAP state assignment. The algorithm then attempts to reduce the total number of functional units, so as to evenly distribute the operations of the same type into all of the available states of the range, producing better implementations against the rest of the heuristics, with an increased run time, however.

Constructive scheduling doesn't do any lookahead into future assignment of operations into the same control step, and this may lead to sub-optimal implementations. After an initial schedule is done by any of the above scheduling algorithms, then iteratively re-scheduling sequences of operations can maximally reduce the cost functions [4]. This is suitable for dataflow-oriented designs with linear control. For control-intensive designs, the use of loop pipelining [5] and loop folding [6] have been reported as scheduling tasks in the bibliography.

### 2.2.3 *Allocation and Binding Tasks*

Allocation determines the type of resource storage and functional units, selected from the library of components, to be assigned for each data object and operation of the input program. Allocation also calculates the number of resources of each type that are needed to implement every operation or data variable. Binding assigns operations, data variables, data structures, and data transfers onto functional units, storage elements (registers or memory blocks), and interconnections, respectively. Also binding makes sure that the design's functionality does not change by using the selected library components.

There are three categories of solutions to the allocation problem: constructive techniques, decomposition techniques, and iterative approaches. Constructive allocation techniques start with an empty implementation and progressively build the datapath and control parts of the implementation by adding more functional, storage, and interconnection elements while they traverse the CDFG or any other internal graph/representation format. Decomposition techniques divide the allocation problem into a sequence of well-defined independent sub-tasks. Each such subtask is a graph-based theoretical problem which is solved with any of the three well-known graph methods: clique partitioning, the left-edge technique, and the weighted bipartite-matching technique. The task of finding the minimum cliques in the graph, which is the solution for the sub-tasks, is an NP-hard problem, so heuristic approaches [7] are utilized for allocation.

Because the conventional sub-task of storage allocation ignores the side-effects between the storage and interconnections allocation, when using the clique partitioning technique, graph edges are enhanced with weights that represent the effect on interconnection complexity. The left-edge algorithm is applied on the storage allocation problem, and it allocates the minimum number of registers [8]. A weighted, bipartite-matching algorithm is used to solve both the storage and functional unit allocation problems. First a bipartite graph is generated which contains two disjoint sets, e.g., one for variables and one for registers, or one for operations and one for functional units. An edge between one node of the one of the sets and one node of the other represents an allocation of, e.g., a variable to a register. The bipartite-matching algorithm considers the effect of register allocation on the design's interconnection elements, since the edges of the two sets of the graph are weighted [9]. The generated datapaths are improved iteratively, a simple assignment exchange, but using the pairwise exchange of the simulated annealing, or by using a branch-and-bound approach. The latter reallocates groups of elements of different types [10].

### 2.2.4 *History of High-Level Synthesis Tools*

HLS has been an active research field for about three decades. Early approaches of experimental synthesis tools that synthesized small subsets of programming constructs or proprietary modeling formats have emerged since the late 1980s. As an example, an early tool that generated hardware structures from algorithmic code, written in the PASCAL-like, digital system specification language (DSL) is reported in [11]. This synthesis tool performs the circuit compilation in two steps: first step is datapath synthesis which is followed by control synthesis. Examples of other behavioral circuit specification languages of that time, apart from DSL, were DAISY [12], ISPS [13], and MIMOLA [14].

In [15] the circuit to be synthesized is described with a combination of algorithmic and structural level code and then the PARSIFAL tool synthesizes the code into a bit-serial DSP circuit implementation. The PARSIFAL tool is part of a larger E-CAD system called FACE and which included the FACE design representation and design manager core. FACE and PARSIFAL were suitable for DSP pipelined implementations, rather than for a more general behavioral hardware models with hierarchy and complex control.

According to [16] scheduling first determines the propagation delay of each operation and then it assigns all operations into control steps (states) of a finite state machine. List scheduling uses a local priority function to postpone the assignment of operations into states, when resource constraints are violated. On the contrary, force-directed scheduling (FDS) tries to satisfy a global execution deadline (time constraint) as it minimizes the utilized hardware resources (functional units, registers, and busses). The force-directed list scheduling (FDLS) algorithm attempts to implement the fastest schedule while satisfying fixed hardware resource constraints.

The main HLS tasks in [17] include allocation, scheduling, and binding. According to [18] scheduling is finding the sequence of operations to execute in a specific order so as to produce a schedule of control steps with allocated operations in each step of the schedule; allocation defines the required number of functional, storage, and interconnect units; binding assigns operations to functional units, variables, and values to storage elements and the interconnections amongst them to form a complete working circuit that executes the functionality of the source behavioral model.

The V compiler [19] translates sequential descriptions into RTL models using parsing, scheduling, and resource allocation. The source sequential descriptions are written in the V language which includes queues, asynchronous calls, and cycle blocks and it is tuned to a kind of parallel hardware RTL implementations. The V compiler utilizes percolation scheduling [20] in order to achieve the required degree of parallelism by meeting time constraints.

A timing network is generated from the behavioral design in [21] and is annotated with parameters for every different scheduling approach. The scheduling approach in this work attempts to satisfy a given design cycle for a given set of resource constraints, using the timing model parameters. This approach uses an integer linear program (ILP) which minimizes a weighted sum of area and execution time of the implementation. According to the authors, their Symphony tool delivers better area and speed than ADPS [22]. This synthesis technique is suitable for dataflow designs (e.g., DSP blocks) and not for more general complex control flow designs.

The CALLAS synthesis framework [23] transforms algorithmic, behavioral VHDL models into VHDL RTL and gate netlists, under timing constraints. The generated circuit is implemented using a Moore-type finite state machine (FSM), which is consistent with the semantics of the VHDL subset used for the specification code. Formal verification techniques such as equivalence checking, which checks the equivalence between the original VHDL FSM and the synthesized FSM are used in the CALLAS framework by using the symbolic verifier of the circuit verification environment (CVE) system [24].

A methodological approach of designing and developing mixed hardware and software parts of a system known as hardware–software co-design has emerged about the same time as early HLS tools in the 1990s. The most known examples of this technology are reported in the following approaches.

The Ptolemy framework [25] allows for an integrated hardware–software co-design methodology from the specification through to synthesis of hardware and software components, simulation, and evaluation of the implementation. The tools of Ptolemy can synthesize assembly code for a programmable DSP core (e.g., DSP processor), which is built for a synthesis-oriented application. In Ptolemy, an initial model of the entire system is partitioned into the software and hardware parts which are synthesized in combination with their interface synthesis.

The COSYMA hardware–software co-synthesis framework [26] realizes an iterative partitioning process, based on hardware extraction algorithm which is driven by a cost function. The primary target in this work is to minimize customized hardware within microcontrollers but the same time to allow for space exploration of large designs. The specialized co-processors of the embedded system can be synthesized using HLS tools. The specification language is based on C with various extensions. The generated hardware descriptions are in turn ported to the Olympus HLS tool [27]. The presented work included tests and experimental results based on a configuration of an embedded system, which is built around the Sparc microprocessor.

Co-synthesis and hardware–software partitioning are executed in combination with control parallelism transformations in [28]. The hardware–software partition is defined by a set of application-level functions which are implemented with application-specific hardware. The control parallelism is defined by the interaction of the processes of the functional behavior of the specified system. The system behavior is modeled using a set of communicating sequential processes [29]. Each process is then assigned either to hardware or to software implementation.

A hardware–software co-design methodology, which employs synthesis of heterogeneous systems, is presented in [30]. The synthesis process is driven by timing constraints which drive the mapping of tasks onto hardware or software parts so that the performance requirements of the intended system are met. This method is based on using modeling and synthesis of programs written in the HardwareC language. An example application which was used to test the methodology in this work was an Ethernet-based network co-processor.

### *2.2.5 Next Generation High-Level Synthesis Tools*

More advanced methodologies appeared at the late 1990s and they featured enhanced input code sets as well as improved scheduling and other optimization algorithms. The CoWare hardware–software co-design environment [31] is based on a data model that allows the user to specify, simulate, and produce heterogeneous implementations from heterogeneous specification source models. This design approach develops telecommunication systems that contain DSP, control loops, and user interfaces. The synchronous dataflow (SDF) type of algorithms found in a category of DSP applications are synthesized into hardware from languages such as SILAGE [32], DFL [33], and LUSTRE [34]. In contrast to this, DDF algorithms consume and produce tokens that are data-dependent, and thus they allow for complex if-then-else and while loop control constructs. CAD systems that allow for specifying both SDF and DDF algorithms and run scheduling are the DSP-station from Mentor Graphics [35], PTOLEMY [36], GRAPE-II [37], COSSAP from Synopsys, and SPW from the Alta group [38].

C programs that include dynamic memory allocation, pointers, and the functions malloc and free are mapped onto hardware in [39]. The SpC tool which was developed in this work resolves pointer variables at compile time and thus C functional models are synthesized into Verilog hardware models. The synthesis of functions in C, and therefore the resolution of pointers and malloc/free inside of functions, is, however, not included yet in this work, but left as future work.

The different techniques and optimizations described above have been implemented using the SUIF compiler environment [40].

A heuristic for scheduling behavioral specifications that include a lot of conditional control flow is presented in [41]. This heuristic is based on a powerful intermediate design representation called hierarchical conditional dependency graph (HCDG). HCDG allows chaining and multicycling, and it enables advanced techniques such as conditional resource sharing and speculative execution, which are suitable for scheduling conditional behaviors. The HLS techniques in this work were implemented in a prototype graphical interactive tool called CODESIS which used HCDG as its internal design representation. The tool generates VHDL or C code from the HCDG, but no translation of standard programming language code into HCDG is known so far.

A coordinated set of coarse-grain and fine-grain parallelizing HLS transformations on the input design model are discussed in [42]. These transformations are executed in order to deliver synthesis results that don't suffer from the negative effects of complex control constructs in the specification code. All of the HLS techniques in this work were implemented in the SPARK HLS tool, which transforms specifications in a small subset of C into RTL VHDL hardware models. SPARK utilizes both control/dataflow graphs (CDFGs) and an encapsulation of basic design blocks inside hierarchical task graphs (HTGs), which enable coarse-grain code restructuring such as loop transformations and an efficient way to move operations across large pieces of code. Nevertheless, SPARK is not designed to process conditional code where the iteration limits are not known at compile time, such as while loops.

Typical HLS tasks such as scheduling, resource allocation, module binding, module selection, register binding, and clock selection are executed simultaneously in [43] so as to achieve better optimization in design energy, power, and area. The scheduling algorithm utilized in this HLS methodology applies concurrent loop optimization and multicycling and it is driven by resource constraints. The state transition graph (STG) of the design is simulated in order to generate switched capacitance matrices. These matrices are then used to estimate power/energy consumption of the design's datapath. Nevertheless, the input to the HLS tool is not programming language code but a complex proprietary format representing an enhanced CDFG as well as an RTL design library and resource constraints.

An incremental floorplanner is described in [44] which combines an incremental behavioral and physical optimization into HLS. These techniques were integrated into an existing interconnect-aware HLS tool called ISCALP [45]. The new combination was named IFP-HLS (incremental floorplanner high-level synthesis) tool, and it attempts to concurrently improve the design's schedule, resource binding, and floorplan, by integrating high-level and physical design algorithms.

Huang et al. [46] discuss a HLS methodology which is suitable for the design of distributed logic and memory architectures. Beginning with a behavioral description of the system in C, the methodology starts with behavioral profiling in order to extract simulation statistics of computations and references of array data. Then array data are distributed into different partitions. An industrial tool called Cyber [47] was developed which generates a distributed logic/memory micro-architecture RTL model, which is synthesizable with existing RTL synthesizers, and which consists of two or more partitions, depending on the clustering of operations that was applied earlier.

A system specification containing communicating processes is synthesized in [48]. The impact of the operation scheduling is considered globally in the system critical path (as opposed to the individual process critical path), in this work. It is argued by the authors in this work that this methodology allocates the resources where they are mostly needed in the system, which is in the critical paths, and in this way it improves the overall multi-process designed system performance.

The work in [49] contributes towards incorporating memory access management within a HLS design flow. It mainly targets DSP applications but also other streaming applications can be included along with specific performance constraints. The synthesis process is performed on the extended dataflow graph (EDFG) which is based on the signal flow graph. Mutually exclusive scheduling methods [50, 51] are implemented with the EDFG. The graph which is processed by a number of annotations and improvements is then given to the GAUT HLS tool [52] to perform operator selection and allocation, scheduling and binding.

A combined execution of operation decomposition and pattern-matching techniques is targeted to reduce the total circuit area in [53]. The datapath area is reduced by decomposing multicycle operations, so that they are executed on monocycle functional units (FUs that take one clock cycle to execute and deliver their results). A simple formal model that relies on an FSM-based formalism for describing and synthesizing on-chip communication protocols and protocol converters between different bus-based protocols is discussed in [54]. The utilized FSM-based format is at an abstraction level which is low enough so that it can be automatically translated into HDL implementations. The generated HDL models are synthesizable with commercial tools. Synchronous FSMs with bounded counters that communicate via channels are used to model communication protocols. The model devised in this work is validated with an example of communication protocol pairs which included AMBA APB and ASB. These protocols are checked regarding their compatibility, by using the formal model.

The methodology of SystemCoDesigner [55] uses an actor-oriented approach so as to integrate HLS into electronic system level (ESL) design space exploration tools. The design starts with an executable SystemC system model. Then, commercial synthesizers such as Forte's Cynthesizer are used in order to generate hardware implementations of actors from the behavioral model. This aids the design space exploration in finding the best candidate architectures (mixtures of hardware and software modules). After deciding on the chosen solution, the suitable target platform is then synthesized with the implementations of the hardware and software parts. The final step of this methodology is to generate the FPGA-based SoC implementation from the chosen hardware/software solution. Based on the proposed methodology, it seems that SystemCoDesigner method is suitable for stream-based applications, found in areas such as DSP, image filtering, and communications.

A formal approach is followed in [56] which is used to prove that every HLS translation of a source code model produces an RTL model that is functionally equivalent to the one in the behavioral input to the HLS tools. This technique is called translation validation and it has been maturing via its use in the optimizing software compilers. The validating system in this work is called SURYA, it is using the Symplify theorem prover and it was used to validate the SPARK HLS tool. This validation experiment with Symplify discovered two bugs in the SPARK compilations.

The replacement of flip-flop registers with latches is proposed in [57] in order to yield better timing in the implemented designs. The justification for this is that latches are inherently more tolerant to process variations than flip-flops. The related design techniques were integrated into a tool called HLS-1. HLS-1 translates behavioral VHDL code into a synthesized netlist. Nevertheless, handling a design where registers are implemented with latches instead of edge-triggered flip-flops is generally considered to be cumbersome due to the complicated timing behavior of latches.

## 2.3 Synthesis for Low Power

Many portable and embedded computing systems and applications such as mobile (smart) phones, PDAs, etc., require design for low power and therefore synthesis for low energy is becoming very important in the whole area of VLSI and embedded system design. In any case, and particularly for IoT devices the majority of custom functions that are implemented in special-purpose hardware imply that they consume much less power than microcontroller equivalent programs. Therefore it is mandatory that most of the special functions in IoT devices found in areas of data compression, security, media playing, image/audio (de-)coding must be implemented using advanced synthesis techniques into specialized hardware for low power consumption and increased security gains.

During the last decade industry and academia invested on significant amounts of research regarding VLSI techniques and HLS for low power design. In order to achieve low energy in the results of HLS and system design, new techniques that help to estimate power consumption at the high-level description level are needed, and they will be a great aid in delivering systems with reduced power consumption such as IoT devices running on batteries. In [58], switching activity and power consumption are estimated at the RTL level taking also into account the glitching activity on a number of signals of the datapath and the controller. The spatial locality, the regularity, the operation count, and the ratio of critical path to available time are identified in [59] with the aim to reduce the power consumption of the interconnections. The HLS scheduling, allocation, and binding tasks consider such algorithmic statistics and properties in order to reduce the fanins and fanouts of the interconnect wires. This will result into reducing the complexity and the power consumed on the capacitance of the interconnection buses [60].

The effect of the controller on the power consumption of the datapath is considered in [61]. Pipelining and module selection was proposed in [62] for low power consumption. The activity of the functional units was reduced in [63] by minimizing the transitions of the functional unit's inputs. This was utilized in a scheduling and resource binding algorithm, in order to reduce power consumption. In [64] the DFG is simulated with profiling stimuli, provided by the user, in order to measure the activity of operations and data carriers. Then, the switching activity is reduced by selecting a special module set and schedule. Reducing supply voltage, disabling the clock of idle elements, and architectural tradeoffs were utilized in [65] in order to minimize power consumption within HLS.

The energy consumption of memory subsystem and the communication lines within a multiprocessor system-on-a-chip (MPSoC) is addressed in [66]. This work targets streaming applications such as image and video processing that have regular memory access patterns. The way to realize optimal solutions for MPSoCs is to execute the memory architecture definition and the connectivity synthesis in the same step.

The above research approaches all attempt to target energy efficiency at the device consumption level and improve architectural and device level techniques for reducing it. It is absolutely necessary to transition from traditional RTL design techniques into HLS-based development so that to automate the design and deliver higher quality of implementations that are suitable for IoT embedded and battery-running devices. Also, by using formal and integrated design and verification techniques a great deal of project time is saved so that necessary focus can be shifted from repetitive verification cycles into advanced (micro-)architectural issues of the system, for improved performance and power consumption.

## 2.4 The C-Cubed Hardware Synthesis Flow

So far in this chapter related work in HLS methodologies for embedded IoT systems was reviewed. From this section onwards, a particular, formal HLS methodology is analyzed and explained, which is directly applicable on embedded system design, and it has been developed by the author of this chapter. The techniques of this work include the front-end compilers which are based on formal compiler generators, the Formal Intermediate Format (FIF) which encapsulates in a formal way the attributes of the input algorithms, and the back-end compiler which is built with formal logic programming relations (or facts in the Prolog language terminology).

The FIF[1] was invented and designed by the author of this chapter as a tool and media for the design encapsulation and the HLS transformations in the C-Cubed

[1]The Formal Intermediate Format is patented with patent number: 1006354, 15/4/2009, from the Greek Industrial Property Organization.

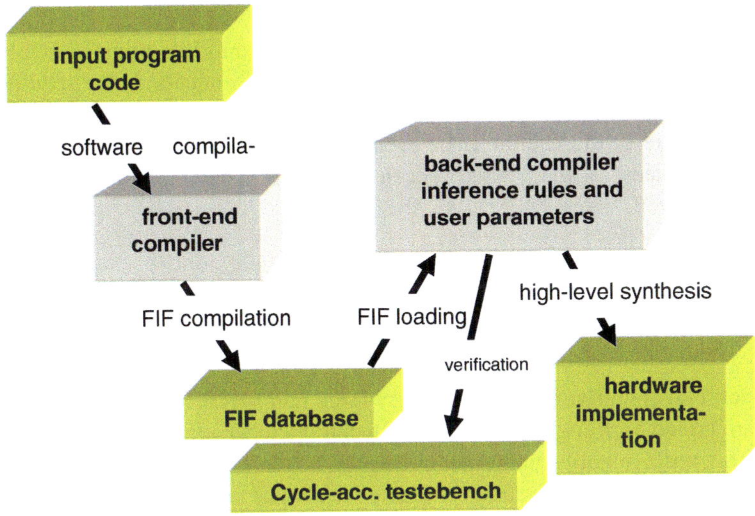


**Fig. 2.1** The C-Cubed hardware synthesis flow and tools

(Custom Co-processor Compilation) hardware compilation tool.[2] A near-complete analysis of FIF syntax and semantics can be found in [67]. The formal methodology discussed here is based on using predicate logic to describe the intermediate representations and transformations of the compilation steps, and the resolution of a set of transformation Horn clauses [68] is used, as the building blocks of the prototype hardware compiler.

The front-end compiler translates the input program code into the FIF's logic statements (logic facts). The inference logic rules of the back-end compiler transform the FIF facts into the hardware implementations. There is one-to-one correspondence between the source specification's subroutines and the generated hardware modules. The source code subroutines can be hierarchical, and this hierarchy is maintained in the generated hardware implementation. Each generated hardware model is an FSM-controlled custom processor (or co-processor, or accelerator), that executes a specific task, described in the source program code. This hardware synthesis flow is depicted in Fig. 2.1. The generated hardware modules specify a standalone function (e.g., DSP filter) and they are coded in the VHDL or the Verilog HDL languages. For verification purposes, and for every hardware module, a fast cycle-accurate testbench, coded in ANSI-C, is generated from the same internal formal model of the FSM and the datapath of the co-processor.

Essentially the front-end compilation resembles software program compilation and the back-end compilation executes formal transformation tasks that are

[2]This hardware compiler method is patented with patent number: 1005308, 5/10/2006, from the Greek Industrial Property Organization.

normally found in HLS tools. This whole compilation flow is a formal transformation process, which converts the source code programs into implementable RTL VHDL hardware accelerator models. If there are function calls in the specification code, then each subprogram call is transformed into an interface event in the generated hardware FSM. The interface event is used so that the "calling" accelerator uses the "services" of the "called" accelerator, as it is depicted in the source code hierarchy as well.

## 2.5 Back-End Compiler Inference Logic Rules

The back-end compiler consists of a very large number of logic rules. These logic rules are coded with logic programming techniques, which are used to implement the HLS transformations and other processes of the back-end compilation phase. As an example, one of the latter processes reads and incorporates the FIF tables' facts into the compiler's internal inference engine of logic predicates and rules [68]. The back-end compiler rules are given as a great number of definite clauses of the following equation:

$$A_0 \leftarrow A_1 \wedge \cdots \wedge \ A_n \text{ (where } n \ \geq 0) \tag{2.1}$$

where $\leftarrow$ is the logical implication symbol ($A \leftarrow B$ means that if $B$ applies then $A$ applies), and $A_0, \ \ldots, A_\mathrm{n}$ are atomic formulas (logic facts) of the equation:

$$\text{predicate\_symbol (Var\_1, Var\_2,} \ldots, \text{ Var\_N)} \tag{2.2}$$

where the positional parameters Var_1,...,Var_N of the above predicate "predicate_symbol" are either variable names (in the case of the back-end compiler inference rules), or constants (in the case of the FIF table statements). The predicate syntax in Eq. (2.2) is typical of the way of the FIF facts and other facts interact with each other, they are organized and they are used internally in the inference engine. Thus, the hardware descriptions are generated as "conclusions" of the inference engine upon the FIF "facts." This is done in a formal way from the input programs by the back-end phase, which turns the overall transformation into a provably correct compilation process. In essence, the FIF file consists of a number of such atomic formulas, which are grouped in the FIF tables. Each such table contains a list of homogeneous facts which describe a certain aspect of the compiled program. For example, all prog_stmt facts for a given subprogram are grouped together in the listing of the program statements table.

In the back-end compiler "conclussions" the correct-by-construction hardware implementations and the cycle-accurate testbench are included, as custom hardware micro-architectures. This along with other features makes the C-Cubed tool very suitable for designing custom hardware blocks of IoT embedded devices and peripherals. Experiments with the tool have shown that it is very suitable for small and big data coding, secure functions and cryptography, DSP and computer graphics, as well as other mathematical blocks, all of which are used in today's embedded systems.

## 2.6 Inference Logic and Back-End Transformations

The inference engine of the back-end compiler consists of a great number of logic rules (like the one in Eq. (2.1)) which conclude on a number of input logic predicate facts and produce another set of logic facts and so on. Eventually, the inference logic rules produce the logic predicates that encapsulate the writing of RTL VHDL hardware co-processor models. These hardware models are directly implementable to any hardware (e.g., ASIC or FPGA) technology, since they are technology and platform-independent. For example, generated RTL models produced in this way from the prototype compiler were synthesized successfully into hardware implementations using the Synopsys DC Ultra, the Xilinx ISE, and the Mentor Graphics Precision software without the need of any manual alterations of the produced RTL VHDL code. In the following Eq. (2.3) an example of such an inference rule is shown:

$$\begin{aligned} &\text{dont\_schedule}\,(\text{Operation1},\ \text{Operation2}) \leftarrow \\ &\quad \text{examine}\,(\text{Operation1},\ \text{Operation2})\,, \\ &\quad \text{predecessor}\,(\text{Operation1},\ \text{Operation2})\,. \end{aligned} \tag{2.3}$$

The meaning of this rule that combines two input logic predicate facts to produce another logic relation (dont_schedule) is that when two operations (Operation1 and Operation2) are examined and the first is a predecessor of the second (in terms of data and control dependencies), then don't schedule them in the same control step. This rule is part of a parallelizing optimizer which is called "PARCS" (meaning: Parallel, Abstract Resource—Constrained Scheduler).

The way that the inference engine rules (predicates relations–productions) work is depicted in Fig. 2.2. The last produced (from its rule) predicate fact is the VHDL RTL writing predicate at the top of the diagram. Right below level 0 of predicate production rule there is a rule at the −1 level, then level −2, and so on. The first predicates that are fed into this engine of production rules belong to level –K, as shown in this figure. Level –K predicate facts include of course the FIF facts that are loaded into the inference engine along with the other predicates of this level.

In this way, the back-end compiler works with inference logic on the basis of predicate relation rules and, therefore, this process is a formal transformation of the FIF source program definitions into the hardware accelerator (implementable) models. Of course in the case of the prototype compiler, there is a very large number of predicates and their relation rules that are defined inside the implementation code of the back-end compiler, but the whole concept of implementing this phase is as shown in Fig. 2.2. The user of the back-end compiler can select certain environment

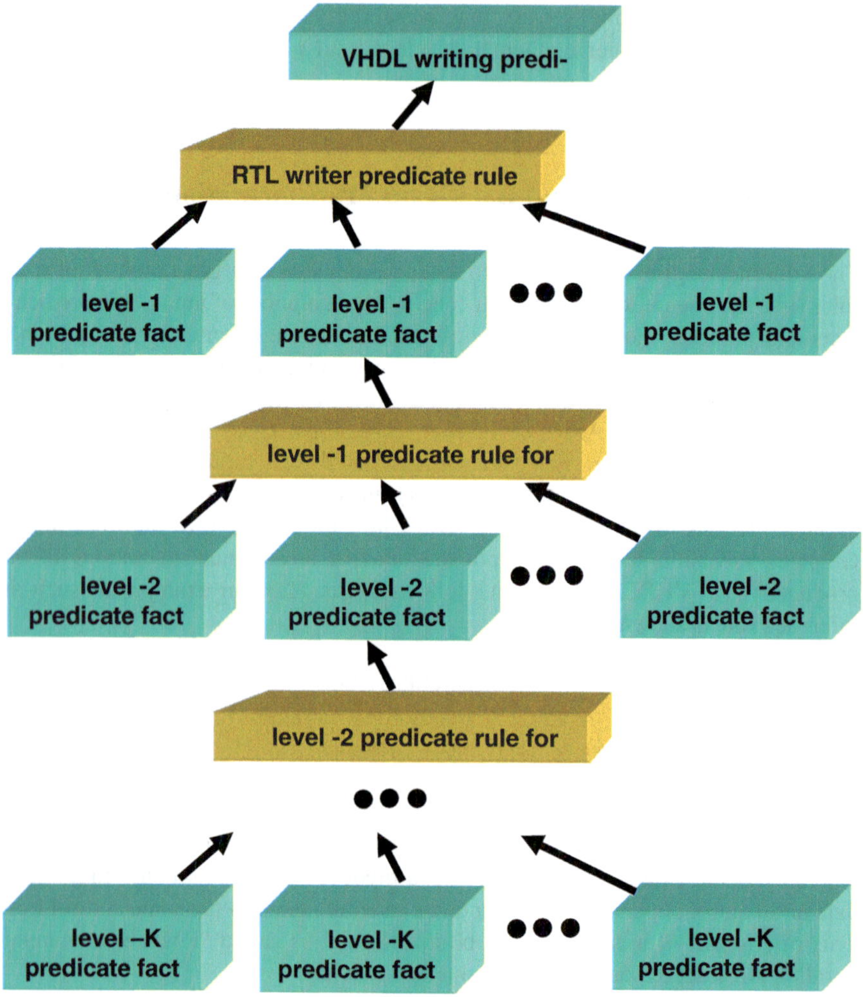


**Fig. 2.2** The back-end inference logic rules structure

command list options as well as build an external memory port parameter file as well as drive the compiler's optimizer with specific resource constraints of the available hardware operators.

The most important of the back-end compilation stages can be seen in Fig. 2.3. The compilation process starts with the loading of the FIF facts into the inference rule engine. After the FIF database is analyzed, the local data object, operation, and initial state lists are built. Then the environment options are read and the temporary lists are updated with the special (communication) operations as well as the predecessor and successor dependency relation lists. After the complete initial schedule is built and concluded, the PARCS optimizer is run on it, and the optimized

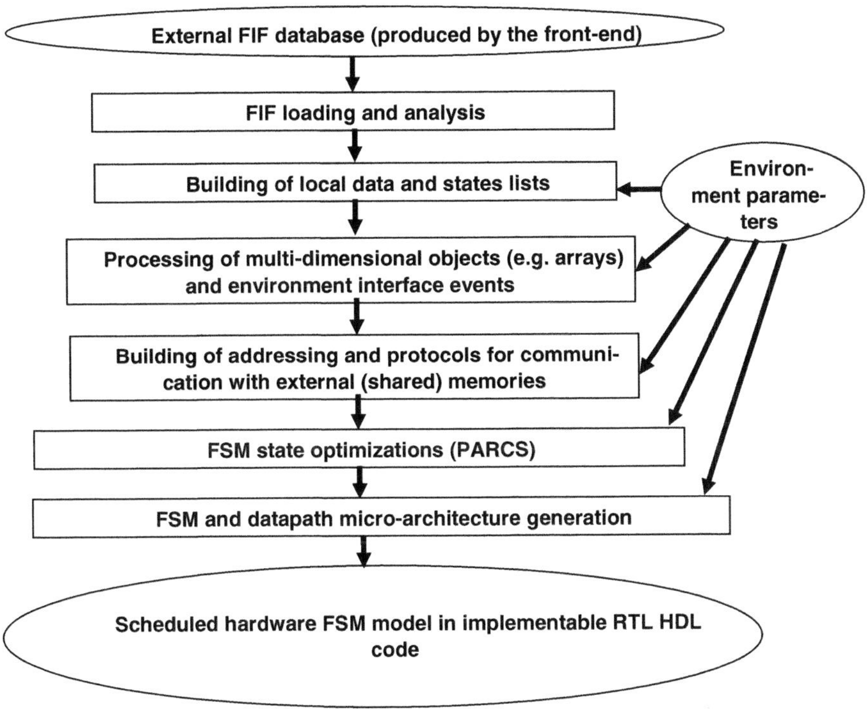


**Fig. 2.3** The processing stages of the back-end compiler

schedule is delivered to the micro-architecture generator. The transformation is concluded with the formation of the FSM and datapath implementation and the writing of the RTL VHDL model for each accelerator that is defined in each subprogram of the source code program.

A separate hardware accelerator model is generated from each subprogram in the system model code. All of the generated hardware models are directly implementable into hardware using commercial CAD tools, such as the Synopsys DC-ultra, the Xilinx ISE, and the Mentor Graphics Precision RTL synthesizers. Also the hierarchy of the source program modules (subprograms) is maintained and the generated accelerators may be hierarchical. This means that an accelerator can invoke the services of another accelerator from within its processing states, and that other accelerator may use the services of yet another accelerator, and so on. In this way, a subprogram call in the source code is translated into an external co-processor interface event of the corresponding hardware accelerator.

## 2.7 The PARCS Optimizer

The PARCS scheduler aggressively attempts to schedule as many as possible operations in the same control step. The only limits to this are the data and control dependencies as well as the optional resource (operator) constraints, which are provided by the user.

The pseudo-code for the main procedures of the PARCS scheduler is shown in Fig. 2.4. All of the predicate rules (like the one in Eq. (2.1)) of PARCS are part of the inference engine of the back-end compiler. A new design to be synthesized is loaded via its FIF into the back-end compiler's inference engine. Hence, the FIF's facts as well as the newly created predicate facts from the so far logic processing "drive" the logic rules of the back-end compiler which generate provably correct hardware architectures. It is worthy to note that although the HLS transformations are implemented with logic predicate rules, the PARCS optimizer is very efficient and fast. In most of benchmark cases that were run through the prototype hardware compiler flow, compilation did not exceed 1–10 min of run time and the results of the compilation were very efficient as explained below.

## 2.8 Generated Hardware Architectures

The back-end stage of micro-architecture generation can be driven by command-line options. One of the options, e.g., is to generate massively parallel architectures. The results of this option are shown in Fig. 2.5. This option generates a single process—FSM VHDL or Verilog description with all the data operations being dependent on

**1.** start with the initial schedule (including the special external port operations)
**2.** Current PARCS state <- 1
**3.** Get the 1st state and make it the current state
**4.** Get the next state
**5.** Examine the next state's operations to find out if there are any dependencies with the current state
**6.** If there are no dependencies then absorb the next state's operations into the current PARCS state; If there are dependencies then finalize the so far absorbed operations into the current PARCS state, store the current PARCS state, PARCS state <- PARCS state + 1; make next state the current state; store the new state's operations into the current PARCS state
**7.** If next state is of conditional type (it is enabled by guarding conditions) then call the conditional (true/false branch) processing predicates, else continue
**8.** If there are more states to process then go to step 4, otherwise finalize the so far operations of the current PARCS state and terminate

**Fig. 2.4** Pseudo-code of the PARCS scheduling algorithm

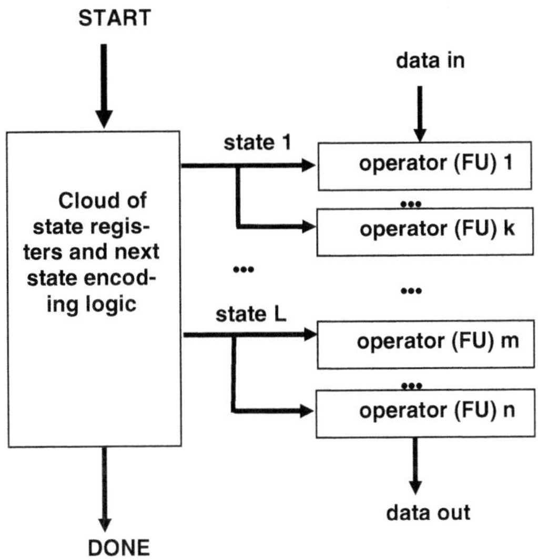


**Fig. 2.5** Massively parallel micro-architecture generation option

different machine states. This implies that every operator is enabled by single wire activation commands that are driven by different state register values. This in turn means that there is a redundancy in the generated hardware, in a way that during part of execution time, a number of state-dedicated operators remain idle. However, this redundancy is balanced by the fact that this option achieves the fastest clock cycle, since the state command encoder, as well as the data multiplexers are replaced by single wire commands which don't exhibit any additional delay, and this option is very suitable to implement on large ASICs with plenty of resources.

Another micro-architecture option is the generation of traditional FSM + datapath-based VHDL/Verilog models. The results of this option are shown in Fig. 2.6. With this option activated the generated VHDL/Verilog models of the hardware accelerators include a next state process as well as signal assignments with multiplexing which correspond to the input data multiplexers of the activated operators. Although this option produces smaller hardware structures (than the massively parallel option), it can exceed the target clock period due to larger delays through the data multiplexers that are used in the datapath of the accelerator.

Using the above micro-architecture options, the user of the CCC HLS tool can select various solutions between the fastest and larger massively parallel micro-architecture, which may be suitable for richer technologies in terms of operators such as large ASICs, and smaller and more economic (in terms of available resources) technologies such as smaller FPGAs.

As it can be seen in Figs. 2.5 and 2.6, the produced co-processors (accelerators) are initiated with the input command signal START. Upon receiving this command the co-processors respond to the controlling environment using the handshake output signal BUSY and right after this, they start processing the input data in order to produce the results. This process may take a number of clock cycles

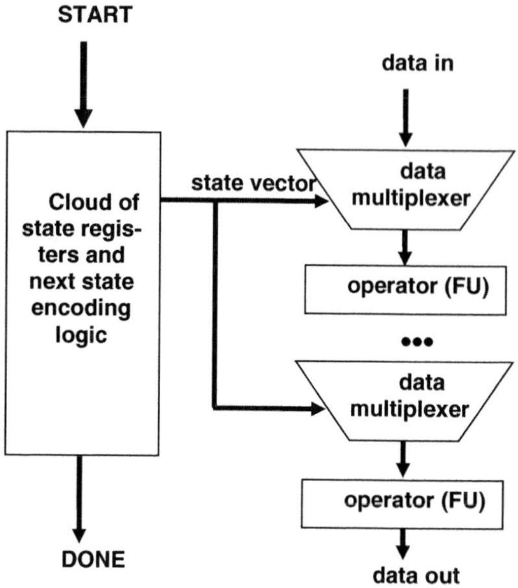


**Fig. 2.6** The traditional FSM + datapath generated micro-architecture option

and it is controlled by a set of states (discrete control steps). When the co-processors complete their processing, they notify their environment with the output signal DONE. In order to conclude the handshake the controlling environment (e.g., a controlling central processing unit) responds with the handshake input RESULTS_READ, to notify the accelerator that the processed result data have been read by the environment. This handshake protocol is also followed when one (higher-level) co-processor calls the services of another (lower-level) co-processor. The handshake is implemented between any number of accelerators (in pairs) using the START/BUSY and DONE/RESULTS_READ signals. Therefore, the set of executing co-processors can also be hierarchical in this way.

Other environment options, passed to the back-end compiler, control the way that the data object resources are used, such as registers and memories. Using a memory port configuration file, the user can determine that certain multi-dimensional data objects, such as arrays and array aggregates, are implemented in external (e.g., central, shared) memories (e.g., system RAM). Otherwise, the default option remains that all data objects are allocated to hardware (e.g., on-chip) registers. All of the related memory communication protocols and hardware ports/signals are automatically generated by the back-end synthesizer, and without the need for any manual editing of the RTL code by the user. Both synchronous and asynchronous memory communication protocol generation are supported.

## 2.9 Generated Hardware Execution Platform

The generated hardware modules can be placed inside the computing environment that they accelerate or can be executed standalone. For every subprogram in the source specification code one co-processor is generated to speed up (accelerate) or just execute the particular system task. The whole system (both hardware and software models) is modeled in algorithmic ADA or C code which can be compiled and executed with the host compiler and linker to run and verify the operation of the whole system at the program code level. In this way, extremely fast verification can be achieved at the algorithmic level. It is evident that such behavioral (high-level) compilation and execution is orders of magnitude faster than conventional RTL simulations. Moreover, now C-Cubed automatically generates cycle-accurate C testbenches from exactly the same FSM information that generates the HDL code. Therefore, using compile and execute these testbenches can be executed on the host computer in a rapid and correct manner.

After the required co-processors are specified, coded in ADA, generated with the prototype hardware compiler, and implemented with commercial back-end tools, they can be downloaded into the target computing system (if the target system includes FPGAs) and executed to accelerate certain system tasks. This process is shown in Fig. 2.7. The accelerators can communicate with each other and with the host computing environment using synchronous handshake signals and connections with the system's handshake logic.

## 2.10 Experimental Results and Conclusions

There have been a great deal of design and verification experiments with the C-Cubed framework. In all the experiments it was found that the quality of the generated HDL modules is very high with increased readability of the code. Among the many experiments, three small benchmarks are analyzed in this paragraph: a computer graphics algorithm, a DSP FIR (finite impulse response) filter, and the classical high-level synthesis benchmark, the second order differential equation approximation solver. Moreover, the statistics of two more benchmarks, an RSA crypto-processor and a complex MPEG engine, are included in the discussion of this chapter.

Table 2.1 shows state reduction statistics with the C-Cubed's optimizer, PARCS. Impressive state reduction is achieved, up to 41 %, even with a complex conditional structure which is found in the line-drawing design (computer graphics benchmark). Due to the formal nature of the C-Cubed synthesizer only high-level source code—level compile and execute verifications are needed. However, in order to prove our argument in practice all synthesized RTL modules were simulated and their behavior matched, as expected, the behavior of the source code programs. Figure 2.8 shows a snapshot of the RTL simulation of the computer graphics hardware module, near the time where the line's pixels are written into the external memory.

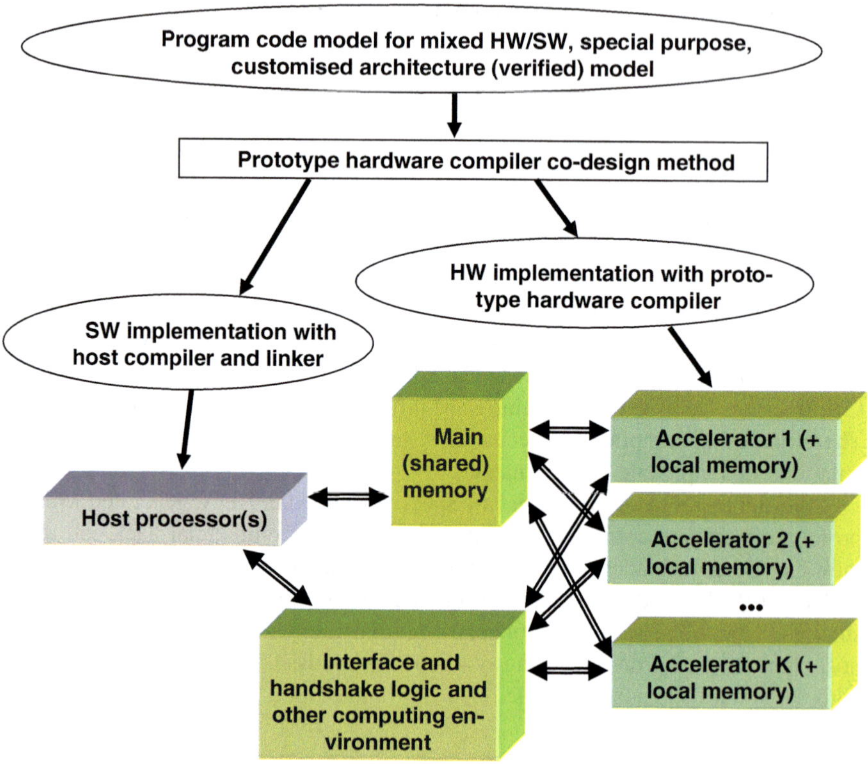


**Fig. 2.7** Host computing environment and hardware execution configuration

**Table 2.1** FSM state statistics before and after optimization with PARCS

| Module name | Initial schedule states | PARCS states | State reduction percentage (%) |
|---|---|---|---|
| DSP FIR filter processor | 17 | 10 | 41 |
| RSA crypto-processor | 16 | 11 | 31 |
| Computer graphics design | 17 | 10 | 41 |
| MPEG top routine (with external memory) | 462 | 343 | 26 |
| Differential equation solver | 20 | 13 | 35 |

The verification flow in the C-Cubed framework is formal and integrated with the synthesis flow. The fact that the input to the tools is executable allows us to compile the source ADA or C and co-simulate with the testbench. Also, we can include commands in the high-level testbench in order to automatically and formally compare simulation outputs of the high-level code with that of the RTL code. The structure and flow of this cross-checking verification is shown in Fig. 2.9.

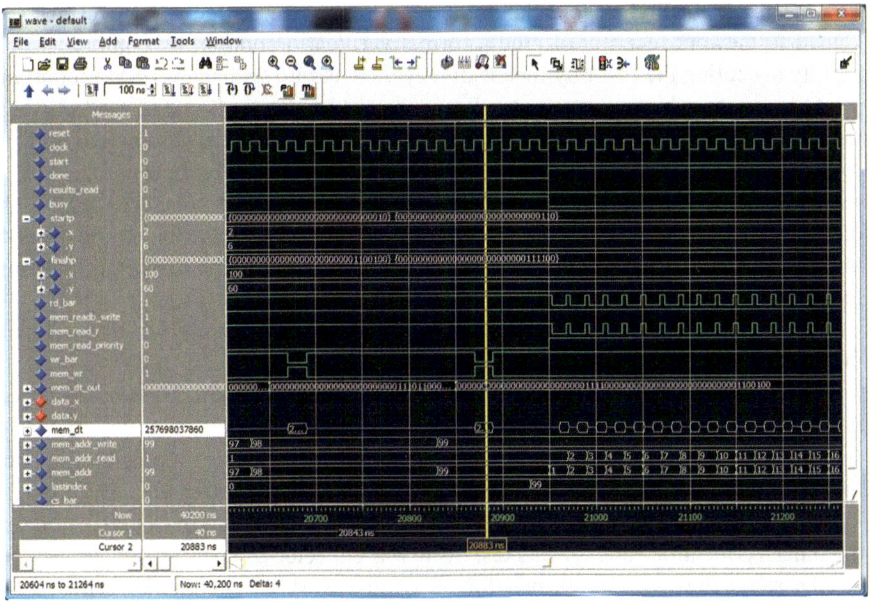


**Fig. 2.8** RTL simulation snapshot of the line-drawing benchmark output

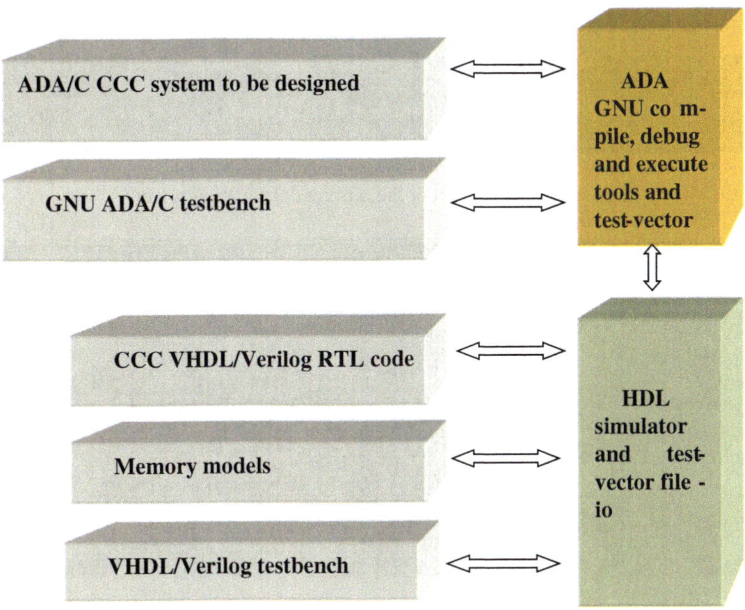


**Fig. 2.9** ADA to VHDL cross-verification flow

Apart from the high-level behavioral testbench (at the source code level) the C-Cubed framework recently offered a formal verification option, by compiling and rapidly executing the FSM by means of a cycle-accurate testbench (FSM model) in the C language which is automatically generated by the back-end compiler using the same internal intelligent FSM models of the optimized hardware modules. The cycle-accurate testbench allows for setting up inputs, reading outputs and registers, resetting the engine, and moving to the next state by pressing corresponding buttons on the keyboard. Thus it is very easy to use and it can increase the confidence of the provably correctness of the generated FSM (co-processor or standalone custom logic).

As an example Fig. 2.10 shown the execution screen of the high-level testbench of the differential equation solver benchmark.

Figure 2.11 shows the RTL simulation of the same benchmark (the result is obviously the same as with the high-level testbench).

The start of the cycle-accurate testbench simulation of this benchmark is shown in Fig. 2.12, where the circuit's inputs are set. After going through the FSM's states the screen in Fig. 2.13 shows reading of the outputs which give the same results as the other verification runs for this test (as it was expected).

The same flow was confirmed for all of this work's benchmarks, but due to limitations in paper length they are omitted.

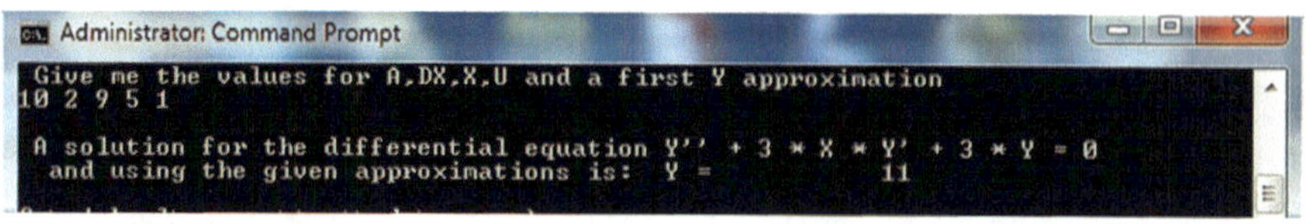


**Fig. 2.10** High-level testbench execution of the differential equation solver

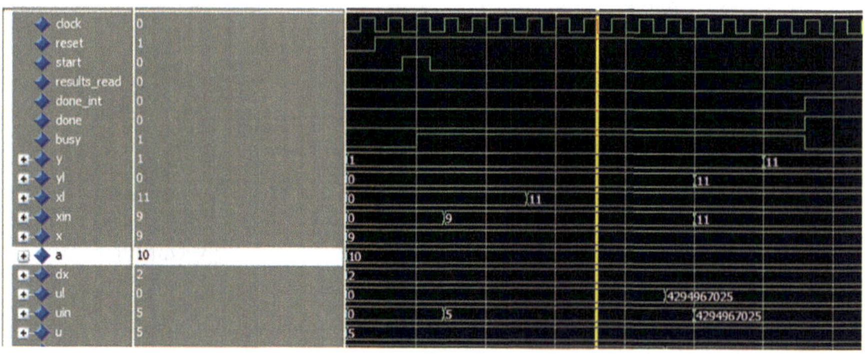


**Fig. 2.11** RTL simulation of the differential equation solver

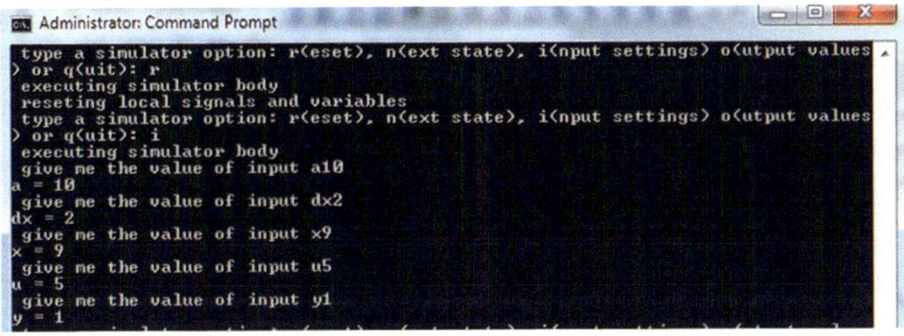


**Fig. 2.12** Setting the inputs of the cycle-accurate diff.eq.solver benchmark

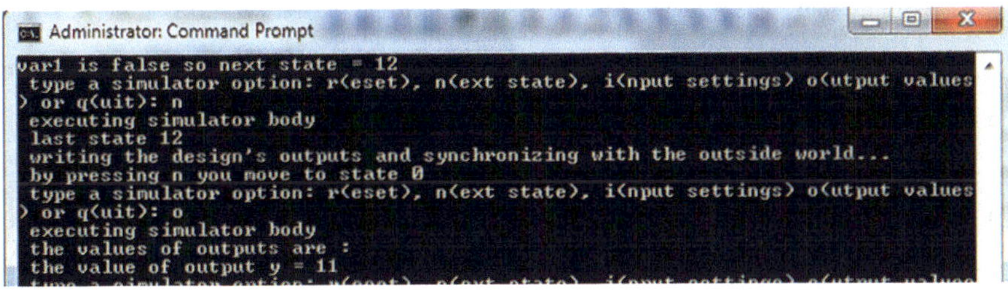


**Fig. 2.13** Reading the outputs of the cycle-accurate diff.eq.solver benchmark

## 2.11 Conclusions and Future Work

The major contribution of this work is an integrated, formal, rapid, and automated methodology and set of tools for the automatic synthesis of custom hardware, which can be used in embedded IoT devices. All of the above characteristics of the presented method make it suitable for short project time and limited project budget for designing custom circuit blocks. It is expected that the IoT industry will be benefited the most by adopting and using such HLS methods, which will make it competitive to the hard international economy.

Due to the nature of the C-Cubed tools implementation they are particularly suitable for future extensions and experiments such as low power design, SystemC testbench generation, and other language input/output interfaces. Also, continuous improvements of the PARCS scheduler and the associated synthesis transformations are envisaged and they are planned for the future.

## References

1. B. Pangrle, D. Gajski, Design tools for intelligent silicon compilation. IEEE Trans. Comput. Aided Des. Integr. Circuits Syst. **6**(6), 1098–1112 (1987)

2. E. Girczyc, R. Buhr, J. Knight, Applicability of a subset of Ada as an algorithmic hardware description language for graph-based hardware compilation. IEEE Trans. Comput. Aided Des. Integr. Circuits Syst. **4**(2), 134–142 (1985)
3. P. Paulin, J. Knight, Algorithms for high-level synthesis. IEEE Des. Test Comput. **6**(6), 18–31 (1989)
4. I. Park, C. Kyung, Fast and near optimal scheduling in automatic data path synthesis, in *Proceedings of the Design Automation Conference (DAC)*, pp. 680–685, San Francisco, CA, 1991
5. N. Park, A. Parker, Sehwa: a software package for synthesis of pipelined data path from behavioral specification. IEEE Trans. Comput. Aided Des. Integr. Circuits Syst. **7**(3), 356–370 (1988)
6. E. Girczyc, Loop winding—a data flow approach to functional pipelining, in *Proceedings of the International Symposium on Circuits and Systems*, pp. 382–385, Philadelphia, PA, May 1987
7. C. Tseng, D. Siewiorek, Automatic synthesis of data path on digital systems. IEEE Trans. Comput. Aided Des. Integr. Circuits Syst. **5**(3), 379–395 (1986)
8. F. Kurdahi, A. Parker, REAL: a program for register allocation, in *Proceedings of the Design Automation Conference (DAC)*, pp. 210–215, Miami Beach, FL, June 1987
9. C. Huang, Y. Chen, Y. Lin, Y. Hsu, Data path allocation based on bipartite weighted matching, in *Proceedings of the Design Automation Conference (DAC)*, pp. 499–504, Orlando, FL, June 1990
10. F. Tsay, Y. Hsu, Data path construction and refinement, in *Digest of Technical Papers, International Conference on Computer-Aided Design (ICCAD)*, pp. 308–311, Santa Clara, CA, November 1990
11. R. Camposano, W. Rosenstiel, Synthesizing circuits from behavioral descriptions. IEEE Trans. Comput. Aided Des. Integr. Circuits Syst. **8**(2), 171–180 (1989)
12. S. Johnson, *Synthesis of Digital Designs from Recursion Equations* (MIT Press, Cambridge, MA, 1984)
13. M. Barbacci, G. Barnes, R. Cattell, D. Siewiorek, The ISPS computer description language. Report CMU-CS-79-137, Department of Computer Science, Carnegie-Mellon University, Pittsburgh, PA, 1979
14. P. Marwedel, The MIMOLA design system: tools for the design of digital processors, in *Proceedings of the 21st Design Automation Conference (DAC)*, pp. 587–593, IEEE Press, Piscataway, NJ, 1984
15. A. Casavant, M. D'Abreu, M. Dragomirecky, D. Duff, J. Jasica, M. Hartman, K. Hwang, W. Smith, A synthesis environment for designing DSP systems. IEEE Des. Test Comput. **6**(2), 35–44 (1989)
16. P. Paulin, J. Knight, Force-directed scheduling for the behavioral synthesis of ASICs. IEEE Trans. Comput. Aided Des. Integr. Circuits Syst. **8**(6), 661–679 (1989)
17. D. Gajski, L. Ramachandran, Introduction to high-level synthesis. IEEE Des. Test Comput. **11**(4), 44–54 (1994)
18. R. Walker, S. Chaudhuri, Introduction to the scheduling problem. IEEE Des. Test Comput. **12**(2), 60–69 (1995)
19. V. Berstis, The V compiler: automatic hardware design. IEEE Des. Test Comput. **6**(2), 8–17 (1989)
20. J. Fisher, Trace Scheduling: a technique for global microcode compaction. IEEE Trans. Comput. **C-30**(7), 478–490 (1981)
21. A. Kuehlmann, R. Bergamaschi, Timing analysis in high-level synthesis, in *Proceedings of the 1992 IEEE/ACM International Conference on Computer-Aided Design (ICCAD '92)*, pp. 349–354, Los Alamitos, CA, 1992
22. C. Papachristou, H. Konuk, A linear program driven scheduling and allocation method followed by an interconnect optimization algorithm, in *Proceedings of the 27th ACM/IEEE Design Automation Conference (DAC)*, Orlando, Florida, USA, pp. 77–83, June 1990

23. J. Biesenack, M. Koster, A. Langmaier, S. Ledeux, S. Marz, M. Payer, M. Pilsl, S. Rumler, H. Soukup, N. Wehn, P. Duzy, The Siemens high-level synthesis system CALLAS. IEEE Trans. Very Large Scale Integr. Syst. **1**(3), 244–253 (1993)
24. T. Filkorn, A method for symbolic verification of synchronous circuits, in *Proceedings of the Comp Hardware Descr Lang and Their Application (CHDL 91)*, pp. 229–239, Marseille, 1991
25. A. Kalavade, E. Lee, A hardware-software codesign methodology for DSP applications. IEEE Des. Test Comput. **10**(3), 16–28 (1993)
26. R. Ernst, J. Henkel, T. Benner, Hardware-software cosynthesis for microcontrollers. IEEE Des. Test Comput. **10**(4), 64–75 (1993)
27. G. De Micheli, D. Ku, F. Mailhot, T. Truong, The Olympus synthesis system. IEEE Des. Test Comput. **7**(5), 37–53 (1990)
28. D. Thomas, J. Adams, H. Schmit, A model and methodology for hardware-software codesign. IEEE Des. Test Comput. **10**(3), 6–15 (1993)
29. C. Hoare, *Communicating Sequential Processes* (Prentice-Hall, Englewood Cliffs, NJ, 1985)
30. R. Gupta, G. De Micheli, Hardware-software cosynthesis for digital systems. IEEE Des. Test Comput. **10**(3), 29–41 (1993)
31. I. Bolsens, H. De Man, B. Lin, K. Van Rompaey, S. Vercauteren, D. Verkest, Hardware/software co-design of digital telecommunication systems. Proc. IEEE **85**(3), 391–418 (1997)
32. D. Genin, P. Hilfinger, J. Rabaey, C. Scheers, H. De Man, DSP specification using the SILAGE language, in *Proceedings of the International Conference on Acoustics Speech Signal Process*, pp. 1056–1060, Albuquerque, NM, 3–6 April 1990
33. P. Willekens et al., Algorithm specification in DSP station using data flow language. DSP Appl. **3**(1), 8–16 (1994)
34. N. Halbwachs, P. Caspi, P. Raymond, D. Pilaud, The synchronous dataflow programming language Lustre. Proc. IEEE **79**(9), 1305–1320 (1991)
35. M. Van Canneyt, Specification, simulation and implementation of a GSM speech codec with DSP station. DSP Multimed. Technol. **3**(5), 6–15 (1994)
36. J. Buck, S. Ha, E. Lee, D. Messerschmitt, PTOLEMY: a framework for simulating and prototyping heterogeneous systems. Int. J. Comput. Simul. **4**, 1–34 (1992)
37. R. Lauwereins, M. Engels, M. Ade, J. Peperstraete, GRAPE-II: a system level prototyping environment for DSP applications. IEEE Comput. **28**(2), 35–43 (1995)
38. M. Rafie et al., Rapid design and prototyping of a direct sequence spread-spectrum ASIC over a wireless link. DSP Multimed. Technol. **3**(6), 6–12 (1994)
39. L. Semeria, K. Sato, G. De Micheli, Synthesis of hardware models in C with pointers and complex data structures. IEEE Trans. VLSI Syst. **9**(6), 743–756 (2001)
40. R. Wilson, R. French, C. Wilson, S. Amarasinghe, J. Anderson, S. Tjiang, S.-W. Liao, C.-W. Tseng, M. Hall, M. Lam, J. Hennessy, Suif: an infrastructure for research on parallelizing and optimizing compilers. ACM SIPLAN Notices **28**(9), 67–70 (1994)
41. A. Kountouris, C. Wolinski, Efficient scheduling of conditional behaviors for high-level synthesis. ACM Trans. Des. Autom. Electron. Syst. **7**(3), 380–412 (2002)
42. S. Gupta, R. Gupta, N. Dutt, A. Nikolau, Coordinated parallelizing compiler optimizations and high-level synthesis. ACM Trans. Des. Autom. Electron. Syst. **9**(4), 441–470 (2004)
43. W. Wang, A. Raghunathan, N. Jha, S. Dey, High-level synthesis of multi-process behavioral descriptions, in *Proceedings of the 16th IEEE International Conference on VLSI Design (VLSI'03)*, ISBN: 0-7695-1868-0, pp. 467–473, 4–8 January 2003
44. Z. Gu, J. Wang, R. Dick, H. Zhou, Incremental exploration of the combined physical and behavioral design space, in *Proceedings of the 42nd Annual Conference on Design. Automation DAC '05*, pp. 208–213, Anaheim, CA, 13–17 June 2005
45. L. Zhong, N. Jha, Interconnect-aware high-level synthesis for low power, in *Proceedings of the IEEE/ACM International Conference Computer-Aided Design*, ISBN: 0-7803-7607-2, pp. 110–117, November 2002
46. C. Huang, S. Ravi, A. Raghunathan, N. Jha, Generation of heterogeneous distributed architectures for memory-intensive applications through high-level synthesis. IEEE Trans. Very Large Scale Integr. **15**(11), 1191–1204 (2007)

47. K. Wakabayashi, C-based synthesis experiences with a behavior synthesizer, "Cyber", in *Proceedings of the Design Automation and Test in Europe Conference*, ISBN: 0-7695-0078-1, pp. 390–393, Munich, 9–12 March 1999
48. W. Wang, T. Tan, J. Luo, Y. Fei, L. Shang, K. Vallerio, L. Zhong, A. Raghunathan, N. Jha, A comprehensive high-level synthesis system for control-flow intensive behaviors, in *Proceedings of the 13th ACM Great Lakes Symposium on VLSI GLSVLSI '03*, ISBN:1-58113-677-3, pp. 11–14, Washington, DC, 28–29 April 2003
49. B. Gal, E. Casseau, S. Huet, Dynamic memory access management for high-performance DSP applications using high-level synthesis. IEEE Trans. Very Large Scale Integr. **16**(11), 1454–1464 (2008)
50. S. Gupta, R. Gupta, N. Dutt, A. Nicolau, Dynamically increasing the scope of code motions during the high-level synthesis of digital circuits, in *Proceedings of the IEEE Conference Computers and Digital Techniques*, ISSN: 1350–2387, vol. 150, no. 5, pp. 330–337, 22 September 2003
51. K. Wakabayashi, H. Tanaka, Global scheduling independent of control dependencies based on condition vectors, in *Proceedings of the 29th ACM/IEEE Design Automation Conference (DAC)*, ISBN: 0-8186-2822-7, pp. 112–115, Anaheim, CA, 8–12 June 1992
52. E. Martin, O. Santieys, J. Philippe, GAUT, an architecture synthesis tool for dedicated signal processors, in *Proceedings of the IEEE International European Design Automation Conference (Euro-DAC)*, pp. 14–19, Hamburg, September 1993
53. M. Molina, R. Ruiz-Sautua, P. Garcia-Repetto, R. Hermida, Frequent-pattern-guided multilevel decomposition of behavioral specifications. IEEE Trans. Comput. Aided Des. Integr. Circuits Syst. **28**(1), 60–73 (2009)
54. K. Avnit, V. D'Silva, A. Sowmya, S. Ramesh, S. Parameswaran, Provably correct on-chip communication: a formal approach to automatic protocol converter synthesis. ACM Trans. Des. Autom. Electron. Syst. **14**(2), 19 (2009)
55. J. Keinert, M. Streubuhr, T. Schlichter, J. Falk, J. Gladigau, C. Haubelt, J. Teich, M. Meredith, SystemCoDesigner—an automatic ESL synthesis approach by design space exploration and behavioral synthesis for streaming applications. ACM Trans. Des. Autom. Electron. Syst. **14**(1), 1 (2009)
56. S. Kundu, S. Lerner, R. Gupta, Translation validation of high-level synthesis. IEEE Trans. Comput. Aided Des. Integr. Circuits Syst. **29**(4), 566–579 (2010)
57. S. Paik, I. Shin, T. Kim, Y. Shin, HLS-l: a high-level synthesis framework for latch-based architectures. IEEE Trans. Comput. Aided Des. Integr. Circuits Syst. **29**(5), 657–670 (2010)
58. A. Raghunathan, S. Dey, N. Jha, Register-transfer level estimation techniques for switching activity and power consumption, in *Digest of Technical Papers, International Conference on Computer-Aided Design (ICCAD)*, ISBN: 0-8186-7597-7, pp. 158–165, San Jose, CA, 10–14 November 1996
59. J. Rabaey, L. Guerra, R. Mehra, Design guidance in the power dimension, in *Proceedings of the 1995 International Conference on Acoustics, Speech, and Signal Processing*, ISBN: 0-7803-2431-5, pp. 2837–2840, Detroit, MI, 9–12 May 1995
60. R. Mehra, J. Rabaey, Exploiting regularity for low-power design, in *Digest of Technical Papers, International Conference on Computer-Aided Design (ICCAD)*, ISBN: 0-8186-7597-7, pp. 166–172, San Jose, CA, November 1996
61. A. Raghunathan, N. Jha, Behavioral synthesis for low power, in *Proceedings of the International Conference on Computer Design (ICCD)*, ISBN: 0-8186-6565-3, pp. 318–322, Cambridge, MA, 10–12 October 1994
62. L. Goodby, A. Orailoglu, P. Chau, Microarchitecture synthesis of performance-constrained low-power VLSI designs, in *Proceedings of the International Conference on Computer Design (ICCD)*, ISBN: 0-8186-6565-3, pp. 323–326, Cambridge, MA, 10–12 October 1994
63. E. Musoll, J. Cortadella, Scheduling and resource binding for low power, in *Proceedings of the Eighth Symposium on System Synthesis*, ISBN: 0-8186-7076-2, pp. 104–109, Cannes, 13–15 September 1995

64. N. Kumar, S. Katkoori, L. Rader, R. Vemuri, Profile-driven behavioral synthesis for low-power VLSI systems. IEEE Des. Test Comput. **12**(3), 70–84 (1995)
65. R. Martin, J. Knight, Power-profiler: optimizing ASICs power consumption at the behavioral level, in *Proceedings of the Design Automation Conference (DAC)*, ISBN: 0-89791-725-1, pp. 42–47, San Francisco, CA, 1995
66. I. Issenin, E. Brockmeyer, B. Durinck, N.D. Dutt, Data-reuse-driven energy-aware cosynthesis of scratch pad memory and hierarchical bus-based communication architecture for multiprocessor streaming applications. IEEE Trans. Comput. Aided Des. Integr. Circuits Syst. **27**(8), 1439–1452 (2008)
67. M. Dossis, Intermediate Predicate Format for design automation tools. J. Next Gener. Inform. Technol. **1**(1), 100–117 (2010)
68. U. Nilsson, J. Maluszynski, *Logic Programming and Prolog*, 2nd edn. (John Wiley & Sons Ltd., Chichester, 1995)

# Chapter 3
# AXIOM: A Flexible Platform for the Smart Home

**Roberto Giorgi, Nicola Bettin, Paolo Gai, Xavier Martorell, and Antonio Rizzo**

## 3.1 Introduction

The Smart Home is an autonomous physical-digital system composed of a network of heterogeneous objects and people that interact with each other [8]. The goal of the Smart Home system is to give to its users a high level of comfort, security, and privacy and in the meantime enable a reduction of energy consumption [17, 25]. According to a recent survey, Smart Homes have a huge potential impact in the next decade, possibly near to that experienced in the past decade by smartphones [16].

To reach this aim the Smart Home system has to be a network of agents able to act in autonomy and to cooperate with each other; moreover, these agents have to exchange information with sensors, actuators, and with human beings in a natural user-interface. The interaction between the human user and its home should be ensured from everywhere inside and outside the home, through local

R. Giorgi (✉)
Department of Information Engineering and Mathematics, University of Siena, Siena, Italy
e-mail: giorgi@dii.unisi.it

N. Bettin
VIMAR SpA, Marostica, Italy
e-mail: nicola.bettin@vimar.it

P. Gai
Evidence SrL, Pisa, Italy
e-mail: pj@evidence.eu.com

X. Martorell
Computer Architecture Department, Barcelona Supercomputing Center, Barcelona, Spain
e-mail: xavier.martorell@bsc.es

A. Rizzo
Department of Cognitive Sciences, University of Siena, Siena, Italy
e-mail: rizzo@unisi.it



G. Keramidas et al. (eds.), *Components and Services for IoT Platforms*,
DOI 10.1007/978-3-319-42304-3_3

and remote connection. Furthermore, another fundamental feature of the agents is their capability to manage a flexible number as well as different types of nodes in the network. The Axiom project (Agile, eXtensible, fast I/O Module) consists in realizing new hardware/software architectures able to meet the features required for the agents of the Smart Home system and for the cyber-physical system (CPS) [2, 11, 24]. This paper presents a possible hardware and software platform that is designed with several principles that may nicely fit the needs of the Internet of Things (IoT) domain [8, 22], e.g., the need of an energy efficient system, including reconfigurable hardware and based on an open-source and open-hardware architecture.

The hardware architecture proposed in this project is based on a family of single board computer (SBC), that could work independently or in a cluster by means of a dedicated fast-interconnect (not interfering with the internet connection) to create a flexible multi-boards system. Each SBCs is based on a heterogeneous multicore composed of a dual core ARM and an FPGA provided in the same chip. The heterogeneous cores allow for a high computational power and in the meantime a low-power consumption [29]. The multi-board solution aims at reaching a higher scalability to provide more computational power when the size of the problem need so [11].

Two key points for the success of the above platform are related to the cost of the technical solutions that are adopted. For these reasons in AXIOM we aim at using an inexpensive but fast interconnect (like SATA cables) for convoying the inter-node traffic due to the application execution. The second pillar of the technical solutions that we are proposing is the use of simple yet powerful programming model such as OmpSs [6], that is open-source, and permits to potentially substitute the proprietary programming toolchain that is needed to take advantage not only of the multiple cores in a board but also of multiple boards and the FPGAs available on all those boards for both task distribution and acceleration, respectively.

One ideal fit for the AXIOM platform in the context of the Smart Home is, for instance, a "Smart-HUB" able to act as a Smart Home server in order to provide a natural center for collecting the more complex computational tasks of the Smart Home. Such Smart Home server is currently one of the more costly components of the Smart Home [8]. The Smart-HUB is a Linux system connected to the local home network and to the internet network by standard protocols and Ethernet connection. It will be composed of AXIOM boards able to process video and audio streams coming from devices/clients inside the Smart Home system and to send commands to clients. In our application case, the Smart-HUB will be focused on processing multimedia streams to extract biometric features and on performing speech recognition to provide voice commands (user-interfaces) to the Smart Home system. The servers could also be used to authenticate the requests from remote clients (via phone, tablet). In this case, the users will have the possibility to get and set the state of the domotic system through internet connection.

The heterogeneous cores on the AXIOM platform will use the FPGA as a coprocessor to process the multimedia data in a power efficient way. The fast-interconnections between the multi-boards and the multi-threaded programming model not only provide the capability to manage a flexible number of clients at the same time as in traditional platforms, but also permit a resilient execution [26, 27].

The issue to define task scheduling and to select in which core the task computation achieve real-time performance and power-efficiency is provided by the AXIOM hardware/software infrastructure [2, 10, 13, 14, 24].

In this contribution we also present some initial results that demonstrate the feasibility of achieving the task distribution across several boards.

The contribution of this chapter are: (1) proposing a more flexible platform based on open-source, open-hardware, reconfigurable, and scalable embedded platform for the Smart Home domain; (2) summarizing the key features that could be useful in such platform for the Smart Home domain; and (3) illustrating some preliminary results of the early simulated prototype of the AXIOM platform.

## 3.2 Smart Home Scenarios

Smart Home means buildings empowered by ICT in the context of the merging ubiquitous computing and the IoT: the generalization in instrumenting buildings with sensors, actuators, CPS allow to collect, filter, and produce more and more information locally, to be further consolidated and managed globally according to business functions and services. A Smart Home is one that uses operational and IT technologies and processes to make it a better performing building—one that delivers lower operating costs, uses less energy, maximizes system and equipment lifetime value, is cyber-secured, and produces measurable value for multiple stake holders [22, 25].

Major challenges in such environments concern cryptography, self-testing, and first of all sensor-networks management. Sensor data brings numerous computational challenges in the context of data collection, storage, and mining. In particular, learning from data produced from a sensor network poses several issues: sensors are distributed; they produce a continuous flow of data, eventually at high speeds; they act in dynamic, time-changing environments; the number of sensors can be very large and dynamic. These issues require the design of efficient solutions for processing data produced by sensor networks.

Figure 3.1 shows different home and office scenarios where AXIOM can help with preventive and interactive maintenance of infrastructures, climate and temperature management. This management can be remotely controlled helping to improve the energy efficiency at home, apartments, and company office buildings. For instance, AXIOM may detect patterns of behavior in a company office building to adapt climate and light switching to the working way of life of the workers.

We are currently considering a wide range of potential uses both for video-surveillance and for Smart Home. They range from dynamic retail demand

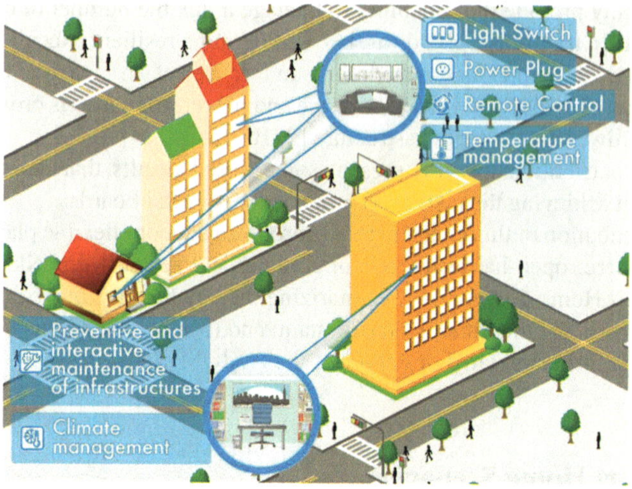


**Fig. 3.1** Smart Home scenarios

forecasting in train/bus station to Smart Marketing in shopping malls for video-surveillance; and from Smart Home comfort to autonomous drone for infrastructure control, for Smart Home.

Another test case for the Smart Home regards comfort at home. Comfort perception and necessities can be different in respect of time of the day/week and to the characteristics of the people actually living that space in that moment. The Smart Home is required to identify and manage the different situations, and to react at the people indications in an easy and smooth way. Networked sensors and actuators are distributed in each room embedded in ordinary appliances. The appliances not only perform their primary "normal" function, but also collect different kinds of information, ranging from presence detection, temperature, humidity, window and door opening, air quality, and audio. The objective of the Smart Home comfort autopilot is to minimize power consumption and to guarantee people's comfort and well-being, without giving the impression of reducing people freedom and capacity of control.

The modular approach explored by AXIOM is particularly well-suited for tackling such challenging scenarios, as it addresses the issues derived from their computational complexity, distributed nature, and need for synchronization among processes. Moreover, we are considering some representative benchmarks to test drive the design of the software stack that two partners already explored in the ERA project [21, 23].

## 3.3 The AXIOM Platform

The AXIOM platform is architected on the following pillars (see also Figs. 3.2 and 3.3):

P1 FPGA, i.e., large programmable logic for acceleration of functions, soft-IPs, implementing specific AXIOM support for interconnects and scaling.
P2 General purpose cores, to support the OS and for running parts that make little sense on the other accelerators.
P3 High-speed, cheap interconnects to permit scalability and deverticalize the technology, e.g., for toolchains.
P4 Open-source software stack
P5 Lower-speed interface for the cyber-physical world, such as Arduino [5] connectors, USB, Ethernet, and WiFi

Below we illustrate those pillars more in details.

[P1] In the first phase we will adopt one of the existing solutions such as the Xilinx Zynq [29] (Zynq is a chip-family, the chip can include a dual ARM Cortex-A9@1 GHz, 4@6.25 Gbps to 16@12.5 Gbps transceivers, low-power programmable logic from 28 k to 444 k logic cells + 240 to 3020 KB BRAM + 80 to 2020 18x25 DSP slices, PCI express, DDR3 memory controller, 2 USB, 2 Gbe, 2

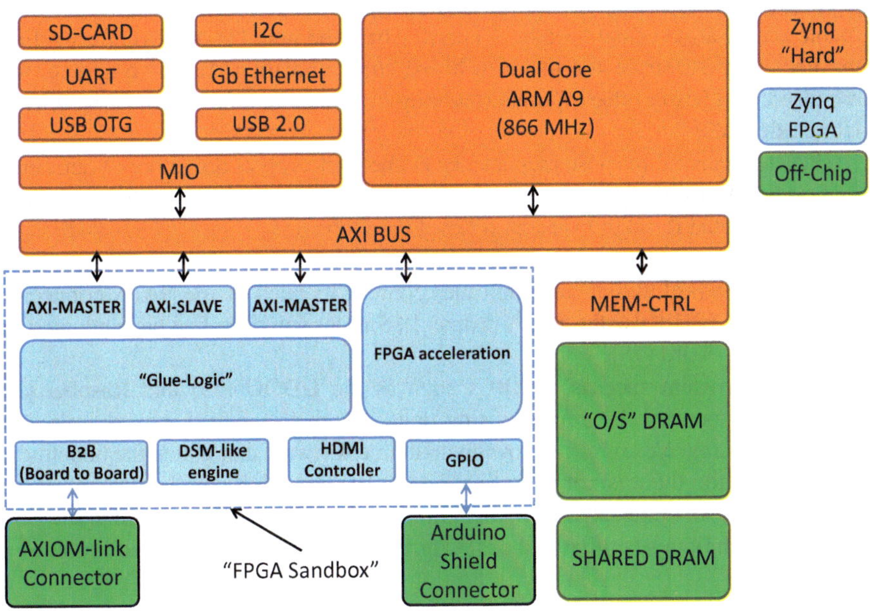


**Fig. 3.2** The detailed architecture of the single-node, relying on a Xilinx Zynq chip or a similar SoC

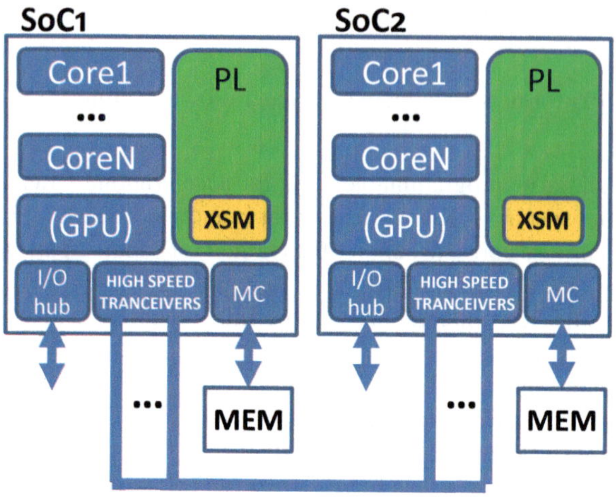


**Fig. 3.3** AXIOM scalable architecture. An instance consisting of two boards, each one based on the same system-on-chip (SoC). GPU is an optional component. *MC* memory controller, *PL* programmable logic, *XSM* eXtended shared memory

CAN, 2SDIO, 2 UART, 2 SPI, 2 I2C, 4x32b GPIO, security features, 2 ADC@12bit 1Msps). The reason for the FPGA is to become the central hearth of the board making possible to integrate all the features, providing customized and reconfigurable acceleration of the specific scenario where the board is deployed, providing the necessary substrate for board-to-board communication. In our roadmap, we are also considering other options that may be available soon such as the Xilinx Ultrascale+ [28].

[P2] The general purpose (superscalar) cores can support a number of activities such as the operating system (system tasks) but also sequential tasks that may need to be accelerated through Instruction Level Parallelism.

[P3] To keep the cost low we are initially oriented to use the FPGA transceivers and use a standard and cheap (multiple) connectors such as the SATA connector (without necessarily use the SATA protocol). Similar solutions had been adopted in the FORMIC board [18].

[P4] The recent success of SBCs such as the UDOO [19] and RaspberryPi further demonstrated the need of using open-source software. Linux has already become a reference example of how open-source software can widen the benefits at any level. While there is not yet a final consensus on which parallel programming model is best, we believe that adopting OmpSs [6] can easy the programmability by providing techniques familiar to the HPC programmer into the embedded computing community.

[P5] In order to interface with the physical world the platform includes support for Arduino connectors for general purpose I/O and other standard interfaces such as the USB, Ethernet, and WiFi. Not less important is the capability of interfacing with sensors and actuators or any other type of external shields as in the Arduino platform.

Not less important is the possibility of bringing together in a single platform all those elements and tackling cross-issues, such as a better real-time scheduling, made possible by the DF-Threads: as the inputs should be available before execution of the DF-Threads, the system can be more predictable too.

## 3.4 Thread Management

Concerning reconfigurable hardware platforms based on FPGA several works have proposed solutions to address the problem of dynamic allocation of tasks to general purpose multicore processors [1], or reconfigurable logic (hardware kernels) [7]. However, such approaches have been successfully explored only on single and multicore super-scalar architectures, so far. In recent systems, the large adoption of many-cores accelerators (i.e., GPUs) has worsen the problem, introducing synchronization issues also among different types of computing elements. Research activity has been always active in this specific context, providing solutions that attempt to efficiently solve the synchronization problem. A more general scheduling unit must be used to efficiently exploit workload distribution, not only among CPU cores, but also on the specific accelerators.

Data-flow threads (DF-Threads [12]) offer a simple and effective solution to address the need of reducing data transfers by moving data where it is requested for a certain computation, expressed by the DF-Threads. While most of computations can be performed in a producer–consumer modality, there is the specific need of accessing mutable shared data. The memory model offered by DF-Threads [12] is encompassing transactional memory [15], which is currently also adopted by manufacturers such as IBM and Intel. DF-Threads are best implemented in hardware through the use of a distributed thread scheduler [13] (DTS). The DTS has to solve the following challenges: (1) when considering the system level, all the available resources and the healthiness of the whole system must be considered by the managing component; (2) at low-level the fine-grain threads coming from the adoption of the data-flow execution model must be distributed across the computing elements (CPUs, FPGAs). This means to understand at runtime what is the best resource assignment (scheduling/mapping on CPU or reconfigurable HW) to a task (or thread), according to multiple goals (e.g., performance/QoS, power consumption minimization, thermal hotspots). The policies should operate effectively both in a single application and a mixed workload scenario. The scheduler can be further extended to enable it distributing fine-grain threads across the different computing elements. To not limit the performance of the system and to avoid the introduction of bottlenecks, the DTS needs to be accelerated in hardware, by mapping its structure into the FPGA as shown in Fig. 3.3 by the XSM block. Standard high-speed and low-latency interconnections (e.g., PCIe 3.0) may provide enough bandwidth to not become the bottleneck of the evaluation platform. The aim of the AXIOM project is also towards an energy efficient improvement of the performance of applications, along with benefits in terms of modular scalability of the platform. In the next sections we will describe the first experiments that enabled us to have more confidence with this approach (see Sect. 3.6.3).

## 3.5 The OmpSs Programming Model

The OmpSs programming model [6] has been introduced with the aim of simplifying the programming of large computing system encompassing heterogeneous multicore processing elements.

In fact programming models such as MPI are quite difficult for developers that are not used to parallel programming and porting legacy code is not an easy task. On the other side, shared-memory systems (such as the nowadays common multicores) are easier to program as they rely on hardware-based transparent mechanisms in order to make sure that the machine can exploit the multiple resource, such as hardware coherence protocols.

AXIOM leverages OmpSs, a task dataflow programming model that includes heterogeneous execution support as well as data and task dependency management [4] and has significantly influenced the recently appeared OpenMP 4.0 specification. In OmpSs, tasks are generated in the context of a team of threads that run in parallel. OmpSs provides an initial team of threads as specified by the user upon starting the application.

Tasks are defined as portions of code enclosed in the *task directive*, or as user-defined functions, also annotated as tasks, as follows:

```
#pragma omp task [clause-list]
   { structured-work |
     function-declaration |
     function-definition }
```

A task is created when the code reaches the task construct, or a call is made to a function annotated as a task. The task construct allows to specify, among others, the clauses *in*, *out*, and *inout*. Their syntax is

```
in (data-reference-list)
out (data-reference-list)
inout (data-reference-list)
```

The information provided is used to derive dependencies among tasks at runtime, and schedule/fire a task. Tasks are fired when their inputs are ready and their outputs can be generated.

Dependencies are expressed by means of data-reference-lists. A data-reference in such a list can contain either a single variable identifier, or also references to subobjects. References to subobjects include array element references (e.g., *a[4]*), array sections (*a[3:6]*), field references (*a.b*), and elaborated shaping expressions (*[10][20] p*). The latter means the rectangular area starting at address *p*, with a shape of 10 rows and 20 columns.

OmpSs is based on two main components: (1) The Mercurium compiler gets C/C++ and FORTRAN code, annotated with the task directives presented above, and transforms the sequential code into parallel code with calls to the Nanos++ runtime system; and (2) The Nanos++ runtime system gets the information generated by the compiler about the parallel tasks to be run, manages the task dependences and schedules them on the available resources, when those tasks are ready. Nanos++ supports the execution of tasks in remote nodes, and heterogeneous accelerators.

At the lower level, the AXIOM project will investigate and implement the OmpSs programming model on top of the following intra- and inter-node technologies:

- **Intra-node:** The most important target here is FPGA programmability support.
  - OmpSs@FPGA, for easy exploiting of the FPGA acceleration;
- **Inter-node:** In this case two different approaches can be addressed based on the performance requirement, although they can be integrated in the same scenarios, to work with different memory address spaces.
  - OmpSs@cluster, for efficient parallel programming hiding message-passing complexities;
  - OmpSs on a DSM-like paradigm, for easy parallelization of legacy code.

Figure 3.4 shows the overall view of OmpSs@FPGA and OmpSs@cluster execution context in a multi-board system. Each FPGA-based node will be addressed by the OmpSs@FPGA support meanwhile the OmpSs@cluster will help to transparently program all the multi-node system.

Figure 3.5 shows the overall view of a DSM system where OmpSs@FPGA would have the same intra-node influence and OmpSs@cluster will appear like a single intra-node OmpSs running over a transparent DSM system.

### *3.5.1 OmpSs@FPGA*

The OmpSs@FPGA ecosystem consists of the infrastructure for compilation instrumentation and execution from source code written in C/C++ to ARM binary and FPGA bitstream for Zynq. The compilation infrastructure provides support to (1) generate ARM binary code from OmpSs code, that can run in the ARM-based SMP of the Zynq All-Programmable SoC, (2) extract the kernel of the part of the application to be accelerated into the FPGA, and (3) automatically generate a bitstream that includes the IP cores of the accelerator(s), the DMA engine IPs, and

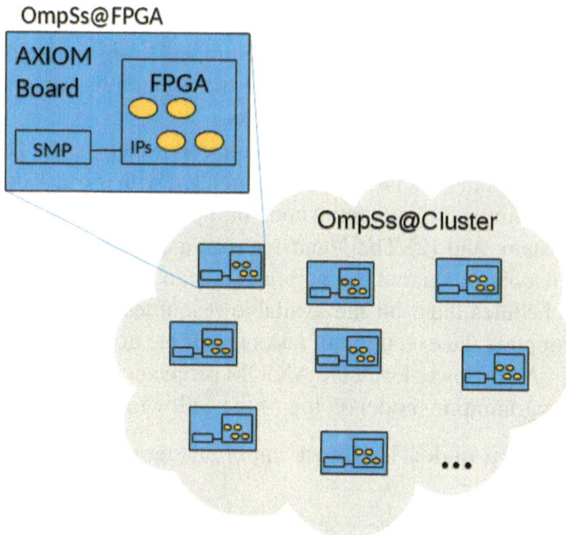


**Fig. 3.4** General view of OmpSs@FPGA and OmpSs@Cluster execution context

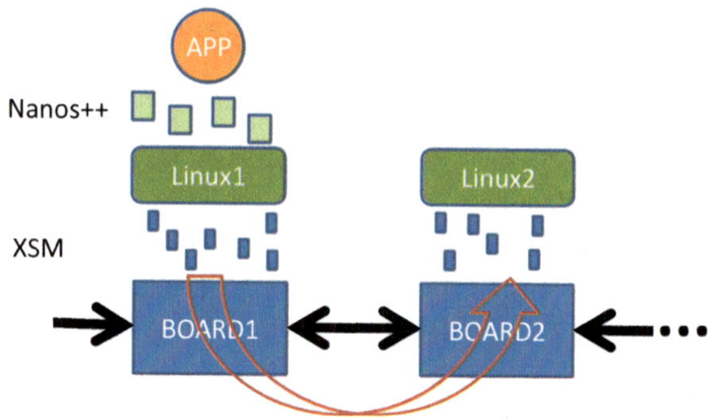


**Fig. 3.5** The application can be launched from a single board and the coarse-grained thread can then rely on the XSM fine-grained threads (called DF-Threads). The DF-Threads are then managed and distributed across boards through the XSM layer

the necessary interconnection. In addition, the ARM binary can be instrumented to generate traces to be analyzed offline with the Paraver tool [20].

The runtime infrastructure should allow heterogeneous tasking on any combination of SMPs and accelerators, depending on the availability of the resources and the target devices.

Figure 3.6 shows the high level compilation flow using our OmpSs@FPGA ecosystem. The OmpSs code is passed through the source-to-source compiler Mercurium [9], that includes a specialized `FPGA` compilation phase to process annotated `FPGA` tasks. For each of those tasks, it generates two C codes. One of them is a `Vivado HLS`[1] annotated code for the bitstream generation branch ("accelerator codes" box in Fig. 3.6). The other is an intermediate host source code with OmpSs runtime (Nanos++) calls that is generated for the software generation

[1]Source to HDL Xilinx tool.

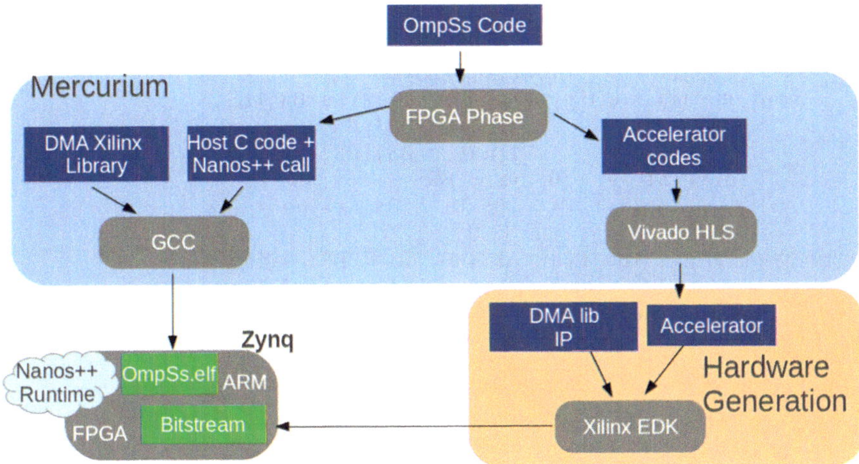


**Fig. 3.6** OmpSs@FPGA ecosystem compilation flow

branch ("Host C code + Nanos++ runtime call" box in Fig. 3.6). Both the hardware and the software generation branches are transparent to the programmer.

Figure 3.7 shows a matrix multiply example that has been annotated with OmpSs directives. This code shows a parallel tiled matrix multiply where each of the tiles is a task. Each of those tasks has two input dependences and an inout dependence that will be managed at runtime by Nanos++. Those tasks will be able to be scheduled/fired to an `SMP` or `FPGA`, as it is annotated in the target device directive, depending on the resource availability. The `copy_deps` clause associated with the `target` directive hints the Nanos++ runtime to copy the data associated with the input and output dependences to/from the device when necessary.

### 3.5.2 *OmpSs@cluster*

OmpSs@cluster is the OmpSs flavor that provides support for a single address space over a cluster of SMP nodes with accelerators. In this environment, the Nanos++ runtime system supports a master-worker execution scheme. One of the nodes of the cluster acts as the master node, where the application starts. In the rest of nodes where the application is executed, worker processes just wait for work to be provided by the master.

In this environment, the data copies generated either by the `in`, `out`, and `inout` task clauses are executed over the network connection across nodes, to bring data to the appropriated node where the tasks are to be executed.

Following the Nanos++ design, *cluster threads* are the components that allow the execution of tasks on worker nodes. These threads do not execute tasks themselves.

```
#pragma omp target device(fpga, smp) copy_deps
#pragma omp task in(a[0:BS*BS-1], b[0:BS*BS-1]) \
                 inout(c[0:BS*BS-1])
void matrix_multiply(int BS, float a[BS][BS],
                     float b[BS][BS],
                     float c[BS][BS]) {
    for (int ia = 0; ia < BS; ++ia)
        for (int ib = 0; ib < BS; ++ib) {
            float sum = 0;
            for (int id = 0; id < BS; ++id)
                sum += a[ia][id] * b[id][ib];
            c[ia][ib] = sum;
        }
}

...
int main( int argc, char * argv[] ){
   int BS = ...
 ...
   for (i=0; i < NB; i++) {
     for (j=0; j < NB; j++) {
       for (k=0; k < NB; k++) {
         matrix_multiply(BS, A[i][k], B[k][j], C[i][j]);
       }
     }
   }
 #pragma omp taskwait
  ...
}
```

**Fig. 3.7** OmpSs directives on matrix multiplication

They are in charge of sending work descriptors to their associated nodes and notifying when these have completed their execution. One cluster thread can take care of providing work to several worker nodes. In the current implementation, cluster threads are created only on the master node of the execution. Slave nodes cannot issue tasks for remote execution and thus they do not need to spawn cluster threads.

In Nanos++, the device specific code has to provide specific methods to be able to transfer data from the host address space to the device address space, and the other way around. The memory coherence model required by OmpSs is implemented by two generic subsystems, the *data directory* and the *data cache*, explained below.

Figure 3.8 shows how the different Nanos++ subsystems are organized to manage the memory of the whole cluster. The master node is the responsible for keeping the memory coherent with the OmpSs memory coherence model, and also for offering the OmpSs single address space view. The master node memory is what OmpSs considers the *host memory* or *host address space*, and it is the only address space exposed to the application. The memory of each worker node is treated as a private device memory and is managed by the master node.

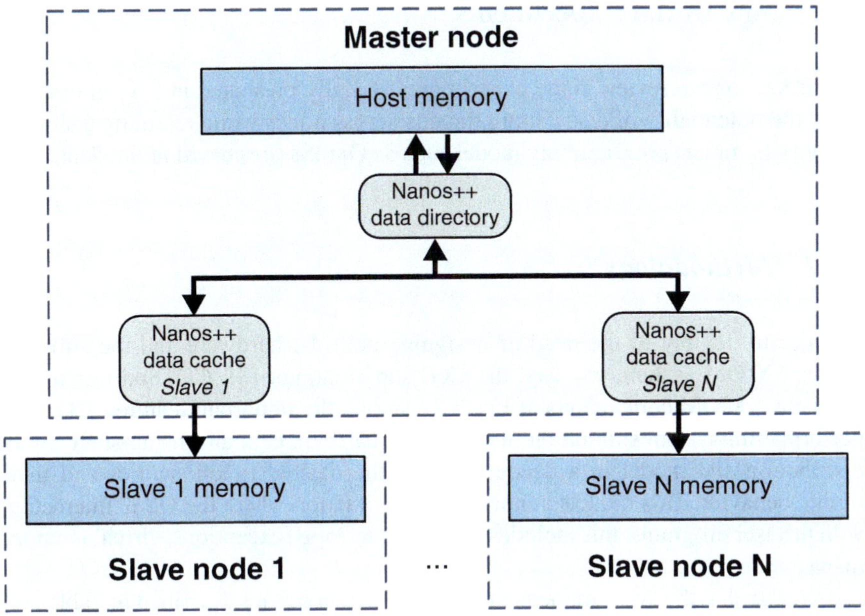


**Fig. 3.8** Nanos++ distributed memory management organization

The *data cache* component manages the operations needed at the master node to transfer data to and from worker memories. There is one data cache for each address space present on the system. Operations performed in a data cache include allocating memory chunks, freeing them, and transferring data from their managed address spaces to the host address space and the other way around. Data caches also keep the mapping of host memory addresses to their private memory addresses. Memory transfer operations are implemented using network transfers. Allocation and free operations are handled locally at the master node.

A memory reference may have several copies of its contents on different address spaces of the system. To maintain the coherence of the memory, the master node uses the *data directory*. It contains the information of where the last produced values of a memory reference are located. With it, the system can determine which transfer operations must perform to execute a task in any node of the system. Also, each task execution updates the information of the data directory to reflect the newly produced data.

The implementation of the network subsystem is currently based on the active messages provided by the GASNet communications library. In the context of AXIOM, we will adapt the networking on the communications library provided for the Zynq platform.

## 3.6 Some Initial Experiments

In this section we review some experiments partially presented in [11] in order to show the potentiality of distributing threads across nodes while retaining a shared-memory (simpler) programming model such as OmpSs (discussed in the Sect. 3.5).

### 3.6.1 Methodology

In order to flexibly fit the need of designing both the hardware and the software of the AXIOM system, we used the COTSon simulator [3]. COTSon can model the main AXIOM components of Fig. 3.3. Among the important features, COTSon performs full-system simulation: the designer can run, e.g., an off-the-shelf Linux distribution and model in a decoupled way the desired functionalities and their timing behavior. This models a more realistic situation where the OS is interacting with the user programs: this includes also any interrupts, exceptions, virtual memory management.

In particular, the key parameters of the modeled cores are described in Table 3.1.

Additionally, the simulator has been extended to support DF-Threads [12]. This means that the simulator is also modeling the DTS [13], which is implemented on the programmable logic through the block XSM of Fig. 3.3.

As of the interconnects among SoCs we are currently exploring several options as offered by the latest technologies. In the COTSon simulator we are performing limit study experiments assuming that we can achieve enough bandwidth and low latency at a reasonable cost. This part is explored in detail within the AXIOM project, but will not be illustrated here.

**Table 3.1** Multicore architectural parameters

| Parameter | Description |
|---|---|
| SoC | 4-cores connected by a shared-bus, IO-hub, MC, high-speed transceivers |
| Core | 1 GHz, in-order superscalar |
| Branch predictor | Two-level (history length=14bits, pattern-history table=16 kB, 8-cycle misprediction penalty) |
| L1 cache | Private I-cache 32 KB, private D-cache 32 KB, 2 ways, 3-cycle latency |
| L2 cache | Private 512 KB, 4 ways, 5-cycle latency |
| L3 cache | Shared 4 GB, 4 ways, 20-cycle latency |
| Coherence protocol | MOESI |
| Main memory | 1 GB, 100 cycles latency |
| I-L1-TLB, D-L1-TLB | 64 entries, full-associative, 1-cycle latency |
| L2-TLB | 512 entries, direct access, 1-cycle latency |
| Write/Read queues | 200 Bytes each, 1-cycle latency |

### 3.6.2 Matrix Multiplication Benchmark

We selected the matrix multiplication kernel to test the performance evaluation infrastructure and to verify the feasibility of supporting DF-Threads on the AXIOM platform.

The matrix multiplication benchmark has the following characteristics:

- Blocked matrix multiplication using the classical three nested loops algorithm.
- Square matrices of size $n \times n$, where $n = 250, 320, 400$.
- Block size $b = 10$ and $b = 25$.

Since the number of operations is $O(n^3)$, the size $n$ of the matrix has been chosen in such a way that the cubed size of each number of the size sequence is approximately the double of the cubed size of the previous number, i.e., $320^3 \approx 2 \times 250^3$, and so on. In the second test we used also the size 800 while focusing instead on the block size.

The DF-Threads are generated in such a way that each thread performs the dot-products of each block, therefore we can expect a number of threads equal to $n/b$.

### 3.6.3 Experiments

In order to verify the feasibility of running programs not only on the single board but also on multiple boards thanks to the adopted programming model (OmpSs) and the underlying runtime system a key point is being able to execute the generated DF-Threads across the boards. This implies that the memory which is local to each node (see Fig. 3.2) has to be managed in such a way that it appears as shared.

The XSM block of Fig. 3.2 serves to that goal by bookkeeping the DF-Threads and by appropriately moving the data where is needed.

We performed two tests to verify that the proposed paradigm can permit the distribution of the threads while varying matrix size and while varying block size. We increase the number of SoCs (for simplicity we refer to the single SoC as if it were a board) and we want to verify if the speedup t1/tN (being t1 the time to execute the program on a single SoC and tN the time to execute the program on N SoCs) is close to the ideally linear speedup (Figs. 3.9 and 3.10).

As we can see in Fig. 3.9, as the number of SoCs is increased from 1 to 2 and then 4, the scalability is particularly good (almost ideal) for higher matrix sizes (e.g., 320).

We explored the sensitivity to the thread granularity. A larger block size generates longer DF-Threads to process a matrix block. In Fig. 3.10, we analyzed the situation for three matrix sizes ($n = 400$, $n = 800$) and while the block size $b$ is equal to 10 and 25. When the block size is decreased, more threads are generated, therefore the parallelism of the computation is higher. In both cases the feasibility of the approach for distributing threads across the boards is confirmed.

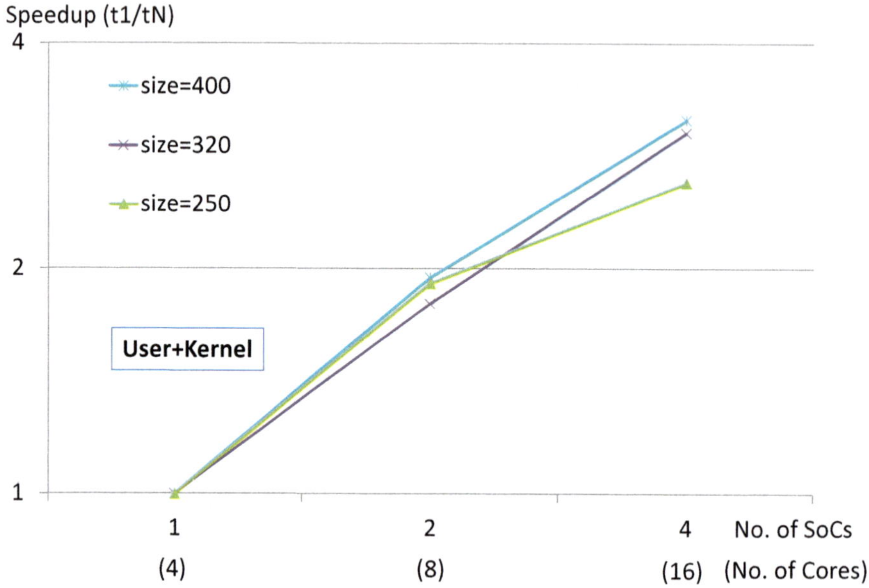


**Fig. 3.9** Strong scaling for benchmark "Dense Matrix Multiplication" (Square Matrices). Matrix size varies: 250, 320, 400. Block size is constant and equal to 10. The time used for calculating the speedup accounts for both the user time and the kernel time

## 3.7 Conclusions

In this contribution, we have shown the potentials of the AXIOM hardware/software platform for the Smart Home. We believe that such platform can offer several advantages as it offers many of the features that are requested in such scenario for both IOT applications and smart CPS.

The AXIOM platform consists of a small SBC which aims at easy programmability through the OmpSs open-source framework, which in turns works on the Linux operating system and an open-hardware platform.

In connection with the fast and inexpensive interconnection method and the XSM components that we are developing, we aim at scaling the performance across several boards so that the system can be tailored or upgraded to the changing needs of the user without the need of upgrading the software.

The initial experiments show that the distribution of the threads across the boards through the proposed model is feasible.

**Acknowledgements** We thankfully acknowledge the support of the European Union H2020 program through the AXIOM project (grant ICT-01-2014 GA 645496), the Spanish Government, through the Severo Ochoa program (grant SEV-2011-00067) the Spanish Ministry of Science and Technology (TIN2012-34557), and the Generalitat de Catalunya (MPEXPAR, 2014-SGR-1051).

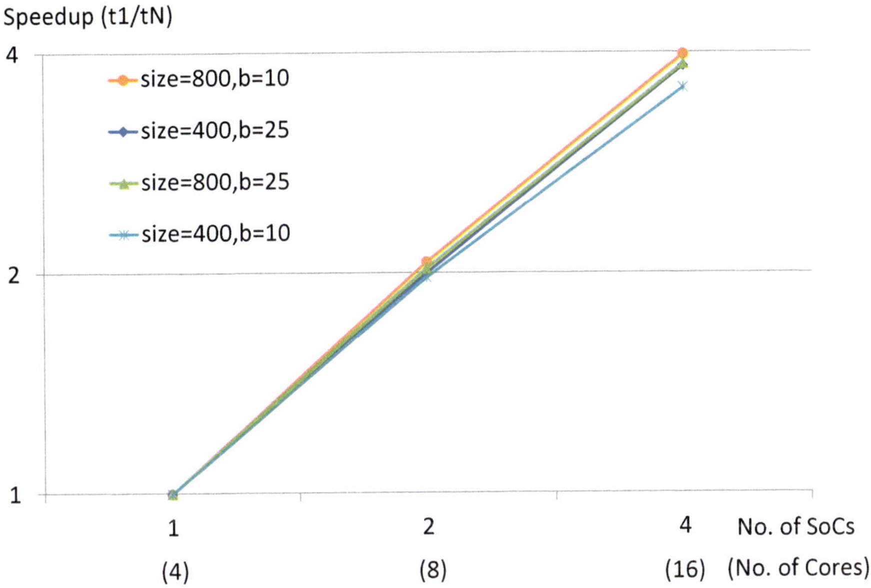


**Fig. 3.10** Sensitivity of the scaling to the block size of the matrix multiplication benchmark (size $= 400, 800$ − block-size $= 10, 25$). When decreasing the block size, more threads are generated thus increasing the parallelism of the application

# References


1. W. Ahmed, M. Shafique, L. Bauer, J. Karlsruhe, Adaptive resource management for simultaneous multitasking in mixed-grained reconfigurable multi-core processors, in *2011 Proceedings of the 9th International Conference on Hardware/Software Codesign and System Synthesis (CODES+ISSS)*, pp. 365–374 (2011)
2. C. Alvarez et al., The AXIOM software layers, in *Digital System Design (DSD)* (2015)
3. E. Argollo, A. Falcón, P. Faraboschi, M. Monchiero, D. Ortega, COTSon: infrastructure for full system simulation. SIGOPS Oper. Syst. Rev. **43**(1), 52–61 (2009). doi:http://doi.acm.org/10.1145/1496909.1496921
4. E. Ayguade, R.M. Badia, D. Cabrera, A. Duran, M. Gonzalez, F. Igual, D. Jimenez, J. Labarta, X. Martorell, R. Mayo, J.M. Perez, E.S. Quintana-Orti, A proposal to extend the openmp tasking model for heterogeneous architectures, in *International Workshop on OpenMP (IWOMP)*, vol. 5568, pp. 154–167 (2009)
5. M. Banzi, *Getting Started with Arduino* (Make Books - Imprint of: O'Reilly Media, Sebastopol, CA, 2008)
6. J. Bueno, L. Martinell, A. Duran, M. Farreras, X. Martorell, R. Badia, E. Ayguade, J. Labarta, Productive cluster programming with OmpSs, in *Euro-Par 2011 Parallel Processing*, pp. 555–566 (2011)
7. J. Clemente, V. Rana, D. Sciuto, I. Beretta, D. Atienza, A hybrid mapping-scheduling technique for dynamically reconfigurable hardware, in *2011 International Conference on Field Programmable Logic and Applications (FPL)*, pp. 177–180 (2011). doi:10.1109/FPL.2011.40
8. D.J. Cook, A.S. Crandall, B.L. Thomas, N.C. Krishnan, CASAS: a smart home in a box. Computer **46**(7), 62–69 (2013). doi:http://doi.ieeecomputersociety.org/10.1109/MC.2012.328

9. R. Ferrer, S. Royuela, D. Caballero, A. Duran, X. Martorell, E. Ayguade, Mercurium: Design decisions for a S2S compiler, in *Proceedings of the Cetus Users and Compiler Infrastructure Workshop in conjunction with PACT* (2011)
10. R. Giorgi, Accelerating haskell on a dataflow architecture: a case study including transactional memory, in *Computer Engineering and Applications (CEA)*, pp. 91–100 (2015)
11. R. Giorgi, Scalable embedded systems: towards the convergence of high-performance and embedded computing, in *Proceedings of the 13th IEEE/IFIP International Conference on Embedded and Ubiquitous Computing (EUC 2015)* (2015)
12. R. Giorgi, P. Faraboschi, An introduction to DF-Threads and their execution model, in *IEEE MPP*, pp. 60–65 (2014). doi:10.1109/SBAC-PADW.2014.30
13. R. Giorgi, A. Scionti, A scalable thread scheduling co-processor based on data-flow principles. ELSEVIER Futur. Gener. Comput. Syst. **53**, 100–108 (2015). doi:10.1016/j.future.2014.12.014. http://www.sciencedirect.com/science/article/pii/S0167739X1400274X
14. R. Giorgi et al., TERAFLUX: harnessing dataflow in next generation teradevices. Microprocess. Microsyst. **38**(8, Part B), 976–990 (2014)
15. M. Herlihy, J.E.B. Moss, Transactional memory: architectural support for lock-free data structures, in *Proceedings of the 20th Annual International Symposium on Computer Architecture, ISCA '93* (ACM, New York, 1993), pp. 289–300. doi:10.1145/165123.165164
16. Intel: Could Smart Homes Be as Commonplace as Smartphones by 2025. http://download.intel.com/newsroom/kits/iot/pdfs/IntelSmartHomeSurveyBackgrounder.pdf. Last accessed on Feb 2016
17. N. Komninos, E. Philippou, A. Pitsillides, Survey in smart grid and smart home security: issues, challenges and countermeasures. IEEE Commun. Surv. Tutor. **16**(4), 1933–1954 (2014). doi:10.1109/COMST.2014.2320093
18. S. Lyberis, G. Kalokerinos, M. Lygerakis, V. Papaefstathiou, D. Tsaliagkos, M. Katevenis, D. Pnevmatikatos, D. Nikolopoulos, Formic: cost-efficient and scalable prototyping of manycore architectures, in *2012 IEEE 20th Annual International Symposium on Field-Programmable Custom Computing Machines (FCCM)* (IEEE, New York, 2012), pp. 61–64
19. E. Palazzetti, *Getting Started with UDOO* (Packt Publishing, Birmingham, 2015)
20. V. Pillet, J. Labarta, T. Cortes, S. Girona, PARAVER: a tool to visualise and analyze parallel code, in *Proceedings of WoTUG-18: Transputer and Occam Developments* (1995)
21. N. Puzovic, S. McKee, R. Eres, A. Zaks, P. Gai, S. Wong, R. Giorgi, A multi-pronged approach to benchmark characterization, in *IEEE CLUSTER*, pp. 1–4 (2010). doi:10.1109/CLUSTERWKSP.2010.5613090
22. S. Ray, Y. Jin, A. Raychowdhury, The changing computing paradigm with internet of things: a tutorial introduction. IEEE Des. Test **33**(2), 76–96 (2016). doi:10.1109/MDAT.2016.2526612
23. A. Scionti, S. Kavvadias, R. Giorgi, Dynamic power reduction in self-adaptive embedded systems through benchmark analysis, in *IEEE Mediterranean Conference on Embedded Computing (MECO)*, pp. 62–65 (2014)
24. D. Theodoropoulos et al., The AXIOM project (agile, extensible, fast I/O module), in *SAMOS* (2015)
25. O. Vermesan, P. Friess, P. Guillemin, S. Gusmeroli, H. Sundmaeker, A. Bassi, I.S. Jubert, M. Mazura, M. Harrison et al. Internet of things strategic research roadmap, in *Internet of Things-Global Technological and Societal Trends*, pp. 9–52 (2011)
26. S. Weis, A. Garbade, J. Wolf, B. Fechner, A. Mendelson, R. Giorgi, T. Ungerer, A fault detection and recovery architecture for a teradevice dataflow system, in, *IEEE Design for Manufacturability (DFM)*, pp. 38–44 (2011)
27. S. Weis et al., Architectural support for fault tolerance in a teradevice dataflow system. Springer Int. J. Parallel Prog. **44**(2), 208–232 (2016)
28. Xilinx Inc.: Xilinx UltraScale Architecture, http://www.xilinx.com/support/documentation/white_papers/wp435-Xilinx-UltraScale.pdf. Last accessed on Feb 2016
29. Xilinx Inc.: Zynq Series, http://www.xilinx.com/content/xilinx/en/products/silicon-devices/soc/zynq-7000.html. Last accessed on Feb 2016

# Part II
# Simulation, Modeling and Programming Frameworks for IoT

# Chapter 4
# Internet of Things Simulation Using OMNeT++ and Hardware in the Loop

**Philipp Wehner and Diana Göhringer**

## 4.1 Introduction

While the internet was primarily used by human beings a couple of years ago, a trend shows up, where "intelligent things" use the network infrastructure to communicate with each other and to exchange data. The aim is to replace the personal computer, as known in its common form, by something that supports humans by their daily activities and procedures, without getting perceived directly [1]. Problematic in this context are the different demands of individual network nodes regarding the network infrastructure. Efficient communication is affected by different protocols that are used [2]. To face this challenge, standardization can help to create preferably huge subnetworks that use a consistent procedure to communicate. But this method is only meaningful as long as it does not produce high costs for individual network participants and complexity of the standard is in a balanced relation to the obtainable benefit. This is not at least related to the energy that is consumed for the standard-compliant communication. The connection of basic sensors, e.g., window or door contact switches in the smart home, would be reduced to absurdity, when a high-cost communication medium, as, for example, Ethernet is used. High acquisition and operating cost would be the result. Hence, utilizing a common interface for communication is meaningful only when requirements of the individual network participants are compatible with the targeted standard. At this, it should get clear that value has to be attached to the interconnection of subnetworks via gateways. To enable the communication across the subnetworks, gateways must be deployed where standardization is not a means to an end. As the complexity of this consideration, especially for huge networks, is

P. Wehner (✉) • D. Göhringer
Application-Specific Multi-Core Architectures (MCA) Group, Ruhr-University Bochum, Bochum, Germany
e-mail: philipp.wehner@rub.de; diana.goehringer@rub.de



G. Keramidas et al. (eds.), *Components and Services for IoT Platforms*,
DOI 10.1007/978-3-319-42304-3_4

not negligible, new technologies have to be developed that allow the exploration of this issue. Particularly to avoid problematic cases which are difficult to investigate in a real world environment, simulation techniques need to be developed that allow a safe but accurate observation.

Subject of this paper is the introduction of a concept that starts by developing a basic gateway. Over the course of time, it can be transferred to a simulator for the Internet of Things (IoT). To achieve this, the network simulator OMNeT++ is introduced that can be utilized for the research on the communication infrastructure. But the target of the paper is not only the simulation of Ethernet networks; it rather introduces a technique to simulate different communication standards in the IoT. It further should be made clear that the presented concept is not limited to an enclosed simulation. If real hardware components, e.g., sensors and actuators are available, they can be integrated in the simulation environment. This Hardware in the Loop (HiL) can essentially increase quality and significance of the results, as the output is not based on probably inaccurate models of the periphery. Furthermore, HiL enables the analysis of components that are not available at an early stage of the development cycle. By using the proposed simulation environment, virtual devices can exist among their physical counterparts. A framework that allows the easy extensibility of IoT networks is of high importance and HiL can enable a rapid prototyping. But there are drawbacks. Especially in large networks, the synchronization between nodes is an important aspect. The presented work enables further analysis on synchronization techniques between the physical devices and the HiL environment. Metrics regarding power consumption [3] and performance [4] can also be considered and analyzed, allowing easy measurements in the entire environment.

The paper is structured in the following manner: Sect. 4.2 introduces the network simulator OMNeT++. Section 4.3 gives an overview of related work. In example of the EU project "RADIO," introduced in Sect. 4.4, the requirements of mobile robot platforms to the communication infrastructure are presented and it is shown, how to deal with these challenges. The concept of the IoT simulation can be found in Sect. 4.5. It starts with an introduction to the idea, including the consideration of HiL and is transferred to a case study, presented in Sect. 4.6. The paper concludes with Sect. 4.7.

## 4.2 OMNeT++

OMNeT++ allows the discrete event simulation of computer networks and other distributed systems [5]. With the target to model large networks, the C++ based project tries to fill the gap between research-oriented simulation software and expensive commercial alternatives, such as OPNET, which is now part of Riverbed Technology [6]. OMNeT++ allows the hierarchical organization of simulation models. The number of layers is hereby not limited. This modular structure supports clarity and facilitates the work with the simulation system. Processes within the virtual network can be visualized via a graphical user interface. This includes the illustration of network graphics, data flow charts, and the possibility of observing objects and variables during simulation. In OMNeT++, the structure of the model is defined using network description (NED) files. Here, parameters are defined for the individual modules and also their connection to other modules and submodules is specified. To efficiently manage large networks, definitions can be split into separate NED files. It is also possible to define parameters for the respective topologies. In this point, OMNeT++ is distinguished from other graphical editors, as it is not limited to fixed topologies. The topology is rather modifiable even during runtime. The user optionally can edit the NED files either via a graphical user interface or based on text files. OMNeT++ includes a library that can be used for the C++ based programming of the modules. It has to be differentiated between two possible implementations: On the one hand, *co-routine based programming* generates a thread for each module where the program code is executed. The respective thread gets the control from the simulation kernel, when a message arrives. In this case, a thread usually never returns; an infinite loop within the module continuously handles incoming and outgoing messages. On the other side, *event-processing* calls the module-function with the message as a parameter. Here, the function immediately returns after processing. Event-processing is feasible especially for extensive simulations, as not every module needs to have its own CPU stack and memory requirements are reduced. In the context of this paper, OMNeT++ is used, as it has the following benefits compared to the related work:

- OMNeT++ is extensible. C++ program code can be used to describe functionality of the respective modules. Furthermore it is possible to integrate OMNeT++ in larger applications by replacing its user interfaces.
- OMNeT++ already contains protocol models like TCP/IP, ATM, and Ethernet.
- The graphical user interface *Tkenv* supports modelling of complex networks.
- OMNeT++ was developed for the simulation of extensive networks.
- As OMNeT++ is open source, functionality of every part of the simulation can be comprehended.

## 4.3 Related Work

The bidirectional communication within home automation networks results in high costs, since each node needs to send and receive data. The benefit of this approach is the higher reliability of the network, as each node can acknowledge packets. In [7], the authors present a technique that is based on less-expensive, unidirectional nodes. A sequence of packets is sent with a constant inter-packet time, guaranteeing that at least one transmitted packet reaches its destination node. The approach is evaluated using an OMNeT++ based simulation. It satisfactorily shows that OMNeT++ can be used for home automation related networks.

Not at least when considering healthcare under the IoT paradigm, the importance of suitable simulation capabilities becomes clear. The authors of [8] present *SimIoT*, a toolkit that can be used to enable experiments on the interactions within an IoT scenario. As monitoring health is an important aspect in the RADIO project, presented in Sect. 4.4, SimIoT is relevant in context of the OMNeT++ based approach presented in this paper.

In context of wireless sensor networks, several approaches exist to simulate these complex environments. OBIWAN [9] is one of those simulators, focusing on cross-layer dependencies and the challenges of large networks that are not sufficiently handled by many of the available simulation tools. The authors present a case study with several thousand nodes, demonstrating the usage of their simulator in large networks. OBIWAN is based on OMNeT++.

There are ambitions to create standards for the interconnection of smart home devices. When a single device is used to manage the communication within the smart home, the whole network is vulnerable. The "Thread Group," as an example, tries to overcome this issue by creating a networking protocol that encourages the development of the IoT [10]. Amongst others, the proposed standard is simple to use, power efficient, and has no single point of failure.

In this context, the Mediola gateway must be mentioned [11]. Its great advantage is the combination of different smart home products to flexibly automate the living environment. The consumer can control the different technologies using a smartphone app. However, the Mediola gateway exactly represents the single point of failure that was mentioned above.

This paper by contrast presents a framework that also provides gateway functionality, but in a decentralized manner. The simulation allows the translation between different communication technologies, but is not dependent on a single device.

A strategic research roadmap for the IoT is presented in [12]. The authors mention the challenges of numerous heterogeneous components and emphasize the need for innovative design frameworks, inspired by HiL. However, HiL is often used in the context of automotive applications [13–15], but has only minor importance in the IoT simulation.

To the best of the authors' knowledge, there is no flexible simulation environment for the IoT that combines decentralized gateway functionality with HiL, enabling the design and analyses of large networks.

## 4.4 Robots in Assisted Living Environments

As demography changes and the population of elderly grow, new paradigms have to be invented to face the challenges of the increasing life expectation. Dealing with chronic diseases results in the need for long term care and innovative technologies to overcome this issue. The classical institutional care hits the brick wall under these challenges as several problems occur, not at least resulting in high costs. Modern information and communication technology (ICT) offers new solution approaches.

The project "RADIO," funded by the European Union's Horizon 2020 research and innovation program, shows up a way to support elderly in their homely environment [16]. RADIO, as short for "Robots in Assisted Living Environments: Unobtrusive, efficient, reliable, and modular solutions for independent ageing" is based on the four main dimensions *user acceptance*, *integrated and power-aware data collection—transmission—processing*, *user interfaces*, and *architecture*. It is the triggering event of the approach presented in this paper and targets the integration of smart home technology and robotic in ways that the user perceives the technical equipment as a natural component, as part of the daily life. Hence, a robot gradually becomes as ordinary as, for example, a radio. To succeed in this challenge, novel approaches to unobtrusiveness and acceptance need to be developed and transferred to a system that satisfies the need for integrated smart home technology together with assistant robot platforms. Instead of utilizing discrete sensing equipment, RADIO rather uses a robot as a mobile sensor for health monitoring. This robot, however, presents special challenges to the ICT as it is mobile.

In the context of this paper, the information technology needs to be capable of handling difficulties regarding the wireless connection. Concepts must be developed that deal with disturbances and disconnection, where data gets delivered delayed and possibly from a completely different spatial location. For this purpose, it is necessary, that suitable data buffers are available and that their integration in the network can be analyzed. Hence, a framework for the simulation of IoT networks, as conceptual introduced in this paper, is beneficial insofar, as complex processes can be broken down. Concepts for problem solving can be developed and transferred to the final application.

## 4.5 Concept

OMNeT++ enables the simulation of large networks and allows the integration of HiL by design. The main challenge of connecting physical devices is its integration to the scheduling mechanism of the simulation environment. OMNeT++ therefore provides a real time scheduler that can be extended by the user. The socket example that comes with the installation of OMNeT++ demonstrates how external applications can get access to the simulation environment: An extension of the real time scheduler opens a TCP/IP socket listening on port 4242. The user can connect to this server with a web browser. The respective HTTP request is visualized in the GUI and shows how data is transferred via the internet to the web server and then sent back to the client. This example sufficiently demonstrates how the interaction between HiL, in form of the client requesting a website, and the scheduler is performed. It also shows that this integration uses a single entry point, in form of the extended scheduler, to connect the two environments, the physical and the virtual world. According to this, the question comes up on how multiple hardware devices can be added to the simulation at different entry points. In case of IoT devices, that could mean that different sensors and actuators need to connect to different nodes in OMNeT++. The socket example does not provide information to face this challenge, as it only shows how a single external device, the web client, can be integrated to the simulation environment. To understand this issue it must become clear that OMNeT++ gets informed about the modified scheduler by an entry in the initialization file of the respective simulation. As this entry is a *general* item, the drawback is that this information affects the entire simulation and hence it is not possible to provide a second scheduler for a second HiL device this way. This naturally makes sense as only one scheduler can handle the simulation process. It is therefore required to develop a mechanism that allows the integration of multiple external devices, on different network nodes, using only one extended version of the scheduler. The approach presented in this paper is based on callback functions that are consecutively invoked by the scheduler. Hence, each node, connecting physical devices to the simulation environment, must provide a method that informs the scheduler about relevant events that need to be included in the simulation flow. The following subsections show how this mechanism is realized in detail.

### 4.5.1 Modified Scheduler

The modified scheduler, `GatewayRTScheduler`, presented in this paper, is derived from the OMNeT++ `cScheduler` class. It provides a method `setInterface()` that enables the user to add an interface to external hardware. The virtual function `getNextEvent()` of `cScheduler` is replaced by an implementation that consecutively calls these interface functions to check if an event occurs. The function `setInterface()` therefore adds all HiL interfaces to a vector, containing a reference to the calling module, a `cMessage` object used to identify the respective event and the actual callback function.

### 4.5.2 HiL Interfaces

The interfaces to the external hardware devices are provided in the form of a shared library. In OMNeT++, a shared library is created by invoking `opp_makemake` with a `--make-so` flag. It mainly consists of the two functions: `virtual void handleMessage(cMessage *msg)` and `static cMessage *interfaceFunction(cModule *module, cMessage *notificationMsg, simtime_t t, cSimulation *sim)`, whereas `msg` is a pointer to the message that must be processed, `module` is a pointer to the module that contains the callback function, `t` is the current simulation time, and `sim` is a pointer to the simulation manager object. `interfaceFunction()` is the callback function that is assigned to `setInterface()` of the gateway real time scheduler class. Its task is to check whether an event of the respective HiL device needs to get considered by the simulation. The task of `handleMessage()`

can be described as follows: The function is called every time a message is present at the respective module. This message can be (a) an external message coming from a distant module, (b) the so-called *self message* that is sent and received within the same module, or (c) the message that was handed over to the scheduler to inform the appropriate module about HiL events. Hence, `handleMessage()` consists of the following distinction of cases:

```
if (msg->isSelfMessage()) {
    // handle self message
} else if (msg == hilEvent) {
    // handle HiL event
} else {
    // handle message from distant module
}
```

### 4.5.3 *Messages*

To face the challenge of different communication standards within a large network, messages of network attendees are translated to an intermediate representation (IR) that is independent from the actual protocol it represents. As this IR should not add an additional payload to the communication mechanism, it must be as lean as possible. In the context of this paper, the IR consists of concatenated integer representations of all relevant input data. Figure 4.1 visualizes this behavior.

The fictitious message coming from the smart home switch consists of a number, identifying the device and its current state (on = 1/off = 0). This message could be transferred to the IR using, for example, the following XML description:

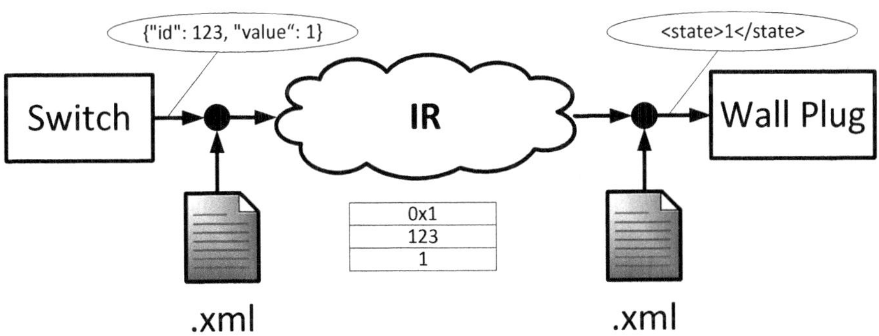


**Fig. 4.1** Message translation example

```
<ir>
    <item pos='0'>
        <fixed>0x1</fixed>
    </item>
    <item pos='1'>
        <variable>
            <uint8>id</uint8>
        </variable>
    </item>
    <item pos='2'>
        <bool>
            <variable>state</variable>
        </bool>
    </item>
</ir>
```

In this case, the intermediate representation of the incoming message would consist of the values 0×1, 123, and 1, representing a unique identifier for the category of smart home switches (0×1), the id of the switch that takes action (123), and the current state (1 = on). The idea is that this IR is only based on variables that can be stored in one or more registers. Therefore, no complex datatypes, for example, "string," are allowed. The respective size that has to be reserved in the IR can be specified by the XML description that is responsible for the incoming message.

In example of Fig. 4.1, the switch should be connected to a wall plug. Considering that this actuator has a communication technique different from the switch, an additional translation mechanism must be included to transfer the messages from the IR to the actual target language. This can exemplarily be done by providing the following XML translation file:

```
<ir>
    <item name='state'>2</item>
</ir>
```

It says that the IR contains the value for the "state" variable at position 2. Using this information, the destination node can translate the information of the IR to the format that must be transferred to the wall plug interface.

The presented format of the IR was selected as OMNeT++ supports this structure in its message files. For the case study, presented in the next section, the following object is used:

```
message GatewayMsg
{
    int source;
    int destination;
    int payload[];
}
```

The integer variables `source` and `destination` contain the ids of the sender and receiver node respectively. `Payload` is an array that consists of the values of the IR.

OMNeT++ provides functions for its dynamic allocation. Adding a new entry is done as follows:

```
int oldSize = msg->getPayloadArraySize();
msg->setPayloadArraySize(oldSize+1);
msg->setPayload(oldSize, newEntry);
```

## 4.6 Case Study

The case study shows how HiL can be added to OMNeT++ using the modified scheduler. There are different techniques on how real hardware can interface with the simulation environment. Using vendor-specific USB dongles, e.g., for Z-Wave communication is one option to enable the data transfer between the respective node and the actual IoT simulator. For the first version of the framework presented in this paper, simple TCP/IP sockets are used as they provide a high degree of flexibility. Therefore, two Raspberry Pis, acting as interfaces, are used to manage the communication on the sender and receiver side. Each Raspberry Pi has a connection to the OMNeT++ simulation environment via Ethernet. The case study implements the concept visualized in Fig. 4.2. Therefore, an EnOcean energy harvester [17] is connected to the first Raspberry Pi using the so-called EnOcean Pi add-on module [18]. It enables the Raspberry Pi to read incoming EnOcean messages via Universal Asynchronous Receiver Transmitter (UART). The Raspberry Pi connects to the socket, opened by the EnOcean interface that was added to OMNeT++. On the receiver side, a Z-Wave module, RaZberry [19], connects the second Raspberry Pi and a Z-Wave wall plug [20]. When the EnOcean switch gets pushed, a message is transferred via Ethernet to the OMNeT++ simulation where it is wrapped into the message object presented in the previous section. The respective switch-state is added to the payload array and then sent to the receiver node, where the message is transferred to the wall plug using the second Raspberry Pi.

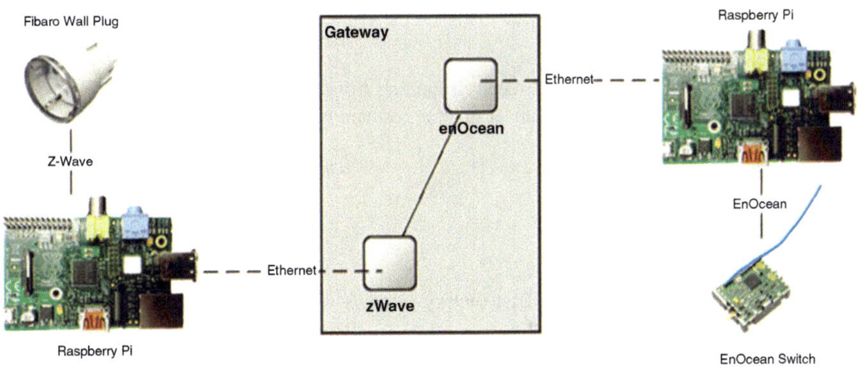


**Fig. 4.2** OMNeT++ simulation acting as gateway between HiL devices

The case study currently does not include the implementation of the XML translation mechanism. In future releases of the presented approach, the data exchange between the sensors and actuators and the actual IoT simulation framework will be improved and further analyses results will be presented.

## 4.7 Conclusion

This work in progress paper provides the basis for an IoT simulation environment including HiL. The network simulator OMNeT++ is used to model the infrastructure between virtual and physical nodes. By extending its scheduler, extendibility with further HiL devices is enabled. Although it is too early to present further simulation results, the authors believe that the presented approach can help to encourage the research on IoT platforms to face the upcoming challenges, including the simulation and analysis of distributed and ubiquitous computing. As OMNeT++ was designed to model large networks, it is suitable for the simulation of future smart environments. The key to success is hereby the consideration of physical hardware in the virtual environment, as a simulator is not a universal solution and might fail in cases where real world data is required to gain feasible results.

The presented approach moreover features the functionality of a gateway, where the translation between different communication standards is not limited to a specific device. By using the intermediate representation, a decentralized mechanism is developed that transfers relevant data to the respective node where it is translated to the final communication standard.

In future, effort will be invested to rework the data flow from multiple sources to multiple destinations in complex networks. In example of a smart home environment, it will be demonstrated how the simulation of not available sensors and actuators can be handled by the presented framework. Also security aspects will be taken into account. Currently, no specific OMNeT++ library, like iNet, is used to model the network. In future versions of the presented approach this will become an essential part of the IoT simulator.

**Acknowledgments** This project has received funding from the European Union's Horizon 2020 research and innovation programme under grant agreement No 643892.

## References

1. M. Weiser, The computer for the 21st century, https://www.ics.uci.edu/~corps/phaseii/Weiser-Computer21stCentury-SciAm.pdf
2. A. Al-Fuqaha, M. Guizani, M. Mohammadi, M. Aledhari, M. Ayyash, Internet of things: a survey on enabling technologies, protocols and applications. IEEE Commun. Surv. Tutorials 17 (2015)

3. S. Schürmans, G. Onnebrink, R. Leupers, G. Ascheid, X. Chen, ESL power estimation using virtual platforms with black box processor models. Proceedings of the 3rd workshop on virtual prototyping of parallel and embedded systems (ViPES), Samos, Greece, July 2015
4. P. Ittershagen, P. Hartmann, K. Grüttner, W. Nebel, A workload extraction framework for software performance model generation. Proceedings of the 7th workshop on rapid simulation and performance evaluation: methods and tools (RAPIDO), Amsterdam, The Netherlands, Jan 2015
5. A. Varga, The OMNeT++ discrete event simulation system. Proceedings of the 2001 European simulation multiconference (ESM), Prague, Czech Republic, June 2001
6. Riverbed application and network performance management solutions, http://www.riverbed.com/products/performance-management-control/opnet.html
7. P. Parsch, A. Masrur, W. Hardt, Designing reliable home-automation networks based on unidirectional nodes. Proceedings of the 9th IEEE international symposium on industrial embedded systems (SIES), Pisa, Italy, June 2014
8. S. Sotiriadis, N. Bessis, E. Asimakopoulou, N. Mustafee, Towards simulating the internet of things. Proceedings of the 28th international conference on advanced information networking and applications workshops (WAINA), Victoria, BC, May 2014
9. E. Egea-López, F. Ponce-Marín, J. Vales-Alonso, A.S. Martínez-Sala, J. García-Haro, OBIWAN: wireless sensor networks with OMNET++. Proceedings of the 2006 IEEE Mediterranean electrotechnical conference (MELECON), Malaga, Spain, May 2006
10. Thread Group, http://threadgroup.org/
11. mediola – connected living AG, http://www.mediola.eu/
12. O. Vermesan, P. Friess, P. Guillemin, S. Gusmeroli, H. Sundmaeker, A. Bassi, I. Soler Jubert, M. Mazura, M. Harrison, M. Eisenhauer, P. Doody, Internet of things strategic research roadmap. Cluster of European Research Projects on the Internet of Things, http://www.internet-of-things-research.eu/pdf/IoT_Cluster_Strategic_Research_Agenda_2011.pdf
13. D. Michalek, C. Gehsat, R. Trapp, T. Bertram, Hardware-in-the-loop-simulation of a vehicle climate controller with a combined HVAC and passenger compartment model. Proceedings of the 2005 IEEE/ASME international conference on advanced intelligent mechatronics, Monterey, CA, July 2005
14. B. Aksun Güvenç, L. Güvenç, S. Karaman, Robust yaw stability controller design and hardware-in-the-loop testing for a road vehicle. IEEE Trans. Veh. Technol. **58**(2), 555–571 (2008)
15. P. Wehner, F. Schwiegelshohn, D. Göhringer, M. Hübner, Development of driver assistance systems using virtual hardware-in-the-loop. Proceedings of the 14th international symposium on integrated circuits (ISIC), Singapore, Dec 2014
16. EU project RADIO, Robots in assisted living environments: unobtrusive, efficient, reliable and modular solutions for independent ageing, http://www.radio-project.eu/
17. EnOcean GmbH, ECO 200 energy converter for motion energy harvesting, https://www.enocean.com/en/enocean_modules/eco-200/
18. EnOcean GmbH, EnOcean Pi transforms raspberry Pi into a wireless gateway, https://www.enocean.com/en/enocean-pi/
19. RaZberry, http://razberry.z-wave.me
20. FIBAR GROUP, FIBARO wall plug, http://www.fibaro.com/en/the-fibaro-system/wall-plug

# Chapter 5
# Towards Self-Adaptive IoT Applications: Requirements and Adaptivity Patterns for a Fall-Detection Ambient Assisting Living Application

**Sofia Meacham**

## 5.1 Introduction

According to the World Health Organization [1] approximately 28–35 % of people aged 65 and over fall each year increasing to 32–42 % for those over 70 years of age. The situation is getting worse due to the situation that elderly people often have to stay alone for long periods of time either in their own home environments or in care homes. In this context, automatic fall-detection systems can enable triggering of an alert (manual or automatic) in an emergency situation, thus enabling help when it is required, reducing deaths from falls and consequently increasing the personal feeling of security of elderly people.

There are several available fall-detection systems, each of which addresses some of the requirements, both for indoor [2, 3] and outdoor environments [4]. However, the self-adaptivity in these systems is still in a primitive form.

On the other hand, a pattern-based approach to software engineering is an established and well-known method that has benefited many application areas and start from Christopher Alexander, a building architect, and his attempt to capture solutions to recurring problems [5]. Patterns allow us to benefit from previous experience. Their use is being explored in several complex systems areas such as Systems of Systems (SoS) [6]. However, their use for adaptivity is still limited due to the fact that areas such as self-aware IoT are at their early stages.

In this chapter, we will explore a pattern-based approach for a fall-detection system, with emphasis on their adaptivity requirements.

Specifically, in Sect. 5.2, the requirements to system design for the fall-detection case study will be presented. The section will start with an overview of the case study (Sect. 5.2.1), and then describe the requirements gathering and

S. Meacham (✉)
Bournemouth University, Fern Barrow, Poole, Dorset BH12 5BB, UK
e-mail: smeacham@bournemouth.ac.uk



G. Keramidas et al. (eds.), *Components and Services for IoT Platforms*,
DOI 10.1007/978-3-319-42304-3_5

analysis processes (Sect. 5.2.2) as well as the resulting high-level system design (Sect. 5.2.3). The section will finish by identifying the adaptivity requirements (Sect. 5.2.4). In Sect. 5.3, the proposed pattern-based approach will be presented with dedicated subsections for the patterns that have been applied (Sects. 5.3.1.1, 5.3.1.2, and 5.3.1.3). The resulting system design and implementation will be presented in Sect. 5.3.2. In Sect. 5.4, problems that have identified from the above requirements/design/implementation path will be analysed along with proposed solutions. In Sect. 5.5, the conclusions and directions for further research will be discussed.

## 5.2 Requirements to System Design: The Fall-Detection Application

### *5.2.1 Case Study Overview*

This case study was set in collaboration between Bournemouth University and the Technological Educational Institute of Western Greece (TWG), the Embedded System Design and Application Laboratory (ESDA lab) of the Computer Engineering and Informatics Department. Specifically, the TWG/ESDA Lab has long-lasting expertise in IoT applications and assisted living systems. The example of their ambient assisted living (AAL) laboratory in Patras provided significant input to this work.

Specifically, in this case study, we are considering the requirements gathering and analysis, system design and implementation for an indoor/outdoor fall-detection AAL system for a UK care home.

The system will automatically detect various elements of fall-detection such as the elderly person moving out of the care home or the garden area, detection of no movement for an extended period of time (8 h) or a sudden acceleration such as a fall.

General medical information needs to be kept using electronic means so that patients, their relatives and medical professionals (doctors, nurses and carers) can access/update and consistently maintain. Security and privacy issues should be maintained for medical data and each user of the system should have different privileges in using the stored information. The data should be handled according to the relevant UK data protection act and ethics rules and considerations for medical information.

The medical professionals (nurses, carers) should have access to the system for observing the level and recharging the batteries of the fall-detection system.

Also, the elderly person should be able to have a choice of communicating directly with corresponding carers, nurses, relatives in addition to the automatic alarms.

In the case of a false automatic alarm, the elderly person should be able to designate that the alarm was false through an appropriately designed interface.

Last but not least, a major requirement for the future would be to add to the system 'intelligent' behaviour wherever appropriate. For example, in case a fall is detected, monitoring mechanisms should be increased in order to obtain more information about the criticality of the incident and the patient's medical condition.

## *5.2.2 Requirements Gathering and Analysis*

### 5.2.2.1 Overview of Modelling Languages

Throughout this project, model-based systems engineering (MBSE) was applied as an approach to the design and development of a number of systems. In MBSE models take a central role, not only for analysis of these systems but also for their construction. According to INCOSE, the adoption of MBSE has several advantages [7] such as improved communication between stakeholders, team members through diagrammatic model representations; improved quality through early identification of problems and fewer errors at the integration stage; increased productivity through reusability of existing models and reduced risk through improved estimates and on-going requirements validation and verification. Overall, it has been said to increase productivity and efficiency in the design and development mainly of complex systems.

Specifically in this project, we used a combination of SysML and UML modelling languages for MBSE within an Eclipse/Papyrus environment [8] to cover most cases. SysML was mainly adopted due to its suitability for modelling a wide variety of systems [9]. It was the result of a UML RFP recommendation for system engineers and has been adopted by OMG since 2006. It offers system engineers several noteworthy improvements over UML. SysML reduces the software-centric restrictions of UML and adds more diagram types such as block definition diagrams (BDD), internal block diagrams, parametric and requirements diagrams. Due to the above additions, SysML is able to model a wide range of systems such as hardware, software, information and processes [10]. On the other hand, we still used UML diagrams wherever it was more appropriate for the project.

### 5.2.2.2 Requirements Engineering Methods

For requirements gathering and analysis, we started defining our requirements using Volere templates [11]. Volere templates are a widespread method for the whole requirements engineering process. In our project, we used them in order to provide a systematic way to formulate the requirements documents. As a second step, we moved on to modelling and diagrammatic techniques, among which was SysML requirements diagrams and use cases. The choice of combining SysML

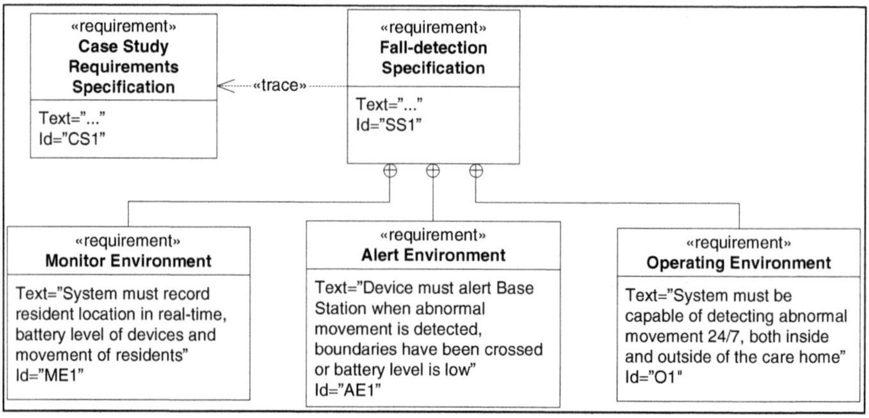


**Fig. 5.1** High-level SysML requirements diagram

requirements diagrams with the traditional use case modelling was deliberate and offered a variety of views to the system requirements. Use cases provide the actor-based description of the system [12] whereas requirements diagrams give a diagrammatic view that connects requirements with block diagram implementations, offering a path to traceability/verification as well as providing relationships between requirements.

In Fig. 5.1, we can see a high-level requirements diagram that consists of three main categories of requirements: *Monitor Environment*, which is used to monitor movement, location and battery levels; *Alert Environment*, which creates an alert for five cases (manual alert, fall-detection, no movement detection, out of range and low battery) and *Operating Environment*, which can be indoor, outdoor and 24/7 operating system.

### *5.2.3 System Design*

In Fig. 5.2, the corresponding high-level SysML block diagram of the system is presented. A one-to-one correspondence between the requirements in the previous diagram and the blocks in this diagram is apparent. In addition to this, the 'flow' of events is depicted. The *Monitor* block monitors the system (*Operating Environment* block) and when something abnormal is detected it raises a *Trigger* to the *Alert* block. The *Alert* block (depending on the specific *Trigger* that was raised) decides on the required adaptation and sends an *Adapt* signal to the *Operating Environment* to perform the changes.

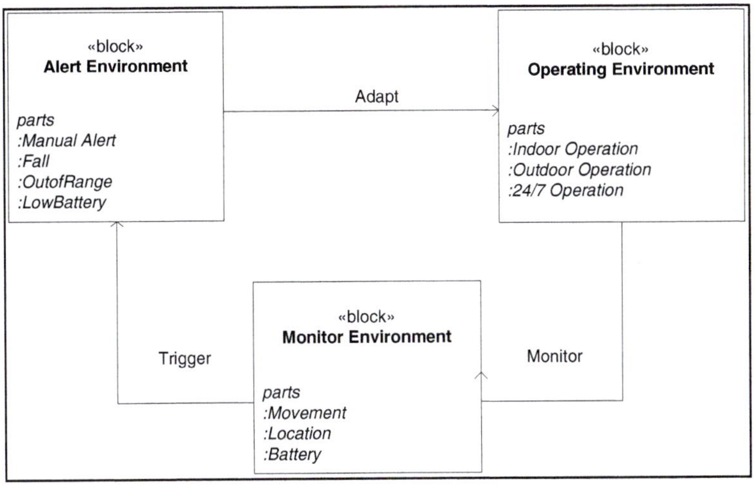


**Fig. 5.2** High-level SysML block definition diagram (BDD)

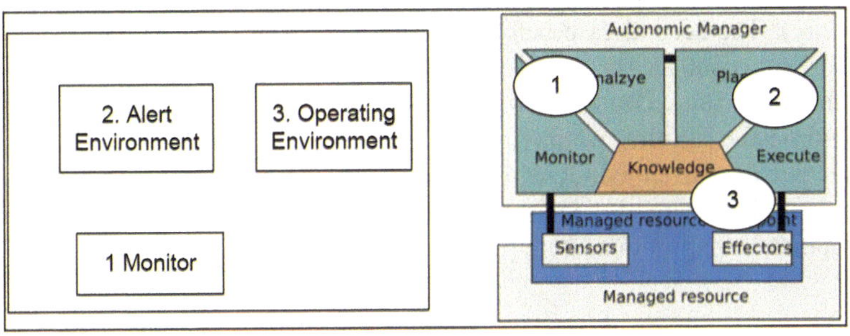


**Fig. 5.3** System design mapping to MAPE-K model

## *5.2.4 Adaptivity Requirements*

From the above block diagram, the similarity between our system and the IBM MAPE-K control loop model [13] for adaptivity is apparent.

In Fig. 5.3, the correspondence between the two models is depicted. The MAPE-K model consists of a monitor that observes the system and the environment (the *Monitor Environment* in our case), an analyser that analyses the monitoring results, a planning component that decides on an appropriate adaptation action (*Alert Environment*) and an execution component that executes the chosen adaptation action (*Alert Environment,* changes in the *Operating Environment*).

Adaptivity is a very important aspect in our fall-detection system and in IoT applications in general.

There are two types of adaptation: design-time adaptation and run-time adaptation. The design-time adaptation is the simplest form of adaptation and consists of mechanisms that are known at design time. For example, our current fall-detection system is designed with design-time adaptation. If a fall, no movement, manual alert or low battery is detected, then it triggers an alert, and the system responds according to the predetermined way. For design-time adaptation, case-based reasoning (which consists of a set of if-then-else statements) can predetermine the system's reaction to different situations.

On the other hand, run-time adaptation consists of the dynamic adaptation of the system according to situations that are discovered at run-time. For example, when a fall is detected, a medical professional might wish to add more sensors to collect data for the patient's health. Which sensors to activate could be given as an option to the medical professional or could be predetermined in a case-based reasoning way. In any case, the system should be able to detect that it should change itself and specifically, the monitoring mechanism should, for example, move from monitoring only movement to monitoring blood pressure and heart rate.

Also, there are different levels and abilities for run-time adaptation that range from cases that can be predicted and the adaptation possibility is inherent in the system (that can be classified as design-time adaptivity) to cases where unpredictable events occur and the system is able to handle them (this is the real run-time adaptivity). The second case belongs to the field of artificial intelligence and its methods and is outside the main scope of this paper.

In the following sections, we will present a pattern-based approach to modelling design-time and run-time adaptivity.

Note that the current fall-detection systems and the system we have developed for our case study belong to the design-time adaptation type. The run-time adaptation development is part of our future plans.

## 5.3 Proposed Pattern-Based Approach: Current IoT

Christopher Alexander describes problems and their solutions using an expression he called a pattern [5]: '*Each pattern describes a problem that occurs over and over again in our environment and then describes the core of the solution to that problem* ***in such a way that you can use this solution a million times over without ever doing it the same way twice***'. Motivated by the work of Alexander, software engineers observed similar recurring problems and solutions throughout the design of software systems [14]. Patterns are described by text representations accompanied by diagrams.

In our project, we focused on adaptivity patterns due to their high-importance and we applied a combination of patterns defined in the master thesis of Ramirez [15] and are referred to software systems, and in the COMPASS FP7 EU project deliverable [6] and are referred to SoS. These patterns were originally developed for software systems due to their inherent flexibility. However, IoT applications have similar requirements and the technological advances have led into more flexible hardware architectures such as wearable sensors (that can be activated/deactivated) and reconfigurable hardware.

For modelling the patterns for our application, SysML block diagrams were chosen for compatibility with the rest of our system design. Note that most of the patterns were originally described in UML.

## *5.3.1 Patterns Applied*

### 5.3.1.1 Adaptation Detection Pattern

In Fig. 5.4, we can see the *Adaptation Detection Pattern* [15] applied to our case study. This consists of an *Observer* block, which has the responsibility to 'observe' the data that are being produced by the *Sensor.* These *Data* are stored so that they get interpreted by the *Observer*. The *Observer* can either work on the stored data or get data directly from the *Sensor*. The amount of data that are normally being produced by a *Sensor* (such as motion sensor) is quite huge and storing them is the easiest way to allow their processing. In the system, we have several *Observers*, one for each *Sensor* dealing with fall-detection, no movement detection, out of range, low battery and manual alert which are part of our fall-detection system

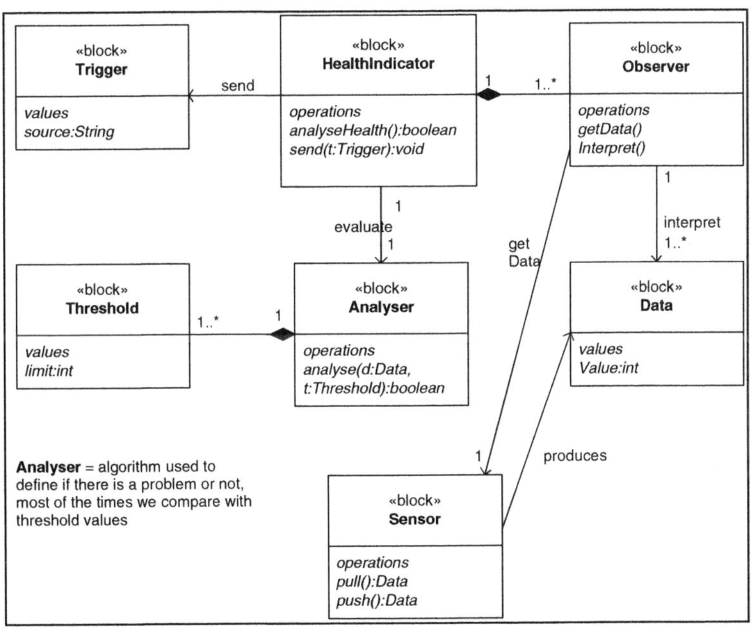


**Fig. 5.4** Adaptation detection pattern

alert cases. The *HealthIndicator*[1] block is responsible for analysing information collected by the *Observer*, calculating if there is anything abnormal in the data by usually comparing the data with some *Threshold* values. If there is something abnormal, the *HealthIndicator* block sends a *Trigger* to the *Trigger* block. The *Trigger* that is produced is the appropriate one for the case that occurred from the aforementioned alert cases (fall-detection, no movement, etc.).This diagram is parametric as regards to the *Trigger–Sensor–Observer* that it refers to and as regards to the *Analyser* algorithm that is being used for decisions. In most cases, we have a physical sensor collecting data, an algorithm that just compares with threshold values and a trigger that is raised appropriately. It is very important to note that the monitoring mechanism that is used in this pattern is fixed and it is polling information that is produced by a sensor. If we change the monitoring mechanism or the algorithm that makes decisions, this pattern no longer applies.

#### 5.3.1.2 Case-Based Reasoning Pattern

In Fig. 5.5, we can see the *Case-based Reasoning Pattern* [15] applied to our case study. The case-based reasoning is the simplest reconfiguration mechanism and is applied when the criteria for reconfiguration is not complex and can be expressed through *if-then-else* statements. The central block for this pattern application is the *InferenceEngine* which reacts according to a raised *Trigger* (fall-detection, no movement, low battery, etc.) and applies the corresponding set of *Rules*. The *Rules* that apply tie together *Trigger* and *Decision*. In the original pattern description, the system can store the *Triggers* and make the *Rules* to learn. However, in our case

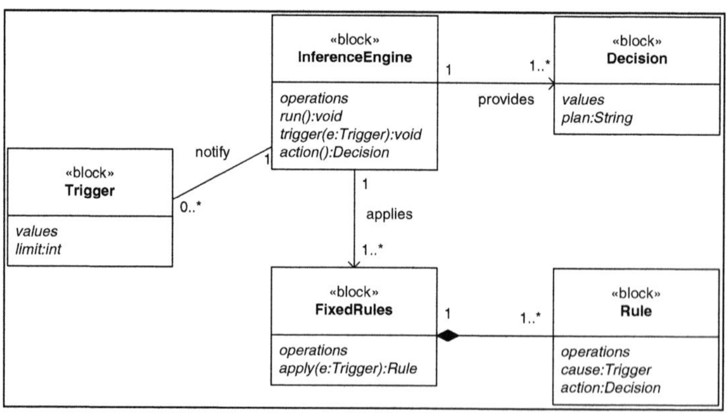


**Fig. 5.5** Case-based reasoning pattern

[1] Note that the name *HealthIndicator* came from the original pattern used and not from our case study health-care application and it refers to the health of the observed data.

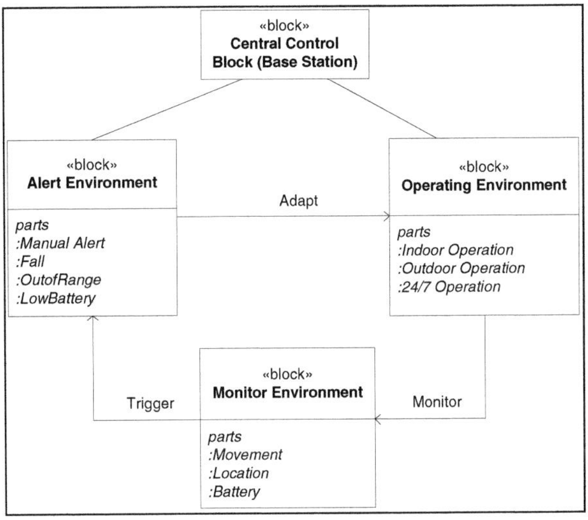


**Fig. 5.6** Centralised architecture pattern

study, we decided not to add any learning capabilities as that would increase the risk of mistakes/failures which is particularly important for critical applications such as fall-detection.

#### 5.3.1.3 Centralised Architecture Pattern

It is important to note that this pattern is not directly related to adaptivity. However, its application is vital for the resulting system design and implementation.

In Fig. 5.6, we can see the *Centralised Architecture Pattern* applied to our case study. The original pattern proposed in [6] was referring to centralised architectures for SoS and contained a central block that would control everything. In our case, the central block is a *Base Station* that communicates with the major blocks in the system. The *Alert* and *Operating Environment* blocks directly communicate with the *Base Station* in order to handle alerts and perform adaptations. The *Monitor* block doesn't seem to require communications with the *Base Station* in the current version of our system as the monitoring mechanisms are static (polling information from sensors).

### *5.3.2 System Design/Implementation Using the Proposed Pattern-Based Approach*

In Fig. 5.7, the corresponding low-level SysML BDD of the system is presented and gives us an idea of the building blocks that are required to perform the basic operations of the system.

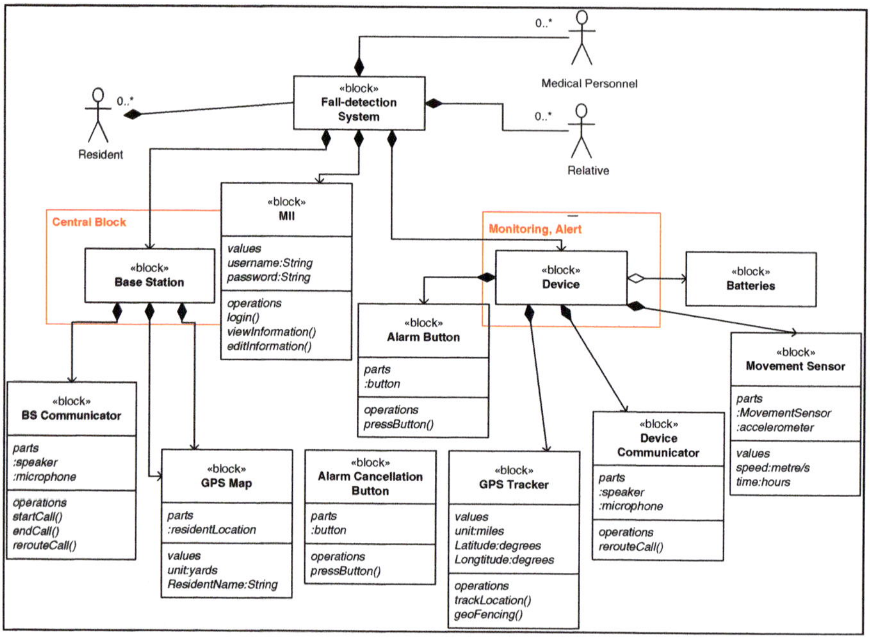


**Fig. 5.7** Low-level SysML block definition diagram (BDD)

As a whole, three parts make the *Fall-detection System* function: a *Base Station*, *Devices* and a *Medical Information Interface (MII)*. These directly related to the centralised architecture pattern diagram of Fig. 5.6, where the *Central Control* block (Fig. 5.6) corresponds to the *Base Station* block (Fig. 5.7), the *Monitoring/Alert Environment* blocks (Fig. 5.6) correspond to the *Device* block (Fig. 5.7). The *Operating Environment* block of Fig. 5.6 corresponds to the whole diagram of Fig. 5.7.

The physical device is a wearable necklace, pendant or wristband assigned to a single resident in a care home. This device will contain a *Movement Sensor*, *Batteries*, *Device Communicator*, *GPS Tracker* and an *Alarm Button* for manual alerts. The *Movement Sensor* block uses an accelerometer and a movement sensor to monitor and detect abnormal movement. When abnormal movement is detected, an alert is automatically sent to the *Base Station* block for the medical personnel to respond. The *GPS Tracker* monitors the location of the device in real time and this information is updated on the *Base Station*. The *Device Communicator* allows conversation between medical personnel and the elderly relative. It consists of both a microphone to speak through and a speaker to listen. The *Device* will contain *Batteries* which are rechargeable. When a battery is below 5 %, an alert is sent to the *Base Station* to inform the medical personnel that the battery needs changing.

The *Base Station* will act as a central communication point for all devices. The *Base Station* will have a *BS Communicator* which also consists of a microphone and a speaker. The *Base Station* will also allow the medical personnel to view the location of any resident. The *Base Station* will also allow the medical personnel to cancel the alarm if it is not deemed an emergency situation or a situation has been dealt with.

The *Medical Information Interface* (*MII*) stores all medical information of each resident and can be accessed by all actors: access is controlled by an appropriate access control system. For example, the relatives shouldn't be able to modify medical records. All the corresponding UK laws should apply.

Most popular systems in the market such as Alert 1 [16], Life Call [17] currently address only some of the above features. In particular, outdoor fall-detection is an area that is missing from most current fall-detection systems.

## 5.4 Problems with Current Fall-Detection Systems and Pattern-Based Run-Time Adaptation: The Future IoT

There are several problems with our case study and the current fall-detection systems that are in the market.

First of all, a centralised architecture that assigns all system control to a single block, the *Base Station* in our case, will inherently cause problems. By centralising control, we are creating a single point of failure for the system. If the *Base Station* fails, then the whole system will fail. Consequently, a more decentralised approach is required.

For the decentralisation problem, patterns that have been proposed in [18] and in [15] should be taken into account. For example, the *Decentralised Reconfiguration Pattern* could be applied to enhance the operation of the *Base Station*. This pattern allows for a decentralised approach to be adopted for client–server architectures. In our case, the *Base Station* acts as a server to adaptation requests and a decentralisation of its operation would remove it from its current single point of failure operation.

Another problem that arises from our case study is the use of polling sensor information as a monitoring mechanism. Polling for information is a continuous monitoring mechanism that has been used in computing architecture for more than twenty years and has always been accused for its high cost in time and processing power. It can be replaced with event-based mechanisms where appropriate. Event-based mechanisms, however, have the overhead of the extra logic required to create and handle the events.

There are several patterns that could be applied to improve and replace at times the polling monitoring. For example, in our case study, the polling of sensor information should be applied to the real-time fall-detection part of the system, but for collecting general health-related patient information such as blood pressure or blood sugar levels that would be useful for general patient monitoring, passive monitoring could be applied. With the term passive monitoring, we mean the case when notifications are sent when a change occurs. The most appropriate pattern for this type of monitoring is the *Content-based Routing Pattern*. With the content-based routing pattern, we route messages across a distributed monitoring infrastructure based on the content of the message. Clients such as medical professionals or relatives subscribe to feeds that they are interested in (e.g. monitoring blood sugar levels in diabetic patients) and they get notifications when there is information generated. This way, they don't have to monitor all feeds and unnecessary traffic is avoided.

Another monitoring mechanism that needs to be explored is the application of the *Sensor-factory Pattern*. This pattern is still based on the polling from sensors type of communication; however, this adds a new dimension of flexibility by enabling the dynamic addition of new sensors when and where required. For example, in an emergency situation, we might want to activate more sensors in order to monitor the patient more closely before the emergency services are in place.

From the above scenarios, it is apparent that there is an increasing trend to dynamically change monitoring mechanisms at run-time in order to adapt to changing environmental conditions and requirements, and consequently alternate appropriately between the adaptivity detection, sensor-factory and content-based routing monitoring patterns.

The ability to dynamically change monitoring mechanisms is provided by another pattern, the *Reflective Pattern*. It is applicable in situations where we cannot directly observe the internal properties of a component, due to visibility constraints such as encapsulation or security purposes, and when we want to change the monitoring mechanism at run-time without affecting the corresponding client. Two intermediate objects are inserted for this purpose: one is the *Metaobject* that contains information about an object; second is the *Proxy* that, by using the *Metaobject* provides a transparent mechanism to intercede during an object's normal behaviour and introduce additional behaviour at run-time. It has originated from (and can be supported by) programming languages with reflective capabilities such as Java, PHP and Python leading to the use of self-improving software [19].

## 5.5 Conclusions and Future Work

The proposed method has been evaluated as regards to the compliance of the designed system with the original requirements and the quality of the final product.

The pattern-based approach offered the general advantages of reusing existing solutions and tailoring them to the specific application problem. The alternative would have been to use one of the existing frameworks for adaptation such as Rainbow [20]. However, that would give us a more restricted solution that would be difficult to adapt to our requirements, to future requirements and more often than not would be domain specific.

On reflection, the system we designed and developed through this approach was effective as regards to satisfying the fall-detection case study requirements. By applying three adaptive patterns, we included all the necessary functional parts of monitoring, alerting and adapting, in a controlled and methodological way.

Two of the patterns that we used (*Adaptivity Detection*, *Case-based Reasoning*) contained inherent scalability and reusability properties. They were designed to be used repeatedly for more than one monitor/trigger pairs and if-then-else cases, respectively. For example, the adaptivity block was 'reused' for both the case of *movement sensors* and *no movement trigger* pair and for the *battery level sensor* and *low battery trigger* pair. This applies at the design level. At the lower implementation level, every part was a separate block though. However, reusability and scalability at the design level is an invaluable property for designing and developing systems as can be observed by the tremendous amount of work by industries in techniques such as product line engineering and variability modelling [21].

Our future plans to extend this work involve applicability of the run-time adaptivity patterns in the fall-detection IoT application as well as in other application areas. Formal methods should also be explored as regards to adaptivity properties to complement and enhance this work as especially for the IoT application domain, real-time and safety-critical properties are fundamental requirements.

Last but not least, we believe that new frameworks for the future IoT should be built that would be combined and informed with patterns and that will have adaptivity requirements as a central point both for the systems to be designed and the frameworks themselves in order to conform to the changing requirements of the changing IoT applications of the future.

## References

1. World Health Organization, *Ageing and Life Course Unit.* WHO global report on falls prevention in older age (World Health Organization, 2008)
2. M. Mercuri, C. Garripoli, P. Karsmakers, P.J. Soh, G.A. Vandenbosch, C. Pace, P. Leroux, D. Schreurs, Healthcare system for non-invasive fall detection in indoor environment. In *Applications in Electronics Pervading Industry, Environment and Society* (Springer, 2016), pp. 145–152
3. Q. Dong, Y. Yang, W. Hongjun, X. Jian-Hua, Fall alarm and inactivity detection system design and implementation on Raspberry Pi. ICACT 2015, 17th IEEE international conference on advanced communications technology, July 2015, pp. 382–386
4. F. Busching, H. Post, M. Gietzelt, L. Wolf, Fall detection on the road. 2013 IEEE 15th international conference on e-health networking, applications and services (Healthcom), IEEE, Oct 2013, pp. 439–443
5. C. Alexander, *The Timeless Way of Building*, vol. 1 (Oxford University Press, New York, 1979)
6. S. Perry, Final report on SoS architectural models. Document Number: D22.6. COMPASS FP7 EU Project public deliverables (2014). http://www.compass-research.eu/Project/Deliverables/D22.6.pdf. Accessed 9 Feb 2016
7. INCOSE, Systems engineering vision 2020, v.2.03 (2007), http://oldsite.incose.org/ProductsPubs/pdf/SEVision2020_20071003_v2_03.pdf. Accessed 9 Feb 2016
8. Papyrus, *Papyrus* (2016), https://eclipse.org/papyrus/. Accessed 9 Feb 2016

9. T. Bouabana-Tebibel, S.H. Rubin, M. Bennama, Formal modeling with SysML. 2012 IEEE 13th international conference on information reuse and integration (IRI), IEEE, Aug 2012, pp. 340–347
10. L. Apvrille, Y. Roudier, Designing safe and secure embedded and cyber-physical systems with SysML-Sec. In *Model-Driven Engineering and Software Development* (Springer, 2015), pp. 293–308
11. S. Robertson, J. Robertson, *Mastering the requirements process: getting requirements right*, 3rd edn. (Addison-Wesley, Upper Saddle River, NJ, 2013)
12. D. Kulak, E. Guiney, *Use Cases: Requirements in Context* (Addison-Wesley, 2012)
13. J.O. Kephart, D.M. Chess, The vision of autonomic computing. Computer **36**(1), 41–50 (2003)
14. J. Vlissides, R. Helm, R. Johnson, E. Gamma, *Design Patterns: Elements of Reusable Object-Oriented Software* (Addison-Wesley, Reading, MA, 1995), p. 11
15. A. Ramirez, *Design Patterns for Developing Dynamically Adaptive Systems*, 1st edn. (Google Books, 2008)
16. Medical Alert Systems | Medical Alert Services for Seniors - Alert1® (2016), https://www.alert-1.com/. Accessed 9 Feb 2016
17. Medical Alert Systems with FallAlert™ 24/7 Medical Alert Systems (2016), http://lifecall.com/. Accessed 9 Feb 2016
18. D. Weyns, B. Schmerl, V. Grassi, S. Malek, R. Mirandola, C. Prehofer, J. Wuttke, J. Andersson, H. Giese, K.M. Göschka, On patterns for decentralized control in self-adaptive systems. In *Software Engineering for Self-Adaptive Systems II* (Springer, Berlin Heidelberg, 2013), pp. 76–107
19. R.V. Yampolskiy, Analysis of types of self-improving software. In *Artificial General Intelligence* (Springer International Publishing, 2015), pp. 384–393
20. D. Garlan, S.W. Cheng, A.C. Huang, B. Schmerl, P. Steenkiste, Rainbow: architecture-based self-adaptation with reusable infrastructure. Computer **37**(10), 46–54 (2004)
21. S. Meacham, F. Gioulekas, K. Phalp, SysML based design for variability enabling the reusability of legacy systems towards the support of diverse standard compliant implementations or standard updates: the case of IEEE-802.15. 6 standard for e-Health applications. In *Proceedings of the 8th International Conference on Simulation Tools and Techniques* (ICST—Institute for Computer Sciences, Social-Informatics and Telecommunications Engineering, Aug 2015), pp. 284–289

# Chapter 6
# Small Footprint JavaScript Engine

**Minsu Kim, Hyuk-Jin Jeong, and Soo-Mook Moon**

## 6.1 Introduction

Internet-of-things (IoT) represents the network of "things" embedded with sensors and computational units [1]. A thing can communicate with another thing, not only as a server but also as a client, so there are frequent I/O requests via the network or the sensor. For example, consider a smart bulb. When a user pushes a button on their smartphone, the smart bulb will receive this message to turn on or off the bulb accordingly. Also, the smart bulb can be turned on if it detects darkness based on information collected by a sensor, or can be turned off if it senses no movement for a while. This scenario indicates that there would be many *event*s initiated from the sensor or the network, so it would be natural to program the IoT computation in an event-driven manner. In other words, the message sent by the phone or the detection by the sensor is regarded as an event, which is adequately handled by its corresponding *event handlers* previously registered for the event.

JavaScript is a mainstream scripting language, supported by most web browsers. JavaScript is known to be particularly useful for event-driven programming, because JavaScript functions are classified as objects, even defined anonymously, thus being easily registered as event handlers. For example, an event handler for the smartphone message can be registered in the smart bulb as follows:

```
bulb.addEventListener("msgTurnOn",
                      function() { bulb.on(); } );
```

This surprisingly simple JavaScript code registers an anonymous function as an event handler for the event `msgTurnOn`. The anonymous function will be fired

M. Kim (✉) • H.-J. Jeong • S.-M. Moon
Dept. of Electrical and Computer Engineering, Seoul National University, Seoul, 08826 South Korea
e-mail: mskim@altair.snu.ac.kr; hjj@altair.snu.ac.kr; smoon@snu.ac.kr



G. Keramidas et al. (eds.), *Components and Services for IoT Platforms*,
DOI 10.1007/978-3-319-42304-3_6

when msgTurnOn occurs, turning the bulb on by calling the device managing function bulb.on().

```
1   (function process() {
2     var time = 0;
3     bulb.addEventListener(“timer”, function f() {
4       if(!movementSensor.getValue()) {
5         if (time ++ > 10000) bulb.off(); }
6       else { time = 0; bulb.on(); }
7     });
8   })();
```

This JavaScript snippet is written for smart bulbs to be capable of self-management. Scrutinising this code will provide some accounts about what makes JavaScript fascinating and, at the same time, complicated. This program makes the bulb keep track of the movement sensor and turn itself on/off based on the movement information. The function f() will be executed as an event handler for the timer event which fires the function periodically. Assuming that a timer event occurs in every millisecond, f() will count the elapsed time in millisecond scale since the last movement, and if the elapsed time is over 10 s, the bulb will be turned off.

One notable figure in this program is that the variable time is defined in line 2 which is located in the outer function of f(), yet is accessed within f() even after the outer function completes execution at line 8. This is called *closure*. Closures allow easier and secure programming, especially for IoT. However, this provides JavaScript engines more burdens to keep local variables of a function even after its execution is completed.

As shown above, JavaScript has distinctive features such as dynamic typing and closure. To implement those features properly JavaScript engines must deal with complexities. Therefore, this study compares and analyses the small footprint engines focusing on how they addressed those complexities first in Sect. 6.2.

JavaScript offers other advantages for programming the IoT platform. For example, a time-consuming job can be programmed with a registration of an event so that the following computations can proceed right after the event registration, instead of waiting for the job to complete. This leads to *asynchronous*, *non-blocking* event handling, yet on a single thread that JavaScript supports, simplifying the programming and the debugging efforts. In fact, JavaScript-based server framework called Node.js [2] allows efficient I/O event handling, so the IoT platforms based on Node.js are popularly employed [3–5], where the stand-alone JavaScript engine works as a server and a client at the same time. Other advantages of JavaScript include easier programming based on the loading of the source code and ample libraries and APIs obtained from being a mainstream web scripting language.

Despite all those strengths, an obstacle of employing JavaScript for the IoT platform is memory constraints since a number of IoT devices would have scarce memory. For example, typical microcontrollers are equipped with only around 500 KB flash memory and 100 KB RAM (https://www.arduino.cc/), which are not enough for the current mainstream JavaScript engines [e.g. JavaScriptCore

(JSC), a JavaScript engine for WebKit, takes over tens of MB in flash memory with much higher than 100KB of RAM usage]. On the other hand, fast event handling and computation would require high-performance JavaScript. Although there are efforts to develop high-performance, high-memory IoT chips [9], low-performance, low-memory IoT chipsets would be employed more often due to cost requirement. Consequently, JavaScript-based IoT platform requires a small footprint, high-performance JavaScript engine.

In this paper, we performed two jobs. First, we analysed three existing small footprint JavaScript engines: *Espruino* (http://www.espruino.com/), *Quad-wheel* (https://code.google.com/p/quad-wheel/), and *Duktape* (http://duktape.org/). We compared the three engines by assessing their binary size, runtime memory, and performance, as well as their compatibility of JavaScript tests. We found that only Duktape correctly supports the JavaScript features, yet with mediocre performance and memory requirement. The second work we accomplished is enhancing Duktape with six optimisations (which we call Duktape+), which leads to a tangible performance improvement as well as a sharp decrease of the runtime memory.

The rest of this paper is organised as follows. Section 6.2 analyses how the small footprint JavaScript engines work in diverse perspectives, and then compares their performance and compatibility with JavaScript features described in the current standard. Section 6.3 proposes Duktape+, a better design in terms of binary size, runtime memory usage, and performance compared to Duktape. Section 6.4 presents the experimental results to evaluate Duktape+. Section 6.5 provides summary and future work.

## 6.2 Analysis of Small Footprint Engines

This section evaluates three small footprint, open source JavaScript engines: Espruino, Quad-wheel, and Duktape.

Espruino directly interprets a JavaScript code while top-down parsing it. On the other hand, Quad-wheel and Duktape first generate an intermediate representation (IR) code for the stack machine and the pseudo register machine, respectively, which is then interpreted. These execution mechanisms may influence performance. Because efficient management of the JavaScript objects is directly related to compact memory usage, we evaluate runtime data structures employed by those engines. Garbage collection (GC) policies of the engines are also analysed because they also may affect memory usage. Finally, we check the ECMA-262 standard [6] coverage.

### 6.2.1 Runtime Data Structure

JavaScript ensures type flexibility of variables. While this provides the web developers the convenience of coding as the developers have no need to be concerned about variable types, JavaScript engines are required to handle all the objects dynamically. An object is a container of properties each of which is a pair of name and value. JavaScript treats even functions and arrays as objects. It is essential for JavaScript engines to implement this object model correctly. Both data structures and object representations play important roles in memory usage.

Quad-wheel uses a dedicated structure to represent an object. The structure includes the followings: the actual type of the object, a red-black tree for saving properties, an 8-entry array for caching recently used properties, and a reference counter to be used for GC. Meanwhile, Espruino and Duktape basically employ similar data structures. They also have a dedicated structure to represent an object just as Quad-wheel does, however, they connect all of the object nodes by forming a linked list as shown in Fig. 6.1.

Because linked lists provide convenience of monitoring reachable memories, wastes on unnecessary memory can possibly be reduced. Each node is generally comprised of the pointers to other nodes. Consequently, in 16-bit machine environment, which requires smaller pointers, this data structure gets more advantageous than in 32-bit machine environment. However, it has a great performance overhead as a linear search must be conducted every time a query about an object or a property takes places.

Additionally, Duktape selectively added a hash part to frequently used objects. This enabled constant-time accesses to frequently accessed property entries. Improved performance may be expected by this method rather than conducting a linear search throughout the nodes every time.

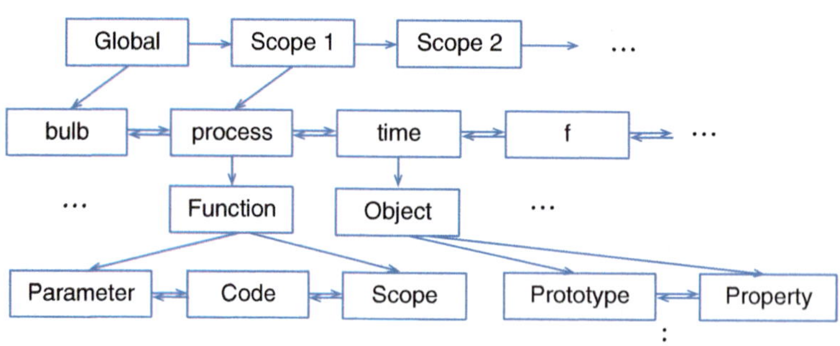


**Fig. 6.1** Runtime data structure of Espruino or Duktape

**Table 6.1** Garbage collection policy

| | JSC | Quad-wheel | Espruino | Duktape |
|---|---|---|---|---|
| Reference counting | Only a few | Yes | Lazy | Yes |
| Mark and sweep | Yes | No | Yes | Yes |

### 6.2.2 *Garbage Collection*

Garbage collection (GC) is a technique introduced for automatic memory managements [7]. The garbage collector attempts to reclaim memory occupied by objects that are no longer necessary. Due to the scarce resources, it is more important to reclaim unreachable memory at the right time.

Reference counting GC assigns a counter to every object. The counter always holds the number of other objects that are referencing its object. When the count becomes zero, memory occupied by the object can be reclaimed because that means there are no more other objects need it. Meanwhile, mark and sweep GC first halts the program during the marking phase where the garbage collector marks all encountering objects while traversing from the root. After the marking phase, memories occupied by unmarked objects can be reclaimed, or swept away.

Table 6.1 summarises garbage collection policies of JavaScript engines. JSC chose to reference count only for a few selected objects because reference counting for every object may produce excessive overhead. On the contrary, all of three small footprint JavaScript engines chose to reference count for all objects. In a memory constrained environment, reclaiming unnecessary memories must be conducted as quickly as possible. While mark and sweep GC policy cannot discover those obsolete memories until the marking phase ends, reference counting GC is capable of catching it right after its reference count becomes zero.

Espruino counts the references only for the directly referenced objects rather than counting for all objects recursively in the tree, in order to avoid tree-traversing overhead. In this case, some child nodes may be remaining in the memory even though their parents do not exist anymore, therefore Espruino also uses mark and sweep GC from time to time to clear those memories. By collecting the memories to be reclaimed into a list and reclaiming them in a batch, Duktape avoids overhead from excessively frequent reclaiming.

### 6.2.3 *ECMA-262 Coverage*

The current JavaScript standard is defined in ECMA-262 specification [6]. We used Test262, a test suite containing thousands of tests, each of which tests a specific item described in the specification. In order to inspect the compatibility of the engines with the specification, a regression test was conducted using Test262 script.

**Table 6.2** ECMA-262 coverage

| | JSC | Quad-wheel | Espruino | Duktape |
|---|---|---|---|---|
| Coverage (%) | 99.6 | <50 | <70 | 99.7 |

**Fig. 6.2** A micro-benchmark

```
function sq(a) {
   return a * a;
};

var b = new Object();
b.x = 0;
b.y = "";
for(i = 0; i < 10000; ++i) {
   b.x = b.x + sq(i);
   b.y = b.y + b.x;
}
```

The result is as shown in Table 6.2. Quad-wheel and Espruino failed to parse common JavaScript codes that Test262 includes for the tests. Therefore we classified some essential components from ECMA-262, and then approximated their compatibilities by analysing how much they have implemented literal, types, expression, and built-in objects. The result implies Quad-wheel or Espruino is not supporting essential features enough. In particular, Quad-wheel has few essential built-in objects or methods such as Date or Math, furthermore it cannot handle regular expressions or Unicode strings. On the other hand, Duktape showed a fairly excellent coverage considering that its percentage was comparable with that of JSC, a robust open source JavaScript engine.

### 6.2.4 *Size and Performance*

To evaluate the engines' memory usage and performance, we have run a micro-benchmark shown in Fig. 6.2. This micro-benchmark repeats 10,000 times of serial operations. The operations include property accesses, a function call, and '+' operations between numbers and strings. Those were selected because firstly they are main components of meaningful semantics in JavaScript code, secondly property access, in particular, was often found a bottleneck of execution. The result is shown in Table 6.3. Duktape+ which will be introduced throughout the rest of this paper was also compared for providing a fair comparison between existing JavaScript engines. It was experimented on an ARM machine (ARMv7 Processor rev 10 v7l @1.2 GHz, 1 GB RAM) and the maximum value of heap size was recorded. It seems that binary size of Quad-wheel is extremely small compared to other engines. However, because the unimplemented requirements would cost around 90 KB additionally, it is hard to tell its superiorities in terms of binary size. Within three engines Duktape takes the smallest heap memory and its binary size is

**Table 6.3** Experimental result of running the micro-benchmark (Fig. 6.2)

| | JSC | Quad-wheel | Espruino | Duktape | Duktape + (proposed) |
|---|---|---|---|---|---|
| Binary size (KB) | 2660 | 93 | 231 | 184 | 187 |
| Heap size (KB) | 16,528 | 688 | 492 | 370 | 356 |
| Time (ms) | 159 | 13,285 | 67,765 | 6573 | 2518 |

**Table 6.4** Results of running programs which create respective objects repeatedly

| | | JSC | Quad-wheel | Espruino | Duktape | Duktape + (proposed) |
|---|---|---|---|---|---|---|
| Object | Heap | 17.8 MB | 2013 KB | 1967 KB | 616 KB | 601 KB |
| | Time (ms) | 19 | 51 | 1151 | 413 | 371 |
| Array | Heap | 17.7 MB | 5675 KB | 3934 KB | 1577 KB | 1400 KB |
| | Time (ms) | 11 | 142 | 3,189 | 848 | 535 |
| Function | Heap | 20.1 MB | 2014 KB | 3934 KB | 3497 KB | 2881 KB |
| | Time (ms) | 15 | 69 | 1232 | 1000 | 878 |
| String | Heap | 17.5 MB | 6732 KB | 983 KB | 1636 KB | 1493 KB |
| | Time (ms) | 16 | 123 | 1062 | 737 | 502 |

quite feasible, yet order of 100 times slower than JSC. Therefore, we decided to take optimisation chances existing in Duktape and propose Duktape+ which has better performance and footprint while not increasing binary size much.

We experimented with a few more micro-benchmarks extending the area of testing. Table 6.4 shows results of running programs which repeatedly create `Object`, `Array`, `Function`, or `String`, respectively. For every test case, Duktape+ has shown better performance and lower footprint than original Duktape or Espruino has. Increase of binary size is only 3 KB from 184 KB originally.

It is interesting that Quad-wheel showed 3–12× better performance than Duktape+, where it appears to be the opposite from the result in Table 6.3. We concluded that this is caused by two reasons. First, the lack of its preparation for built-in objects and methods. Because Quad-wheel does not support full features required by standard, it has nothing much to process for start-up which has more significant effects for shorter programs. Second, Quad-wheel seems to create objects relatively efficiently. Nevertheless, since experiments with Table 6.4 only focus on creating objects, they are less meaningful than those with Table 6.3.

## 6.3 Proposed Engine

Three small footprint JavaScript engines were compared in Sect. 6.2. Out of those, Duktape fairly satisfies ECMA-262 coverage and offers a compact footprint. However, performance issues remain prominent. To solve these issues, this study proposes several optimisations. Sections 6.3.2–6.3.5 explain those in detail.

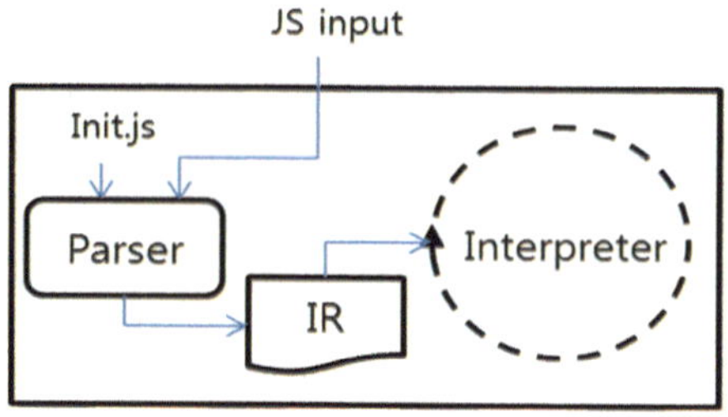


**Fig. 6.3** Overall structure of Duktape

Furthermore, this study proposes lazy built-in objects construction, a means of potentially reducing the footprint. This is explained in Sect. 6.3.6.

### *6.3.1 Overall Structure of Duktape*

Duktape was designed following the general structure of JavaScript interpreters. Figure 6.3 depicts the overall structure of Duktape. When Duktape takes a source code written in JavaScript, a parser produces an intermediate representation (IR) code. Then, the pseudo register machine-based interpreter fetches an IR instruction one by one from the code, executing it by performing pertinent behaviour. Due to memory constraints, it works as a pure interpreter without such techniques as JIT compilation.

When Duktape engine starts, it executes an embedded initialisation code before executing the actual input JavaScript code. The initialisation code is also written in JavaScript and this code initialises objects to prepare the engine for executing the actual JavaScript application. The objects which are initialized by the initialisation code could be implemented by native codes. However, by dynamically creating them, binary size can be reduced while sacrificing the performance because that increases duration of the initialising step.

### *6.3.2 Improved String Concatenation and Join*

Strings are frequently used in JavaScript as well. For the sake of Unicode handling, every string object contains a value called '`clen`'. '`clen`' refers to the length of a string in code point. In the UTF-8 encoding scheme, two different Unicode characters may occupy different number of bytes. Therefore, every string object created by Duktape contains '`blen`' string length in byte, as well as '`clen`' string length in code point.

String concatenation is a JavaScript operation which creates a new string by concatenating two strings. The natural way to implement this operation is building a new string while scanning two input strings linearly. String join also requires a linear scan of two strings as with string concatenation. The point is that, to calculate the `clen` value linear scan of the strings is being repeated. If we calculate `clen` during the linear scan for string operations, we can avoid unnecessary computation. We improved the performance by suggesting a way to calculate `clen` effectively. Moreover, by classifying Unicode strings that require special handling like one described above and the others of no need for it, it is also possible to remove calculations of `clen` because `clen` is equal to `blen` for most strings.

### *6.3.3 Array Putprop Fastpath*

As a result of profiling, a native method called putprop was found a bottleneck of execution. This is an implementation of a requirement, property write, in ECMA-262 specification. This function is passed an object and a property (a pair of key and value), and then it writes the property to the object. For example, the semantics of a JavaScript code `person.age = 30;` is to write a property <age, 30> to an object named as `person`.

Because characteristics of objects differ depending on the types of them, data structures or required variables are different. Accordingly, putprop method passes several steps as shown on the right column in Fig. 6.4. We exploited the discovery that only writing to array part is necessary for array objects. We designed a fast path which only passes the writing to array part, branching from the beginning of the putprop method.

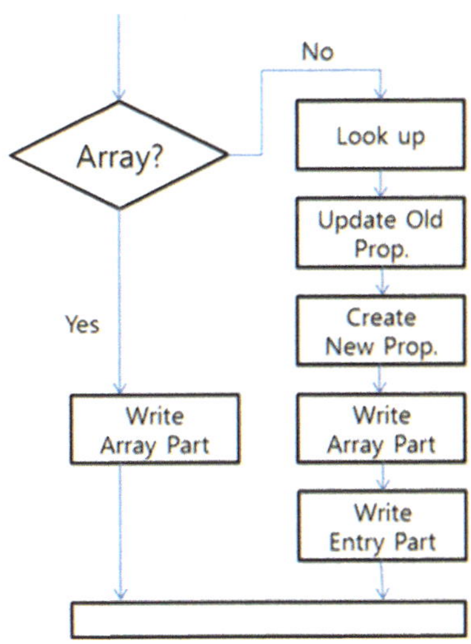


**Fig. 6.4** Array putprop fastpath

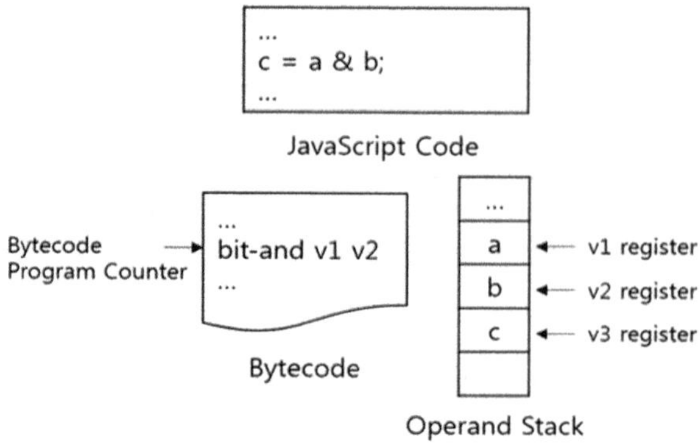


**Fig. 6.5** Bitwise and operation

## 6.3.4 *Bitwise Operation/Compare with Native Stack*

When Duktape is interpreting IR, it delivers operand values through the operand stack. In Fig. 6.5, a JavaScript code `c = a & b` is converted to an IR instruction '`bit-and v1 v2`'. When this instruction is being executed, the actual values of virtual registers v1 and v2 are fetched from the operand stack. The problem is, because the size of the operand stack is variable, the stack can be relocated to a new memory address in case of overflow. If so, the pointer referencing the top of the stack also needs to be modified appropriately. Therefore, every access to the memory pointed by this pointer requires checking if the stack has moved.

By acquiring value of operands using the native stack rather than dynamically allocated memory region, we removed the checking overhead. For the bitwise operations or compare operation, instead of accessing the operand stack, values of the operands are saved as local variables.

## 6.3.5 *Indirect Threading*

Indirect threading is a way of interpretation. It is a well-known threaded code generation technique which invokes subroutines [8]. A dispatch table contains a map between instructions and the addresses of corresponding routines. The underlining idea is to let the interpreter always know where to jump in advance by looking-up the dispatch table. Duktape is using switched interpreting, where the interpreter knows where to jump only after decoding the opcode. These indirect branches in switched interpreting are not efficient due to cache miss or branch prediction issues.

Direct threading omits lookup by putting the addresses directly into IR. Even though direct threading may show a better performance than indirect threading would do, we decided to use indirect threading scheme. Because the size of an address is usually much bigger than that of a typical instruction, the size of address-containing IR can sometimes exceed a considerable extent in scarce memory resources.

### 6.3.6 *Lazy Built-in Objects Construction*

Duktape creates all of built-in objects and methods during the initialisation step as explained before, in Sect. 6.3.1. However, because most of the applications do not use all of the built-in objects, the memories taken for creating useless built-in objects end up being a waste. Therefore, this study suggests lazy built-in objects construction, where creating a built-in object right before when needed. In the initialisation process, only the essential objects for starting program are created. The rest are waiting to be created until the engine makes the decision that they are actually required for running the application.

Each of built-in objects has its flag that indicates if it has been created or not. The initialisation stage is conducted by setting all of them to 0 at the start, with no actual instances of the objects created. Every time a reference to a built-in object is required, it checks the flag finding if the required object has already been created. If there exists the responding instance, it simply directs to the instance. Otherwise, it calls the initialisation routine which constructs the required built-in object and marks its flag. By doing so, unnecessary memory usage is expected to be reduced.

## 6.4 Experimental Results

### 6.4.1 *Environment*

The execution time for running 26 programs included in the prominent Sunspider Benchmark suite was measured for evaluation. The execution was conducted on a Linux system (Ubuntu 12.04) equipped with a 64-bit Intel® Core™ i7-4770 CPU @ 3.40 GHz and 16 GB RAM.

### 6.4.2 *Performance*

Figure 6.6 is a graph showing the geometric mean of the speed-up for all benchmark programs when each optimisation is applied. The rightmost bar named 'All applied' indicates the result when all of six optimisations are applied together. Lazy built-in objects construction made it 1.4 % slower because of time to check if the built-in object has already created or not. Rather, it is faster for start-up time because it does not need to create objects that are never used.

Improved string concatenation and join optimisation enhanced the performance a lot, especially for benchmarks handling strings. Array putprop fastpath optimisation evenly enhanced, particularly benchmarks which have frequent property accesses.

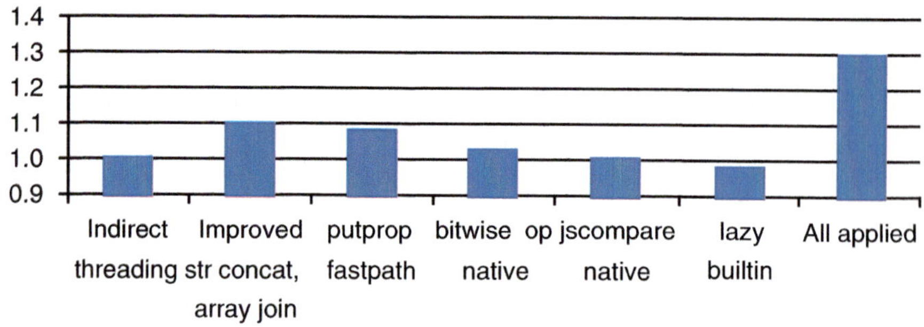


**Fig. 6.6** Speed-up results

**Table 6.5** Built-in objects lazy construction

| | Original | Applied | Diff (%) |
|---|---|---|---|
| Built-in objects memory (KB) | 25.36 | 11.95 | −52.88 |
| Total initial memory (KB) | 44.32 | 30.91 | −30.26 |

Bitwise operation or compare with native stack optimisation also showed a good result for benchmarks specialised for those operations, not for ones that use the value copied to native stack only a time.

### 6.4.3 Footprint

Table 6.5 shows RAM usage until the end of Duktape initialisation. The second row is the total memory consumption right after initialisation, and the first row is that consumed by built-in objects out of total consumption. Built-in objects have quite a big portion seeing that they take 25.36 KB out of total 44.32 KB. Among those 25.36 KB, 52.88 % was not essential at that moment. While running the application, if an uncreated built-in object is turned out necessary, it takes additional memory. To sum up, lazy built-in objects construction led to reduction of RAM usage by up to 30.26 %.

### 6.4.4 Binary Size

It is consequent that the binary size increases when adding a new design. Even so, the total increase of binary size is only 2.8 KB as shown in Table 6.6. As binary size of Duktape is 184 KB originally, the increase is less than 2 % which is almost nothing compared to the merits it brings.

**Table 6.6** Changes of binary size

| Optimisation | Binary increase (Bytes) |
|---|---|
| Indirect threading | 1736 |
| Improved string cat/join | 512 |
| Putprop fastpath | 136 |
| Bitwise op native | −8 |
| Compare native | 16 |
| Lazy built-in | 712 |
| All applied | 2808 |

## 6.5 Summary and Future Work

There is an emerging need for small footprint JavaScript engine for the sake of event-driven programming which is promising in IoT platform. Out of Quad-wheel, Espruino, and Duktape, this paper proposed several optimisation ideas for Duktape as Duktape was distinguishable, but still has mediocre performance and this study implemented the suggested ideas.

By the optimisations suggested through Sect. 6.3, this study has achieved an increase of performance by 30.5 %, a reduction of footprint by 30.3 %. The increase of binary size while applying those optimisations was negligible as it was only 2 % of original size 184 KB.

Such a small footprint JavaScript engine can be exploited as a server-side software platform as Node.js is. Node.js is embedding the V8 JavaScript engine, which is an excessively oversized program to be used on small IoT devices. Therefore, by associating the small footprint engine with libUV which is the I/O library used in Node.js, a prototype of small-sized server-side software platforms can be exhibited.

**Acknowledgments** This research was supported by the Ministry of Trade, Industry and Energy (MOTIE) through the Electronics and Telecommunications Research Institute (ETRI) (Project No. 10045344), Global PhD Fellowship Program through the National Research Foundation of Korea (NRF) funded by the Ministry of Education (NRF-2014H1A2A1019270), and Samsung Electronics.

## References

1. L. Atzori, A. Iera, G. Morabito, The internet of things: A survey, International Journal of Computer and Telecommunications Networking, **54**, 2787–2805 (2010)
2. S. Tilkov, S. Vinoski, Node.js: Using JavaScript to build high-performance network programs. IEEE Internet Comput. **14**(6), 80–83 (2010)
3. T.-M. Grønli, G. Ghinea, M. Younas, *A lightweight architecture for the Web-of-Things*. Mobile Web Information Systems (Springer, Berlin and Heidelberg, 2013), pp. 248–259
4. M. Kovatsch, M. Lanter, S. Duquennoy, Actinium: A restful runtime container for scriptable internet of things applications. 2012 3rd international conference on the internet of things (IOT), IEEE, 2012.

5. L. Tong et al., Interactive surface composition based on arduino in multi-display environments. Proceedings of the ninth ACM international conference on interactive tabletops and surfaces, ACM, 2014.
6. ECMAScript, ECMA International, European Computer Manufacturers Association, ECMAScript Language Specification 5.1 Edition (2011), Accessible at http://www.ecma-international.org/ecma-262/5.1/Ecma-262.pdf
7. R. Jones, R.D. Lins, *Garbage Collection: Algorithms for Automatic Dynamic Memory Management* (Wiley, New York, 1996)
8. J.R. Bell, Threaded code. Commun. ACM **16**(6) (1973)
9. C. Wootton, *Beginning Samsung ARTIK*. Apress (2016)

# Chapter 7
# VirISA: Recruiting Virtualization and Reconfigurable Processor ISA for Malicious Code Injection Protection

**Apostolos P. Fournaris, Georgios Keramidas, Kyriakos Ispoglou, and Nikolaos Voros**

## 7.1 Introduction

Instilling security in the evolving IT world where considerable number of diverse systems are involved, is a difficult multi-domain task, requiring collaborative work on a single target between design experts from different technology fields like fault-tolerance, low-power, or communication/connectivity domain. In all such systems information is becoming increasingly valuable. Such systems are employed for sensitive data storage and transactions (bank account keys, passwords, user habits, etc.). This introduces the need for strong security, privacy, and trust in the future internet computing paradigm. Effectively shielding a system against privacy and security attacks is a well-known uphill battle to be fought in all computer systems from high end computers to embedded system devices.

The extensive toolset of malicious software aiming at computer system compromise is reason enough for supporting the common belief that computer systems cannot be trusted. When malicious code is executed in a computer system then it can replicate itself by exploiting other program or operating system (OS) vulnerabilities in order to inject code segments in running or stored programs. Such attacks, known as code injection attacks, occur when arbitrary code is executed remotely or locally not by its intended user but by an unauthorized entity (user or malicious program). Code injection is the outcome of various system vulnerability exploitations like stack smashing, buffer overflow [1, 2], dangling pointers in memory, and format string attacks [3, 4]. Code injection attackers can cause direct damage to a system

---

A.P. Fournaris (✉) • G. Keramidas • N. Voros
Technological Educational Institute of Western Greece, Embedded System Design and Application Laboratory (ESDA), Antirio, Greece
e-mail: afournaris@teimes.gr

K. Ispoglou
University of Patras, Patras, 26504 Greece

 
G. Keramidas et al. (eds.), *Components and Services for IoT Platforms*,
DOI 10.1007/978-3-319-42304-3_7

by executing infection vectors of a broader attack scheme (e.g., computer worm execution/propagation, privilege rights changes, etc.) on existing programs.

Detecting and prohibiting code injection can be done by various means including the use of static (compile time) and dynamic (run-time) source code analysis for detecting vulnerabilities that can enable code injection attacks [4, 5] as well as process sandboxing where the attacks can be constrained to a controlled environment. However, none of the above methods can cancel a code injection attack after it has been launched. To solve this issue, processor instruction [6, 7] and address space randomization (address obfuscation) [8] techniques have been proposed in an attempt to reduce the code injection attack risk or even render the injected code unusable using hardware means. However, the above approaches cannot protect against all code injection attack types, can be applicable to specific processor types and can be considered reliable only when the OS is considered trusted. To increase the protection shield of such countermeasures, it is necessary to migrate them to a system lower architecture layers like the OS kernel, the BIOS, or even the computer hardware [9, 10].

In this work, we extend the randomization techniques for code injection protection to lower layers in the computer system architecture in an effort to describe a ubiquitous malicious code injection protection mechanism that is not restricted to a specific processor platform, has few prerequisites, and leverages the flexibility provided by the current and future multicore architectures. More specifically, the proposed methodology utilizes a trusted virtual execution environment, denoted as virtualization ISA environment (VirISA), similar to a hypervisor or microvisor, capable of using subsets of the target core instruction set architecture (ISA). The proposed virtualization environment can be employed in order to generate fully isolated virtual machines (VM) that use a randomized subset (or the full set) of the processor ISA. VirISA can also assign such virtual machines to a single core or to core groups in a multicore platform. In order to increase the level of protection, the VirISA environment is loaded before the OS kernel which can be loaded on the environment's VM.

All programs (malicious or not) running on a VirISA supported VM must be compatible with the "specialized" VM ISA. To achieve that, all executable programs when loaded from the system storage area to its memory must be recompiled on the VirISA environment (through appropriate compiler options and preferably following a just-in-time compiling approach) using a randomized ISA subset and executed through a dedicated VM supporting that ISA. The randomly and dynamically selected instruction set (a subset of the regular ISA) cannot be known from the processor external environment (the ISA is located at the heart of each computing system and it is impossible to be identified via reverse-engineering approaches). Therefore, when malicious code is injected in a target process executed through a VirISA VM in the system memory, it will contain, with high probability, instructions that do not dictate to the specific ISA instantiation of the target VM. This will have a profound impact to the injected code expected behavior. As a result, the malicious code will become not-executable or will be executed fairly slowly thus providing distinct evidence of its existence to the user. Such evidence can be identified by a detection mechanism so as to alert the user of possible intrusion and take all necessary measures to mitigate the attack. The description of this mechanism is beyond the scope of this paper.

To the best of our knowledge, this is the first work that proposes the explicit and dynamic reduction of a target processor ISA for increasing its security capacity. The underlying virtualization mechanisms play an important role in our methodology, since they offer a practical mean to instantiate a processor with limited ISA (a processor supporting a subset of its regular instructions) on top of an off-the-shelf processor.

The rest of the paper is organized as follows. In Sect. 7.2, the underlying threat model is described. In Sect. 7.3, background information and related work are outlined. The details of the proposed methodology are presented in Sect. 7.4 and finally, Sect. 7.5 concludes the paper and outlines future work plans.

## 7.2 Threat Model

A computer system (device under attack, DUA) can be the target of a wide variety of attackers both insider and outsider. In our threat model, an attacker target point is a given software code (host program) that is executed as a process in an OS environment through the use of a DUA low architecture layer software acting as a trusted computing base (TCB). A software entity can act as a TCB when its security functionality is trusted not to fail (if it fails the whole security policy collapses). As such, TCB cannot be manipulated by an unauthorized user while it always functions as intended to. We assume that such an entity exists in our proposed approach and uses the BIOS operations as well as all system hardware resources via the regular processor ISA. In the context of the specific work and in order to not pertain to a specific system configuration, the TCB can be a trusted hardware virtualization layer, a trusted boot loader or, in general, any kind of a trusted middleware layer. Trusted Computing Group approaches (including TPM chips) can be further utilized to attest the TCB security reliability.

We assume that the host OS environment cannot be trusted. The OS has no protection mechanism against attacks and can be manipulated, using existing vulnerabilities, by an attacker in order to launch locally or remotely a code injection attack. Target of such attack is the host program process and goal of such attack is the alteration of the host program functionality in order to act as an attack vector of a broader malicious intrusion like malware execution, distribution (worm, Trojan), administrator privileges acquiring and sensitive data eavesdropping (when the host program handles sensitive information). The attacker can execute arbitrary code as an OS process and he can have access to the OS services. Two types of code injection attacks can be identified on the given computer system: attacks manifested during host program execution (run-time attacks) and attacks manifested when the host program is stored in the system physical disk (static attacks). It can be assumed in our threat model that the computer system non-volatile storage elements (physical disk) have trust integrity check mechanisms and all data stored in them are impervious to malicious data changes. So, a code injection attack is limited to run-time attacks. During such an attack, the host program executable code is loaded onto RAM and is executed as a regular process. Since, the OS and its RAM is considered untrusted (they are vulnerable to attacks), an attacker can profile the host program process binary code and statically or dynamically inject malicious code in order to alter the process functionality (the process will operate in a different way than the one it was built for).

Finally, without loss of generality, we assume that an attacker cannot mount an unlimited number of code injection attacks on the target host process before been detected. An evaluation of the tolerance criteria of the attack detection mechanism is left for future work.

## 7.3 Motivation, Background, and Related Work

The goal of the proposed methodology is to enhance the toolbox of system security designers with a new set of shielding techniques. While processors (the ISA of a processor formulates its functionality and its hardware specifications) are designed to be generic enough in order to efficiently execute a large spectrum of applications, our view is quite different. If the capabilities of a processor are intentionally reduced (by dynamically limiting the processor ISA), then a large class of code injection attacks will become obsolete or not applicable. This happens because a malicious software code is typically implemented and compiled using the full ISA of the system. In other words, a malicious injected code is currently designed to attack a specific system.

Our vision is to formulate a run-time execution environment with limited capabilities by explicitly and intentionally reducing the processor ISA. Not surprisingly, someone may argue that by limiting the ISA of a target processor, all applications (malicious or not) may become not "runnable," simply because they may contain specific assembly instructions which are not part of this limited ISA. However, if the limited ISA of a processor is carefully formulated by including instructions that are part of the non-malicious host program/process, but not part of the malicious injected code, then we can end up with a low-level protection scheme that cannot be bypassed.

Consider, for example, the case of an MUL instruction and let assume that the target host process does not contain any. If we intentionally remove the commonly used MUL instruction from the group of supported machine instructions, then all the malicious codes injected in the host process containing this instruction will simply become non "runnable." Another first-class example is the support of any kind of indirect memory addressing modes which are mainly used to implement pointer-like operations. In this case, the host process code can be rewritten (either by the programmer or by employing dynamic binary rewriting techniques) in a way that does not rely on pointer indirections. However, this will be catastrophic for a typical malicious code in which its operation is heavily relied on pointer-based indirections in order to alter and migrate the dynamic execution. As a result, our ultimate goal is to provide a virtual execution environment that is exclusively matched to a given host program process so that code injection attacks can be only possible for system-AND-host-program oriented malicious codes which by principle are very difficult to be realized.

In the computer architecture domain, designing variable instruction set processors can be traced back to the 1990s. An example is the work of De Gloria [11] which can be considered as the ancestor of the explicitly parallel instruction computer and the modern very long instruction word architectures [12, 13]. Another more recent approach is the compiler-level techniques provided by Liu et al. [14] in their attempt to configure the best instruction set to use for executing compiled programs targeting the Cognigine's CGN16100 Network Processor [15].

The notion of hardware or architecture level code injection protection has found fertile ground in research cycles. There are a lot of works that aim at computer architecture instruction set or memory section alterations in order to render injected codes unusable. Randomization plays an important role in the above endeavors. Instruction set randomization (ISR) has been proposed as a generic protection approach against code injection attacks in [6, 7]. ISR is based on encryption (e.g., an XOR operation between an instruction and a random key) of the host process binary in order to randomize/obfuscate the processor instructions. The encrypted process is executed by an augmented emulator that is responsible for instruction decryption before execution. In [6], the authors use a special register for the encryption key. Encryption is based on XOR. When loading an encrypted process into the core, it is decrypted by XORing with the stored encryption key. Similar approach is followed in [7, 16], where the key is provided by a pseudorandom generator (a one-time pad approach) and it is as long as the process code. Both approaches are demonstrated in processor emulators and have limitations due to performance overhead and restricted compatibility with dynamic code execution (e.g., dynamic linked libraries). Unfortunately, XOR encryption has been found lacking and can be compromised [17]. ISR has been further refined by adopting sophisticated encryption functions, like AES [18], and employing efficient software dynamic translation systems [5, 19] as well as introducing key management systems for process key handling [19]. All such attempts manage to generate a software-based virtual execution environment that has a complex and performance intensive operation mechanism due to the use of symmetric key encryption algorithms and associated key management for a large number of encryption keys.

Apart ISR, architecture-level code injection protection follows memory address space randomization or address obfuscation. In the PaX system [20, 21], the code, stack, and heap segment addresses are randomized, leading to a methodology denoted as address space layer randomization (ASLR). ASLR is currently being adopted as a code injection protection mechanism in a wide range of commercial OSs including Linux, Windows, and Mac OS X and has been proposed for Android [22]. In [8], the ASLR methodology has been expanded into an address obfuscation technique where random padding is used. However, in ASLR systems, address space remains fixed thus enabling the attacker to probe the system and gather useful attack information.

Overall, the above approaches are static in nature and do not provide a fully dynamic virtual execution environment that can be arbitrary adjustable to host process characteristics. Software virtualization offers an interesting expansion to the above approaches that can lead to strongly isolated execution environment. However, when viewing a software hypervisor operating on an untrusted OS, then this hypervisor is a target of attack and cannot be trusted [23, 24]. This issue can be solved by hardware virtualization environments that operate on architecture layers lower than the OS kernel (directly on top of the hardware structure). Such structures can act as TCBs and can support a code injection protection mechanism [25].

More specifically, in this work, we envision a dynamic environment that acts as a TCB using state-of-the-art full virtualization frameworks and can support randomized, dedicated ISA for each host process. In such an approach, there is no need for instruction encryption since randomization is provided by the TCB. The above idea can be applicable to many use cases and can be adapted to various application systems, in an effort to provide a trusted environment inside an untrusted device. Critical processes can be executed in dedicated ISA with isolated VMs regardless of the security status and vulnerabilities of the full ISA and host OS. Expanding the proposed idea to support dedicated VMs (not only assembly instructions), our solution can be migrated to various platforms including mobile devices or embedded systems. For example, a prime use case of the proposed approach can be the java virtual machine (JVM) macro-instructions typically existing in all Android systems. In such environments, the run-time system using the Dalvik hypervisor on top of a Linux kernel is able to bundle together (at run-time) groups of Java macro-instructions. Using the proposed approach, a VirISA java-based virtual execution environment can be formulated and restricted java instruction set VMs can be instantiated. In this way, each Android application will run using a dedicated Java instruction subset. Note that such infrastructure already exists in some Android systems. Such systems, primary designed for power reduction, can generate one JVM for each executed application to minimize power consumption. Using the proposed VirISA idea, the Android virtualization mechanism can be extended in order to provide strong security and trust.

## 7.4 Proposed Code Injection Protection Architecture

In the proposed solution, illustrated in Fig. 7.1, the VirISA scheme is structured on a full scale single or multicore system. Apart from restricted ISA VM generation, VirISA can provide VM hardware resource allocation to specific cores in the multicore case. The VirISA virtualization environment can be used in order to provide multiple VirISA VMs that are isolated between them and can communicate as separate entities through the VirISA infrastructure. To achieve that, each VM is equipped with a VirISA software driver capable of providing a secure, trusted

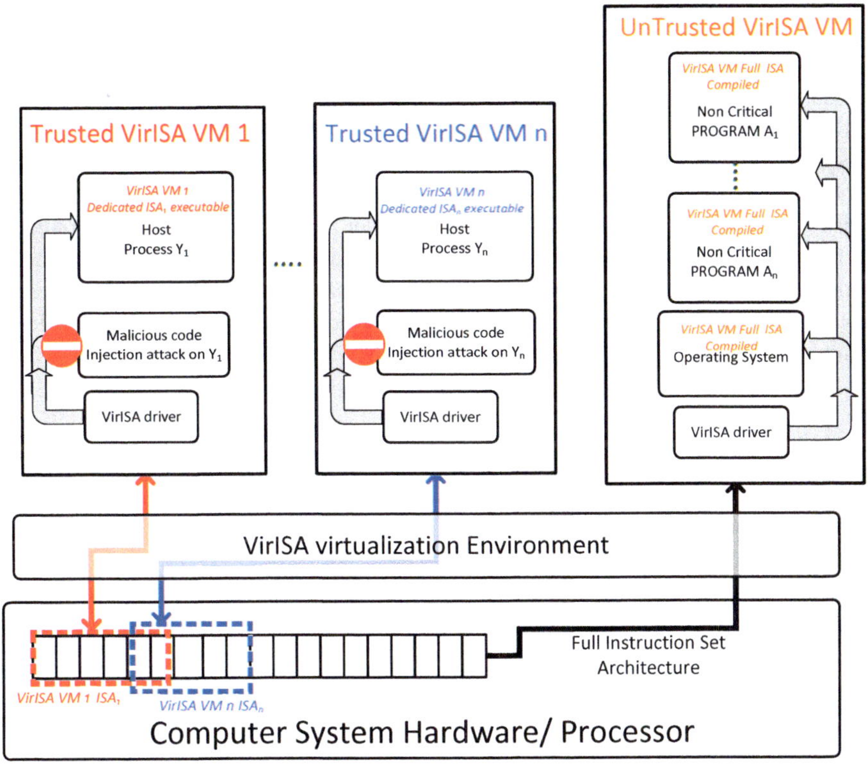


**Fig. 7.1** The proposed VirISA architecture

interface for communication with other VMs. We assume that VirISA virtualization environment can be trusted and it is very difficult to be compromised. Based on the above specification, system operations/processes can be categorized into critical and noncritical ones. Critical operations involve sensitive information activities or crucial system activities for the system functionality and must be executed in a trusted VM. A trusted VM can be generated by the VirISA virtualization environment using a dedicated, restricted ISA and is assigned to a critical operation process that is recoded and, thus, executed on this VM. Noncritical operations may include OS processes and are executed in a separate VM with full ISA support (we assume that the OS kernel is too complex to be recoded in restricted ISA). Any attacker's injecting code to a critical process will be considerably restricted by the ISA of the trusted VirISA VM and will hardly be executable.

To make the proposed security and trust scheme more practical, the VirISA virtualization environment can be preferably employed in a multicore processor so as to assign a VirISA VM to a specific core of the processor and consequently achieve VM performance overhead reduction.

### 7.4.1 VirISA Main Functionality

Assume that a host program executable code X resides in a physical disk or non-volatile memory that adopts the VirISA protection mechanism and an execution request for this program is issued. As Fig. 7.2 shows, the request is initially transferred to the VirISA run-time system. The executable code is read from the disk and is analyzed by the VirISA entity in order to be profiled and transformed to a functionally equivalent executable code Y. The key point is that the Y executable consists of different assembly instructions (actually, a subset of the supported system instructions) as compared to the original code X, which was compiled taking into account the full ISA. The role of the VirISA is to analyze the input code X, to

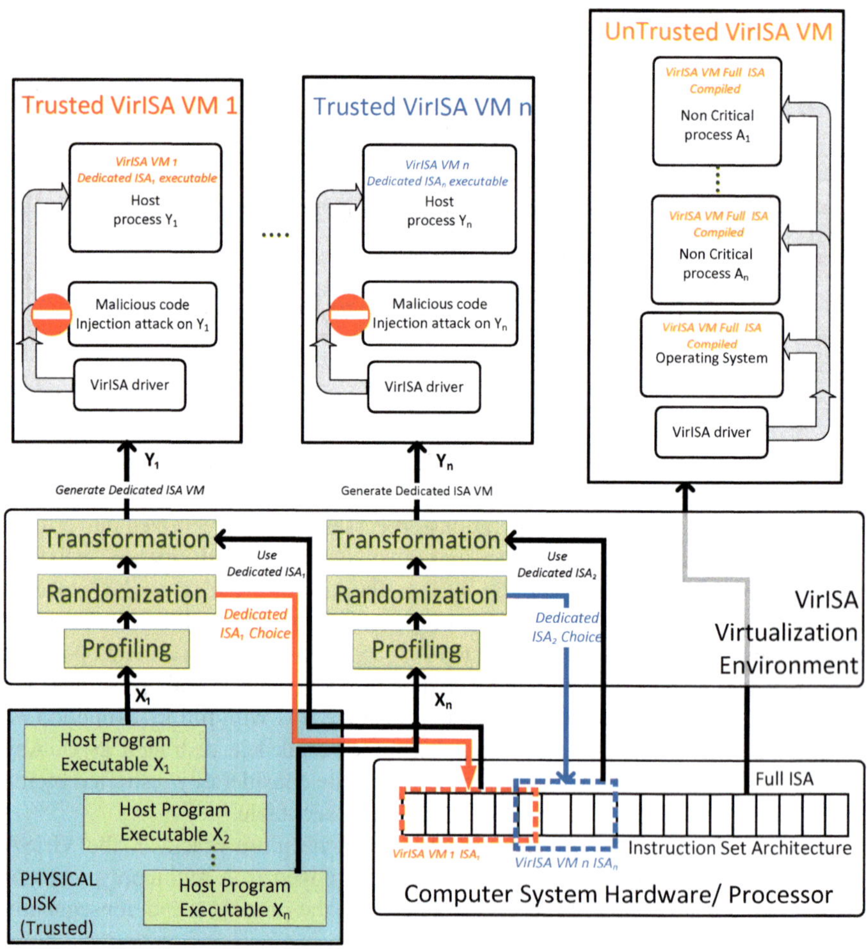


**Fig. 7.2** VirISA functionality example

create a dedicated subset of the processor ISA for X in a random manner and to transform code X into code Y. Note that the functionality of the two executable codes remains exactly the same.

When this step is finished, the VirISA virtualization environment creates a new, dedicated VM (VirISA $VM_Y$) in the system memory that utilizes appropriate system resources and employs a dedicated subset of the regular ISA. Code Y is executed in this VM as a regular process and communicates with the rest of the system locally or remotely through a VirISA software driver that passes all messages to a different VM using the VirISA virtualization environment. This functionality is shown in Fig. 7.2.

Note that the OS runs in a separate VM and although it cannot be trusted (as assumed in our threat model), it remains isolated from the rest of the system through the VirISA virtualization environment. Obviously, this functionality can be extended to support more than one OS, each one running in a separate VM assuming that the VirISA acts as a full virtualization environment. An important characteristic of our solution is that it is generic enough to be employed to a wide variety of single or multicore architectures by appropriately adjusting the VirISA environment to the corresponding target architecture. In case an attacker manages to take advantage of some OS/system vulnerability and acquires the necessary means to launch a code injection attack by accessing the system memory, his only option would be injecting a full ISA malicious code to Y process code (note that process Y is executed and monitored via its dedicated $VM_Y$). However, code Y uses an unknown to the attacker, dedicated subset of the regular ISA that will not be compatible with the injected codes instructions.

## 7.4.2 *VirISA Architecture*

The VirISA virtualization environment is depicted in Fig. 7.3. The proposed architecture consists of four modules, namely the profiler, the transformer, the ISA definition, and the VM generation (instantiation) module. Following the example of the previous section, the host program full ISA executable code X, before loaded into memory, is inserted to the VirISA virtualization environment (as shown in Fig. 7.3).

Initially, the control is transferred to the profiler module. Instructions that can be altered/transformed are identified and are fitted in ISA rule categories (see next subsection for more details). At this step, the code requirements are also specified in terms of critical (unchangeable) operations and code structures that must be retained or not. The consolidated code profile is outputted from the profiler module and is inserted to the ISA definition module. The latter module is responsible for determining a dedicated, restricted ISA (noted as $ISA_Y$ in Fig. 7.3) for code X. The choice of $ISA_Y$ is made through the use of a series of ISA permutation table rules packed appropriately in mutually exclusive categories. The main role of the definition module is to randomly select a series of ISA permutation rules from some

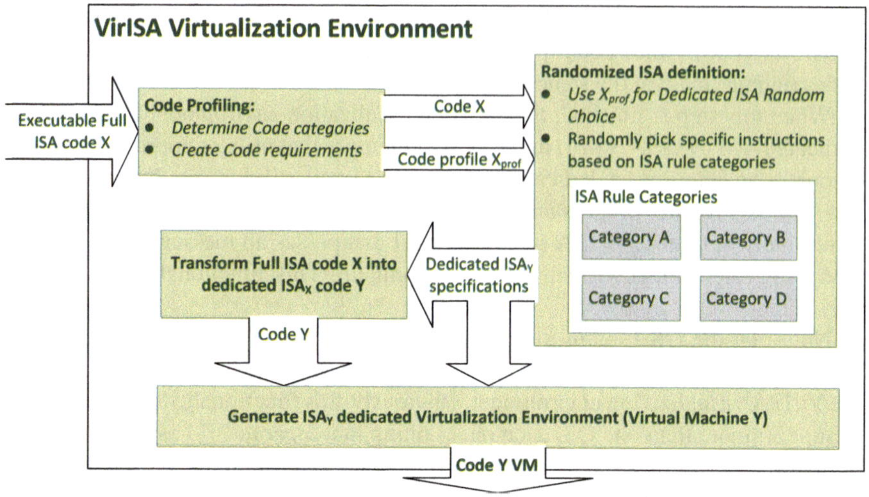


**Fig. 7.3** VirISA virtualization environment architecture

category in an effort to formulate a uniformly random (obviously attacker agnostic) ISA environment.

Note that rules for a single $ISA_Y$ cannot be chosen from all categories since not all category rules can be combined into a functional program code (e.g., removal of both MUL and ADD processor instructions cannot be simultaneously applied to the same executable code). It must be pointed out that each time an instance of code X is inserted to the ISA definition module, a different dedicated ISA is generated. Each version of the newly generated code (taking always the same code X as input) will be assigned with a different random dedicated ISA. However, the exact evaluation of the resulting collision probability requires more investigation and it will be considered in future work.

### *7.4.3 VirISA Permutation Table Rules*

Key point in the overall concept of the VirISA virtualization environment is the formulation of a random, dedicated ISA for an input executable code. As noted, this dedicated ISA will be a subset of the core conventional ISA. This subset will include instructions enabling the execution of a given host program, but preventing the execution of an injected code (the underlying VirISA virtualization environment is responsible for detecting the unsupported instructions/opcodes and issue the corresponding alert messages using some detection mechanism).

At first glance, someone may argue that the target ISA subset can be created by excluding from the regular ISA all the machine-level instructions that are not part of the host executable code. However, since this protection approach can be bypassed by an experienced attacker (e.g., by acquiring and profiling a copy of the host program), we move one step forward and propose a randomly generated ISA which is based on specific permutation rules. In the proposed VirISA scheme, the host executable code is randomly transformed (dynamically rewritten) to another, new executable code. Obviously, the initial (input) and the new (output) executable codes must have exactly the same functionality (i.e., produce exactly the same results), although different machine-level instructions will be included in each code instance.

To enable a set of rules in order to dynamically and randomly create dedicated ISA subsets, we sketch a set of permutation rules targeting the popular x86 instruction set. However, we consider our approach as a generic solution in concept is applicable to any general purpose ISA, e.g., ARM, Sparc, or Alpha ISA. As a case study we explore the x86 ISA which is a fairly rich CISC ISA exhibiting great opportunities for our scheme and has wide appliances in the market (PCs as well as embedded systems; Intel Quark or Atom processors). The main reason for our choice is the fact that x86 specific, functionally equivalent, machine-level instructions are associated with different opcodes when the instruction operands (registers) are different.

Table 7.1 provides some instruction permutation examples, divided in rule categories, to demonstrate the functionality of the proposed VirISA definition module. In the first example case of Table 7.1, our focus point is the `INC` (increase) instruction for which each x86 instruction has different opcodes, defined mainly by the input register (this behavior exists in other instruction categories as well). Among all the `INC` instructions shown in the same row in Table 7.1, in the proposed VirISA approach only one `INC` instruction (e.g., `INC ecx`) will be randomly selected and only this `INC` instruction/opcode will be included in the new dedicated ISA subset. All the other `INC` instructions will be marked as non-valid opcodes. Obviously, this transformation will require recoding (via extra

**Table 7.1** Exemplary permutation tables

| | | | | |
|---|---|---|---|---|
| *Arithmetic instructions category* | | | | |
| 1 | INC EAX | INC ECX | INC EBX | INX EDX |
| 2 | (NOT) ADD | | (NOT) SUB | |
| *Stack instructions category* | | | | |
| 3 | PUSH EAX | PUSH ECX | PUSH EBX | PUSH EDX |
| 4 | POP EAX | POP ECX | POP EBX | POP EDX |
| *Register assignment instructions category* | | | | |
| 5 | PUSH/POP | MOV | | |
| *Addressing mode instructions category* | | | | |
| 6 | (i) | MOV R, [R'] | Register indirect | |
| | (ii) | MOV R, [R'+index] | Register relative | |
| | (iii) | MOV R, [R'*index+offset] | Register relative plus offset | |
| | (iv) | MOV R, [R'+offset*R"] | Scaled index | |

assembly code inserted into the newly generated executable) of all the other INC instructions. These instructions will be either realized via the INC ecx instruction or via some other, dedicated ISA valid, instruction combination (e.g., via an ADD instruction). Note that all INC instructions are heavily used by a typical compiler. The same transformation can be applied to other instruction categories, e.g., DEC (decrease) and PUSH/POP instructions. Another possible transformation is to randomly choose between the ADD and SUB instructions and allow only one of them to be included in the generated dedicated ISA subset (row 2 in Table 7.1).

Another beneficial transformation is to convert all the PUSH/POP instructions to MOV equivalent instructions (or the opposite). Of course, this transformation will require extra code modifications to appropriately align the top-of-the-stack (TOS). However, this kind of conversion will have a significant impact in the generated code, since a large volume of the instructions in a typical application belong to those two categories. Finally, a heavily influential transformation is to randomly use one among the register-based addressing modes (only one of them will be part of the target ISA subset), as can be seen in the last line of Table 7.1.

Overall, we have to point out that the cases illustrated in Table 7.1 are only indicative so as to briefly illustrate the basic idea of our approach. We have created full tables (and the required code transformations) taking into account the majority of x86 assembly instructions. Our analysis revealed that in some cases removing one instruction from the dedicated ISA subset may require a fairly large number of additional assembly instructions in order to produce a functionally equivalent assembly code. For example, this is the case for the XOR (exclusive OR) instruction which requires ten assembly instructions in order to be realized using logical NOT/OR/AND operations. Thus, such kind of transformations will have a great impact in the reported execution time and code size of the newly generated code resulting in trade-offs between performance/code size and security/protection capacity. We intent to fully investigate those trade-offs in a future version of this work.

## 7.5 Conclusions

In this paper, a code injection protection methodology was sketched that uses processor ISA subsets in order to generate dedicated virtual execution environments for each protected process. We proposed the VirISA scheme, acting as a TCB, that is loaded before the OS kernel and is capable to recode a full ISA executable code into a randomized, dedicated, and restricted ISA subset executable code with the same functionality as the original. The VirISA scheme can mount a VM based on the chosen dedicated ISA so as the recoded program can be safely executed. An attacker with no knowledge of the chosen restricted ISA will inject full ISA code to such a process and since the malicious code will include instructions that are not supported by the limited ISA environment it cannot be functional. The proposed approach does not need to employ encryption algorithms for randomizing the ISA nor need to retain encryption keys. The VirISA scheme can be used to any off the shelf single or multicore system and through the ISA restriction methodology can protect against a wide range of code injection attacks. Currently, the proposed solution is at design stage and our future goal is to realize a functional prototype for several use cases (mainly in mobile operating environments) in order to provide actual performance results under several attack scenarios and extract "performance against protection" trade-off patterns.

## References


1. Aleph One. Smashing the stack for fun and profit. Phrack Magazine. **7**(49), (1996)
2. M. Conover, w00w00 on heap overflows, w00w00 Security Team, http://www.w00w00.org/articles.html (1999)
3. Scut's team TECO, Exploiting Format String Vulnerability, http://www.team-teso.net/articles/formatstring, (2001)
4. H. Shahriar, M. Zulkernine, Taxonomy and classification of automatic monitoring of program security vulnerability exploitations. J. Syst. Softw. **84**(2), 250–269 (2011)
5. W. Hu et al., Secure and practical defense against code-injection attacks using software dynamic translation, in *Proceedings of the 2nd International Conference on Virtual Execution Environments,* ACM, New York (2006)
6. G. Kc, A. Keromytis, V. Prevelakis, Countering code-injection attacks with instruction-set randomization, in *Proceedings of the Conference on Computer and Communications Security* (2003)
7. E.G. Barrantes, D.H. Ackley et al., Randomized instruction set emulation to disrupt binary code injection attacks, in *Proceedings of the Conference on Computer and Communications Security* (2003)
8. S. Bhatkar, D. DuVarney, R. Sekar, Address obfuscation: an efficient approach to combat a broad range of memory error exploits, in *Proceedins of USENIX* (2003)
9. R.B. Aussel, J. Sailer, Only hardware-assisted protection can deliver durable secure foundations. IEEE Software Magazine. **28**(2), 57–59 (2011)
10. A. Baldwin, S. Shiu, Hardware encapsulation of security services, in *Proceedings of the European Symposium on Research in Computer Security* (2003)
11. A. De Gloria, VISA: A variable instruction set architecture, ACM SIGARCH Computer Architecture News. **18**(2), 76–84 (1990)
12. K. Hwang, Z. Xu, *Scalable Parallel Computing: Technology, Architecture, Programming* (McGraw-Hill, New York, 1998)
13. J. Sanchez, A. Gonzalez, Instruction scheduling for clustered VLIW architectures, in *Proceedings of the Symposium on System Synthesis* (2000)
14. J. Liu, F. Chow et al., Variable instruction set architecture and its compiler support. IEEE Trans. Comput. **52**(7), 881–895 (2003)
15. C. Corp, Cognigine Corporation, CGN16100 Network Processor User Manual (2002)
16. E.G. Barrantes, D.H. Ackley et al., Randomized instruction set emulation. ACM Trans. Inf. Syst. Secur. **8**(1), 3–40 (2005)
17. A. Sovarel, D. Evans, N. Paul, Where's the FEEB? The effectiveness of instruction set randomization, in *Proceedings of the USENIX Security Symposium* (2005)
18. M. Milenković, A. Milenković, E. Jovanov, Hardware support for code integrity in embedded processors, in *Proceedings of the Conference on Compilers, Architectures and Synthesis for Embedded Systems* (2005)

19. G. Portokalidis, A.D. Keromytis, Fast and practical instruction-set randomization for commodity systems, in *Proceedings of the Annual Computer Security Applications Conference* (2010)
20. Pax Team. PaX. [Online]. Available: http://pax.grsecurity.net (2000)
21. Pax Team. PaX address space layout randomization (ASLR)
22. H. Bojinov, D. Boneh et al., Address space randomization for mobile devices, in *Proceedings of the Conference on Wireless Network Security* (2011)
23. S. Bratus, P.C. Johnson et al., The cake is a lie, in *Proceedings of the Workshop on Virtual Machine Security* (2009)
24. S. Bratus, M.E. Locasto et al., VM-based security overkill, in *Proceedings of the Workshop on New Security Paradigms* (2010)
25. C. Gebhardt, C.I. Dalton, A. Tomlinson, Separating hypervisor trusted computing base supported by hardware, in *Proceeings of the Workshop on Scalable Trusted Computing* (2010)

# Part III
# Opportunities, Challenges and Limits in WSN Deployment for IoT

# Chapter 8
# Deployment Strategies of Wireless Sensor Networks for IoT: Challenges, Trends, and Solutions Based on Novel Tools and HW/SW Platforms

**Gabriel Mujica, Jorge Portilla, and Teresa Riesgo**

## 8.1 Introduction

The research field of the wireless sensor network (WSN) has undergone an important evolution process during the last few years along with the progressive inclusion of the Internet of Things (IoT) as the main foundations for the future of smart and sustainable cities [1]. The intrinsic nature of a WSN-based system as the integration of different technologies related to multi-domain research areas (such as low-power embedded processing architectures, lightweight communication capabilities, industrial technologies for environmental sensing, and energy-aware mechanisms for long-term operability of the application) relies on the implementation of novel and more efficient hardware/software techniques, in order to properly cope with the requirements and constraints that this type of distributed system pose to researchers [2]. In this direction, such innovation process also implies new challenges and more complexity when designing and developing smart applications based on this heterogeneous technology, starting from the requirement stage up until the final operational release of the system. However, the creation and optimization of sensor networks goes beyond robust and low cost platforms, since the node distribution, communication, and dissemination strategies as well as the cooperative processing capabilities of hundreds or even thousands of small devices are critical aspects to assure the actual performance and reliability of the planned application.

This is particularly important during the deployment, commissioning, and in-field validation phases as a fundamental part to analyze and optimize the real behavior of the network with respect to what is expected from the theoretical simulation models. At that stage of the development cycle, wireless communication

G. Mujica (✉) • J. Portilla • T. Riesgo
Centro de Electrónica Industrial, Universidad Politécnica de Madrid - José Gutiérrez Abascal, 2, 28006 Madrid, Spain
e-mail: gabriel.mujica@upm.es



G. Keramidas et al. (eds.), *Components and Services for IoT Platforms*,
DOI 10.1007/978-3-319-42304-3_8

technologies and node distribution mechanisms play a relevant role for the establishment of an overall collaborative platform, so as to integrate a whole correlated system aside from the individual capacity of the sensor nodes. However, the issue of optimizing WSNs has been usually approached from the basis of a more conceptual perspective at the planning phase, in the sense that modeling and simulation techniques are certainly accomplished albeit a gap with the actual on-site deployment strategy is still a noticeable aspect that might have a direct effect on the performance of the sensor network at its operational stage. Therefore, in this chapter a new perspective of WSN deployment strategies targeting IoT smart application scenarios is in-depth discussed, trying to address today's shortfall of methodologies to efficiently carry out the commissioning, configuration and on-site performance optimization tasks, and converging deployment capabilities with runtime enhancement of multi-hop mesh networking mechanisms. New trends and solutions based on hardware–software support tools are introduced so as to provide an overall context for accomplishing effective distributed sensor deployments.

The rest of the chapter is organized as follows, starting with a general overview of state-of-the-art technological trends that contributes to introduce the proposed concept. Then, an on-site deployment toolset is presented to detail the integration of optimization techniques in order to achieve an energy-efficient cohesion of the wireless network as well as the long-term maintainability of the deployed system. Based on such an optimization process in addition to in situ simulation capabilities, a dynamic parametrization of network routing mechanisms is proposed, in combination with the on-site multi-hop mesh communication assessment that is intended to provide a comprehensive decision-making analysis of the type of routing strategy to be adopted in a particular scenario. A discussion regarding experimental results and potential benefits of the proposed system is subsequently clarified, and finally conclusions as well as future perspectives are also analyzed.

## 8.2 Research Trends and State-of-the-Art Approaches

There have been different contributions in the state-of-the-art regarding simulation tools that aim at tackling the problem of system design optimization in terms of sensor coverage, radio propagation, hardware cost, power awareness, and embedded software capabilities. In this way, common research trends regarding deployment techniques for WSNs center the attention in random strategies [3] in which non-uniform, uniform, and grid distributions are combined with coverage optimization analysis [4] to maximize/balance the connectivity of the sensor network to be deployed. Node placement strategies targeting network lifetime and network coverage enhancement have also been addressed in the state of the art, in which optimization methods for energy consumption improvements can be achieved by using constrained and relay-node-based deployment schemes [5, 6]. In this direction, even though several optimization approaches can be distinguished for performing more efficient network distributions, actually most of the current solutions certainly face the problem of deploying WSNs from the point of view of planning techniques, but not as the result of a proper on-site real deployment, configuration, and performance evaluation/enhancement methodology.

On the other hand, there are several contributions in the literature that focus on studying the differences between wireless models and real radio communication, service, and application behavior by means of using WSN testbed deployments, such as the work proposed in [7], in which authors attempt to enhance the simulation engine by analyzing the main mismatches with respect to a real sensor network test case, so that more realistic simulations can be offered prior to the final system deployment. The implementation of WSN testbeds is also a widespread research activity that commonly relies on the indoor deployment of sensor platforms supported by a laboratory testing infrastructure upon which researchers can perform controlled experimental tests, such as Motelab or Indriya [8, 9], where the configuration and monitoring of the nodes is performed by using backchannel-based management architectures. Such support environments are mainly intended to provide users with a more structured scheme for testing hardware and software prototypes before their final implementation into the actual WSN application scenario, though flexibility, scalability, mobility, and modularity are usually penalized. The above-mentioned concerns are even more relevant when referring to the ad-hoc mesh networking essence of the WSNs, whose flexibility, scalability, and versatility promote their suitability for IoT application contexts. In fact, heterogeneous networks based on wireless multi-hop mesh approaches constitute the foundations of current initiatives by the research community for establishing IoT-based technologies for smart city application scenarios. The efficiency of data dissemination techniques among remote points of the networks will be certainly defined by the reliability and effectiveness of adaptable routing mechanisms according to the topology and partitioning strategies to be adopted. Moreover, energy-efficient routing optimization capabilities are a must for the long-term feasibility of the WSN deployment. Based on this, modeling and simulation tools are also proposed in the literature for the performance evaluation of routing protocols prior to their implementation into wireless sensor and communication platforms [10, 11].

### *8.2.1 A New Perspective for On-Site WSN Deployments*

However, during the deployment, configuration, and tuning process of the distributed system, in-field routing protocol implementation analysis are to be allowed for so as to produce a better assessment of the multi-hop data dissemination strategy to be finally applied, specially taking into consideration how specific constraints of the target scenario may affect the overall performance of the wireless network deployment. This would lead to generate on-site optimization approaches in order to guarantee a high level of system adaptability to the target scenario, which results in a more efficient and energy-aware implementation.

In this direction, one of the major issues when progressing within the system development cycle is that there is a lack of well-defined frameworks and tools focused on facing the aforementioned gap between planning/simulation strategies and on-site efficient commissioning methodologies. Moreover, the idea of taking advantage of a comprehensive on-site deployment approach to dynamically optimize or even reconfigure the mesh networking mechanism in situ represents a very challenging perspective that has not been properly covered yet. The new era of emerging technologies is creating synergies among IoT and WSN application contexts that demand the determination of novel methods and toolsets for the optimization and performance enhancement of collaborative wireless distributed systems. This can be indeed realized by means of integrating already defined hardware/software components with new strategies that provide runtime on-site dynamic optimization capabilities for the target mesh network implementation.

## 8.3 General Overview of the Proposed System

The system architecture of the proposed WSN deployment platform encompasses the integration of hardware and software elements in a unique flexible and powerful toolset, so as to provide users with an easy-to-use mechanism for the analysis and optimization of mesh networking implementations in situ. As shown in Fig. 8.1, the general structure of the designed framework is based on the inclusion of an intelligent mobile device as the core element in which deployment optimization

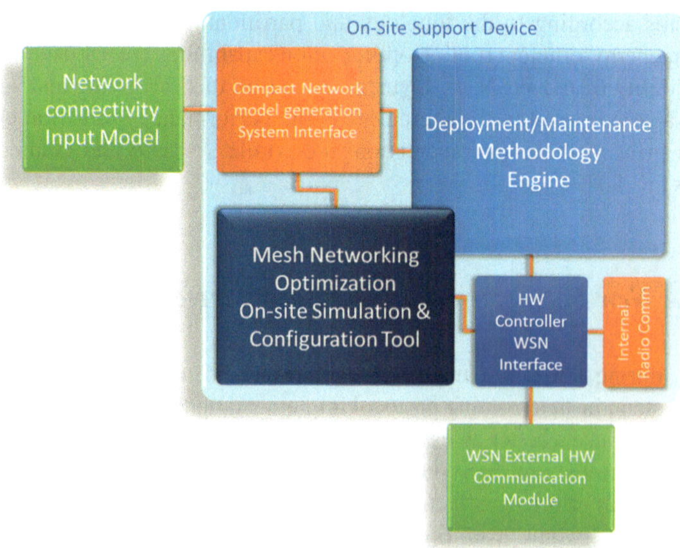


**Fig. 8.1** General view of the proposed system architecture

methodologies as well as on-site simulation capabilities are implemented. This means that the intrinsic nature of the platform as an in-field tool requires that mobility, usability, portability, and robustness have to be properly assured. The system prototype is composed of an Android-OS-based element that allows the integration of WSN technologies by using a host-mode connection approach via a serial interface communication, such as USB. This is particularly important for equipping the smart platform with low-rate wireless communication protocols that are commonly used in WSN implementations, such as IEEE 802.15.4 [12], which is carried out by integrating a modular and highly flexible WSN platform [13] within the system implementation. Although the developed prototype considers such an integration approach that is mainly intended to provide a high degree of adaptability to different WSN hardware elements, the performance analysis, and optimization techniques that will be subsequently presented in this chapter can be particularly applicable in todays' work-in-progress ah-hoc multi-hop mesh networks for IoT scenarios, where IEEE 802.11 [14] modules are used as the wireless communication interface. Therefore, by taking advantage of the available hardware in the smart devices, this on-site deployment and maintenance support toolset implementation is highly usable and portable among such application contexts.

As shown in the system architecture scheme proposed in Fig. 8.1, regardless of the type of communication protocol to be used, the main requirement of the proposed tool is the provision of two input elements. On one hand, a radio propagation and connectivity model among the wireless nodes to be deployed is to be produced as a result of the network simulation stage. This model shall provide information regarding radio transmission configurations in addition to metrics related to radio signal strength and link quality between pairs of points. This will define an overall theoretical correlation map as the starting point to carry out the on-site deployment and commissioning process. Although this model is intended to be downloaded as the output of a planning/simulation engine, the on-site deployment tool also provides a compact runtime model generation based on the experimental optimization of a radio propagation simulation models, considering the one proposed in [15]. In any of the aforementioned cases, the network model serves as the main input to the in-field optimization engine as well as the deployment methodology.

On the other hand, the actual connectivity from the mobile device to the wireless sensor nodes to be deployed is a fundamental aspect to guarantee the runtime execution of configuration and analysis tasks according to the real performance of the deployment. This means that a synergy between simulation capabilities and on-site behavioral data is created so as to produce a more comprehensive optimization process. The developed prototype considers such a level of interaction with the wireless network by using low-abstraction-layers of the communication interface, that is, acting right on top of the medium access control (MAC) layer of the IEEE 802.15.4 or 802.11 modules. Based on this, real metrics related to the network interconnection process can be gathered to evaluate and refine the runtime deployment methodology and enhance the mesh networking array to be finally released. When referring to the low-rate wireless communication interfaces, the parameters that are mainly of interest relate to the radio signal strength indicator (RSSI), link quality indicator (LQI), as well as the link delivery ratio (LDR).

## 8.4 On-Site Deployment and Maintainability Optimization Mechanism

The issue of deploying a WSN in such a way that both the cohesion and the maintainability of the network can be assured poses important challenges to be faced, especially when adding energy efficiency as another variable to the problem analysis. The setup of the sensor nodes not only in terms of the mesh networking communication mechanisms but also in relation with the platform parametrization (hardware/software properties and application tuning) might be unified by means of triggering a collaborative configuration procedure from the mobile on-site deployment tool, in which the target node executes a frame dissemination task at MAC level in order to generate two main actions: network connectivity correlation with its surrounding environment (and thus establishing bidirectional information for routing purposes), as well as platform setup where specific parameters regarding application support services (data transmission rate mask, service provision, synchronization, and power mode configurations) and node properties (such as node weight determination, which refers to the generation of routing metrics and cluster-based configuration entries, as explained in next sections) are distributed among its enclosed area. In both cases, the performance verification, reconfigurability, and maintainability processes could seem to be a subsequent generation of config dissemination triggers among the sensor nodes that compose the wireless mesh networking array. However, this scenario becomes a non-trivial execution task and even less stable scheme when the WSN-based smart system is a conjunction of points in a large-scale deployment, where problems related to distributed processing capabilities, network flooding, and power dropping situations may affect the overall performance and long-term autonomy of the application.

This leads to propose an on-site optimization scheme in which the configurability and maintainability of the multi-hop distributed system can be efficiently assured. The basic idea behind such an optimization process relies on the dynamic calculation of what is considered in this methodology as *critical points* of the mesh network (called *Cps)*, upon which the config dissemination scheme is realized with the most resource+energy effective cost. However, the actual efficiency of this execution process does not only depend on the number of encountered points, but also on how the sequence of dissemination tasks is performed along the Cp's scheme. So let's first analyze the impact of generating an optimized selection of Cps. When encompassing mechanisms such as node discovery and flooding actions across the mesh array, a high level of processing and energy resources is spent in unnecessary redundant points, whose subsequent performance can certainly produce an unbalanced resource consumption profile of the system. Instead, by minimizing the number of specific points of the networks that allows obtaining the same overall dissemination and link correlation scheme, the side effects on the network performance can be also reduced. Then, the optimization of the type and number of Cps also relies on the definition of the execution sequence to be carried out at the deployment and maintenance stage, so as to minimize the level of redundancy when applying the configuration tasks.

### *8.4.1 Modeling of the Deployment and Maintenance Methodology*

In order to be able to accomplish the above-explained process, a proper modeling of the deployment methodology is to be created, so that dynamic and runtime optimization algorithms can be employed accordingly. An important aspect to be highlighted is the inclusion of actual WSN hardware parameters as part of the modeling process, especially in terms of power consumption and radio communication configurations. This is achieved by determining the actions that compose the deployment and dissemination strategy in the form of a problem cost function, which will serve as the main component to compare different possible solutions. The main target is to minimize this cost function within an affordable computational time (considering the nature of the tool as an on-site mobile mechanism). To produce such modeling process the inclusion of the sensor node resource consumption (related to every stage of the discovery, configuration, and dissemination mechanism) is considered into the cost function, which is materialized as the transient energy associated with the corresponding action to be triggered, as follows:

*Node Deployment Sequence, Transient Cost* This cost representation corresponds to the process of deploying or maintaining the WSN in field, which refers to the influence of activating a network setup mode while the deployer/maintainer executes the deployment configuration sequence. As shown in Eq. (8.1), this cost is expressed as the average energy weight of the network right before starting the deployment/maintenance sequence (pre-deployment/installation stage), plus the transitions among critical point's analysis, that is represented as a distance cost factor for every pair-point of the sequence. An efficient generation of the node deployment sequence (*NDoSq*) (based on the optimization algorithms explained subsequently), in addition to the discovery and config dissemination mechanisms, will assure the cohesion of the network according to the provided network correlation model, but with energy/resource cost effectiveness.

$$f_{tr} = \alpha \cdot N + \sum_{i=2}^{p} P_{sb} \cdot N \cdot \delta\,(i-1, i) \qquad (8.1)$$

*Discovery and Link Correlation Scheme Generation Costs* After identifying the target node from the configuration sequence, the on-site deployment support tool triggers a controlled discovery mechanism in which the surrounding area is evaluated, so as to update the link correlation scheme with those pair-points that have not been analyzed/configured yet. In this binding phase the mesh network information is updated (particularly regarding neighbors indexing) both in the involved embedded nodes and in the deployment tool, so that a dynamic recalculation and refinement of the deployment models can be performed in situ. During this procedure, the transactions among the different comprised nodes are computed as the relationship between the radio transmission power configuration and the energy weight of the critical point under commissioning and its associated nodes during the transaction and acknowledgement process. Clear-To-Send-like mechanisms (CTS) for Collision Avoidance (CA) are also considered within the modeling process so as to properly express the transaction slots among the involved nodes.

$$f_{sr}(e) = \left(E_T(e) + \sum\nolimits_{r=1}^{m} E_R(r)\right) + \left(\sum\nolimits_{r=1}^{m} E_a\left(e, r\right)\right) + \left(\sum\nolimits_{r=1}^{m} \left[(m-r)\cdot E_{ck}(r)\cdot \xi + (r-1)\cdot E_s\right]\right) + (n-m-1)\cdot E_{sn} \quad (8.2)$$

Four main elements can be distinguished within the cost expression shown in Eq. (8.2). First, the critical node triggering action that involves the discovery mechanism with the surrounding neighbors. It is represented as the energy resources spent at the source and destination points during the performed transaction. Then, the association stage is expressed as redundancy filtering and pair-point establishment, which includes the energy representation of the recognition execution between the critical and discovered points. Medium access influence is defined as the average power during stand-by and active transitory modes of the critical points' surrounded nodes multiplied by a CA time slot factor. This transaction also allows a bidirectional neighbor table indexing for subsequent routing optimization purposes. Finally, the energy weight of passive elements that are not directly involved in the discovery dissemination process at that point of the deployment sequence is also included as an overall stand-by offset of the rest of the sensor nodes.

*Reconfiguration and Parametrization Cost Based on the Node Correlation Scheme* Once the node association is properly performed from the corresponding node of the deployment sequence under action, the configuration transaction is then triggered between the Cp and the correlated node, so as to establish a runtime parametrization in relation with application level services as well as network settings for communication and multi-hop purposes, but particularly focusing on link metrics and node's weights for the cluster-based routing optimization mechanism. This stage of the deployment process represents a fundamental part for the dynamic updating of the network simulation models based on the on-site gathered data, so that a refinement of the optimization strategy can be accomplished according to a more comprehensive feedback from the deployment tool. Moreover, the result of this strategy indeed enhances the reduction of flooding schemas that significantly affect the overall power and resource consumption of the network platform. As shown in Eq. (8.3), the cost of the configuration transaction is expressed as the energy weight of the Cp and the target setting-point during the cross configuration exchange process, in addition to the stand-by energy offset of the nodes that are not directly involved in the performed action.

$$f_{\mathrm{conf}}\left(e,r\right) = E_{tr}\left(e,r\right) + \left(n-2\right)\cdot E_{sn} \tag{8.3}$$

It is important to highlight that although in the above-declared functions the energy/resource weight of the transactions associated with the process of performing the deployment/maintenance tasks are expressed as overall energetic costs for the sake of simplicity on the representation, every element is certainly broken down into specific parameters within the system implementation of the deployment tool, especially regarding radio power configurations of the hardware module, platform's operational and processing modes, and protocol transmission rate definition. The general representation of the cost function for the proposed methodology is shown in Eq. (8.4), in which the explained underlying transactions are represented, and where $m(i)$ represents the nodes that have not been configured/parameterized yet, whereas $p$ refers to the Cps in the configuration sequence to be carried out.

$$f_{\mathrm{cost}} = f_{tr} + \sum\nolimits_{i=1}^{p}\left[f_{sr}(i) + \sum\nolimits_{r=1}^{m(i)} f_{\mathrm{conf}}\left(i,r\right)\right] \tag{8.4}$$

One of the main advantages of the proposed method is its platform independence nature. The system implementation allows developers to adjust the parameters related to the technology to be used, so as to adapt the provided optimization methodology to the specifics of the target WSN deployment. This, combined with the simulation model refinement, is to guarantee the aim of achieving the most efficient/cost-effective cohesion of the mesh network.

### 8.4.2 Deployment Optimization Mechanism

Based on the definition of the deployment and maintenance methodology scheme as well as the cost functions that allow the evaluation of different strategies to be performed, the next step relies on the ability to dynamically optimize the result of these functions such that the commissioning task can be efficiently carried out with a minimum cost of operation. This process is to be accomplished considering the nature of the on-site deployment tool in terms of time and processing capabilities. Since it is a dynamic optimization method based on both modeling and real data from the actual behavior of the WSN under deployment/maintenance, efficient though affordable runtime optimization mechanisms are to be adopted. The first consideration that could seem to be a solution for the optimization problem is the application of what is named here as *swap-based algorithm*, which refers to the assessment of each solution as a combination of every possible NDoSq, which will lead to compare and obtain the most efficient result for a given deployment scenario. The main problem of this solution is its performance and scalability in large-scale WSNs, since the algorithm makes the problem of optimizing a network composed of $n$ nodes rely on the evaluation of $n!$ possible solutions, which is not time and computationally feasible for the support tool.

As an alternative to guarantee a trade-off between solution efficiency and time/processing effort of the mechanism, the concept of bio-inspired optimization algorithms has been included within the system implementation of the proposed tool. These types of strategies rely on evolutionary computation to generate a solution space based on optimal or suboptimal results, by applying natural evolution and selection principles [16]. In the proposed toolset, a *genetic-based algorithm* has been implemented as part of the runtime optimization process, in which several modifications were integrated in order to adapt the dynamics of this mechanism to the problem of deploying a WSN by using the proposed methodology.

Considering the process of applying this algorithm to the optimization problem, the first step is the generation of candidate solutions that are expressed as a population composed of individuals. Traditionally in genetic algorithms individuals are represented as strings of binary codes, which might simplify the subsequent operations and evaluations that the candidate solution has to undergo. However, in case of the deployment methodology, the genotype of the individual is created as an array of nodes that corresponds to the NDoSq. Apart from this, in order to evaluate the degree of aptitude of a candidate solution, the problem cost function proposed in the modeling process is used in such a way that the higher its value is, the less the aptitude of the deployment optimization sequence will be (that is, the fitness expression is defined as $f_{ap} = 1/f_{\text{cost}}$). However, as explained in the experimental result section, an additional execution factor is included to take into account the time/processing performance of the solution space ($ef_n=N/(f_{\text{cost}}*t_c)$). During the evaluation of the individual, the optimization engine performs the different transactions according to the provided genotype, then based on the redundancy detection strategy the Cps associated with the initial sequence are generated.

In order to produce a more appropriate population in every generation of the genetic evolution, the *selection* process takes place within the optimization algorithm, in which a number of individuals is selected from the population, and from such a group the best one of them is chosen for the subsequent crossover operation. The process is then repeated for a second candidate solution. Both individuals undergo the process of *crossover* that is intended to modify the genotype from one generation to the next one, although allowing for the nature of the WSN deployment optimization problem. This means that the recombination of the genotypes will be executed considering that the result of the segment mapping has to be a sequence of nodes. Thereafter, the *mutation* operation is carried out to guarantee that a premature convergence into local optimal solutions is avoided, so that a level of diversity is maintained throughout the genetic generations. In this case a swapping-like mechanism is adopted to assure a coherent gene alteration. As a result of this process, the created individual is included in a new population that will replace the previous one for the next genetic generation. Then, the aforementioned operations are repeated up until the population is finally completed with new candidate solutions. An additional operation that has been included into the optimization mechanism is the *elitism*, in which a minimum level of aptitude is kept across the generations. In this case this is done by including that individual with the best fitness of the previous generation into the new produced population.

In order to achieve an efficient solution not only in terms of the deployment optimization sequence but also regarding time and computational resources when triggering the proposed methodology, boundaries related to the maximum number of generations as well as converge rate can be dynamically defined during the on-site execution support process. In this case, the convergence parameter is defined as the percentage of candidate solutions with the same value of aptitude. By default this parameter is set to 90%, but it can be reconfigured by the deployer in situ. Therefore, four main elements will determine the performance of the optimization process, that is, the size of the population, the number of generations, the convergence threshold, and also the size of the network that influences the genotype of the candidate solutions, which will impact the complexity of the underlying operations as well.

## 8.5 Wireless Mesh Networking Optimization Approach

The deployment optimization strategy proposed in previous sections pursues a highly efficient mechanism to obtain the cohesion and overall network correlation scheme of the wireless distributed system. But it is noticeable to point out that one of its major impacts is also related to the ability to provide an enhanced multi-hop mesh network optimization approach while carrying on the on-site configuration process. This is particularly important in large-scale WSN implementations where the efficiency associated with routing protocol strategies is a key to the success of the overall operability of the deployment. It is well known that wireless mesh networking approaches represent the main foundations for this promising era of interconnected IoT-based systems and technologies. Therefore, flexible and resource-aware optimization mechanisms aiming at reducing the overhead of the WSN are to be properly addressed. This is certainly one of the main targets of the proposed on-site toolset, that is, the creation of an adequate synergy between deployment and multi-hop dissemination optimization processes.

Taking advantage of the previously described deployment methodology, a novel routing protocol optimization technique is also approached in order to provide deployers with an in-field multi-hop modeling process that allows the runtime configuration and execution of the most suitable and efficient routing strategy for the target application. The underlying idea relies on including different routing optimization strategies within the on-site deployment support tool so that users can simulate and evaluate them based on real network data, so as to perform runtime reconfigurations from the basis of the proposed deployment methodology accordingly. Right after the deployment optimization algorithm run by the deployer, a new model of the mesh network that includes the result of the node's parametrization is created, which then serves as the entry point to generate the routing optimization schema to be evaluated and configured into the sensor nodes.

Since the main target of the developed tool is its applicability to large-scale WSNs, a combination of flat and cluster-based protocol strategies [17, 18] are taken into account so as to produce an optimized scalable routing mechanism for mesh networking. The default strategy within the routing modeling engine is an hierarchical structure optimized from the basis of the work proposed in [19] and composed of four possible types of node states or functional roles, that is, *Undefined* status when the node is not included in the hierarchy, being the default initial status for the nodes of the network before executing the deployment optimization engine; normal or *Cluster-member* status when the node is part of a cluster defined by a *Cluster-head* node; *Gateway* (GW) status when the node may be both part of a cluster or part of two different clusters, thus being responsible for the communication between cluster-heads and serves as a highway of information exchanging; and *Cluster-head* (CH) status when the node is in charge of its corresponding cluster. It generally has a series of cluster-member nodes under its domain, and manages advanced procedures within the cluster regarding the multi-hop communication mechanism. The determination of the different types of states will depend on the weight calculation performed at the previous stage of the optimization process, which will then generate an overall scheme of suitable clustering roles for the sensor nodes. In order to manage and maintain the routing mechanism, every node has to handle two routing information tables regardless of its type of role: a *Neighbor Table* that contains information related to every node within its rage of radio communication, and a *Routing Table* similar to flat routing mechanisms such as AODV [20], in which information regarding next hop and destination node, sequence number and quality metrics of the route can be included. However, in addition to this general structure, CHs have to include two more information entries: the *CH Neighbor table* in which data related to Cluster-head neighbors and the associated Gateway in between are registered, and the *Cluster Table* that comprises the *Cluster-member*s that belong to the cluster as well as their associated neighbors, such that CHs can be completely aware of nodes 2-hops ahead from them. There are cluster-based routing mechanisms proposed in the state of the art that include this 2-hop information in every node of the wireless network [21], whereas the one proposed here manages this capability through the CHs as a result of the deployment and maintenance optimization process, which thus leads to reduce the network overhead and energy/resource cost of the routing strategy. Finally, the GW nodes also include an additional table (called *Gateway Table*) for registering the information about pairs of CHs that use the gateway as the main interconnection point between two network areas. Based on the discovery and config dissemination mechanisms of the proposed deployment and maintenance optimization process together with the underlying weight calculation and comparative evaluation, the cluster-based multi-hop link formation is carried out by dynamically updating the aforementioned routing data management structure.

Another important optimization that lies on including enhanced capabilities into the CHs is related to the runtime routing discovery mechanism. Let's first briefly consider how the establishment of a route between two remote points is performed during the discovery process. While in flat routing strategies such as AODV the source node generates a broadcast with the RREQ (Route Request) control message, in the cluster-based approach the RREQ is sent to the associated CH, which then checks both its neighborhood and its Routing Table for the destination node. If it does not find it, the RREQ message is forwarded to its neighbor cluster-heads through the corresponding gateway, which also checks its Neighbor Table and Routing Table before forwarding the message to the corresponding CH. When a node discovers the destination node in its neighborhood or in its Routing Table (or the destination node itself receives the RREQ message), it sends back a RREP (Route Reply) control message to the previous cluster-head. The optimization comes when the RREP is travelling back to the source route by using the established route during the RREQ transaction, because the CHs can execute a runtime comparative analysis in order to dynamically design an alternative adjacent route based on their *Cluster Tables*, particularly intended to be applied when efficiency in multi-route code dissemination strategies across the distributed system are to be supported.

As shown in Fig. 8.2 the involved CH makes a local evaluation based on the already available routing management information, and if succeed it will notify the nodes involved, add the alternative adjacent route to the RREP message and let the next CH continue developing the optimization process. When the last CH receives the RREP message, it checks whether the alternative path is more suitable or not in terms of the multi-hop communication cost metric and then sends the adjacent solution to the source node. The main advantage is that such strategy relies on already available information from the CHs, so there is no impact in terms of network flooding and packets overhead. This multi-hop optimization mechanism for mesh networking (including route maintenance and integrity verification strategies)

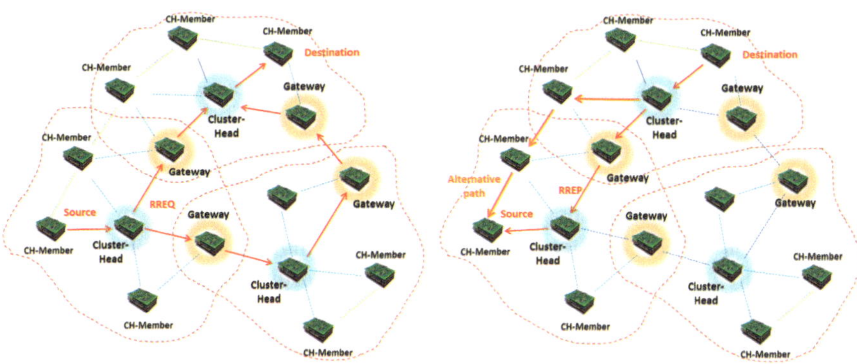


**Fig. 8.2** Route optimization mechanism based on CH dynamic evaluation

has been integrated not only as a fundamental part of the deployment support toolset but also to evaluate its implementation into hardware nodes such as the already mentioned modular platform [13], so that a runtime comparative analysis of the protocol performance with respect to the generated simulation models can be properly carried out, thus assuring a more adapted solution to the target scenario.

## 8.6 System Implementation

As shown in the system architecture of the proposed platform, the implementation of the deployment optimization support toolset has been conceived in such a way that deployers just need to handle a smart device that includes the corresponding wireless protocol for performing the configuration and optimization tasks in situ. In case of WSN deployments based on low-rate data communication capabilities such as IEEE 802.15.4, the interface to the radio module that serves as the link to the nodes under deployment/maintenance is performed by means of a Host-USB-based controller developed into the software implementation, which relies on programming capabilities under the Android-OS environment. Based on the modularity and flexibility that a Java-based system implementation confers, nodes are modeled as entities that include their own WSN properties such as network identification, coordinates and metrics parametrization, power modes and radio communication configuration, whose structure is generated from a generic platform definition. Then, their properties and configurations are set up from the specific definitions of the sensor nodes to be deployed, which are supported by an internal database that gathers the overall information of the runtime deployment activities. This, combined with the network connectivity model, allows determining the initial system correlation among the sensor nodes.

In order to perform a user-friendly interaction with the deployment optimization capabilities, the Google Maps API for Android-OS is combined with graphical interfaces to monitor and control the process of configuring and evaluating the executed strategies in real time. As shown in Fig. 8.3 (left) setting dialogs allow users to configure the parameters of the optimization process, particularly regarding the population size, the number of generations, and the convergence rate. After the execution task is started, users can analyze the behavior of the optimization algorithm and the computed solutions during every generation interval, with the aid of statistical values related to the evolution process as well as the time estimation for the remaining execution. When this execution process finishes, the result of the most optimal solution is represented as the NDoSq to be carried out, together with the list of associated Cps from which the node configuration capabilities can be subsequently triggered. Users can also show related results regarding the cost of the solution and computational time.

Thereafter, the generated results serve as the input model for the routing protocol analysis and optimization engine, in which the implemented strategies for multi-hop networking can also be simulated prior to their parametrization into the sensor

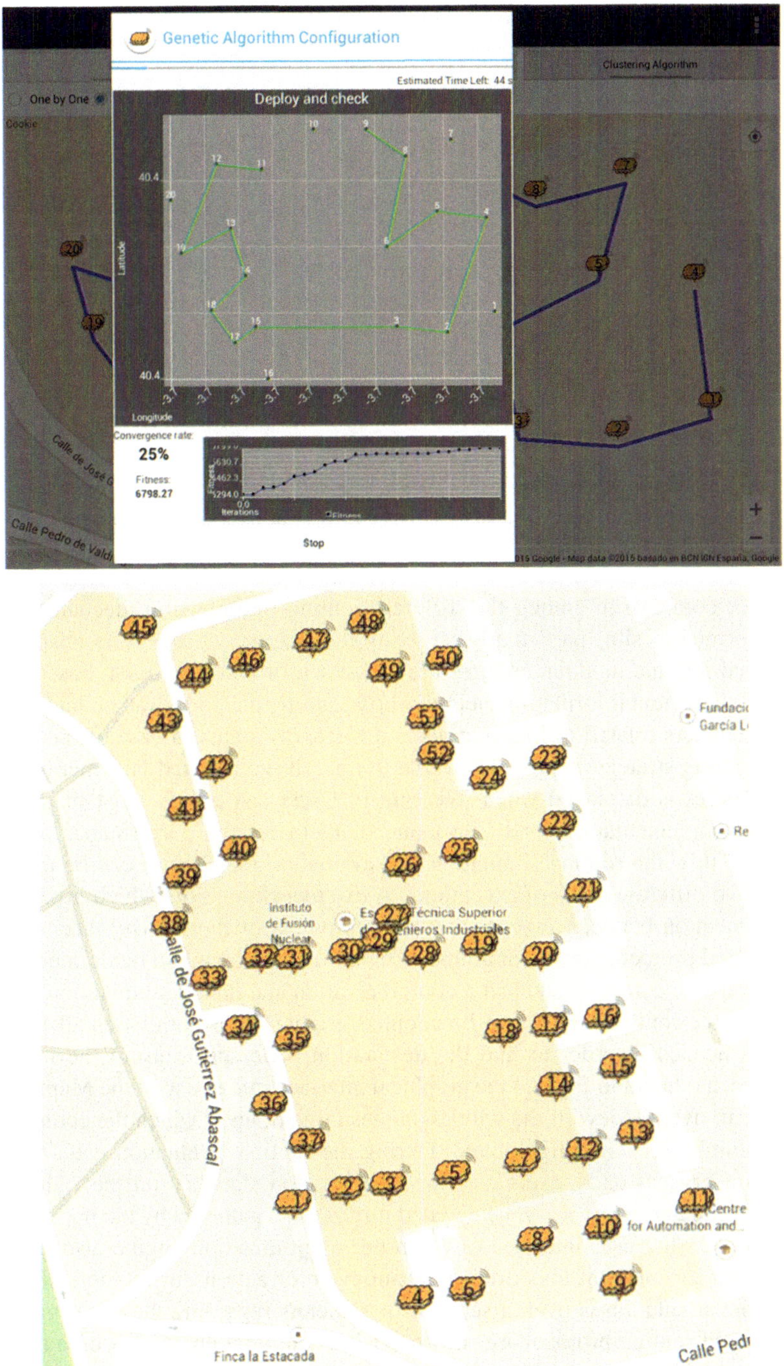


**Fig. 8.3** User interface for the deployment optimization strategies (*left*) and network deployment scenario for the experimental test case (*right*)

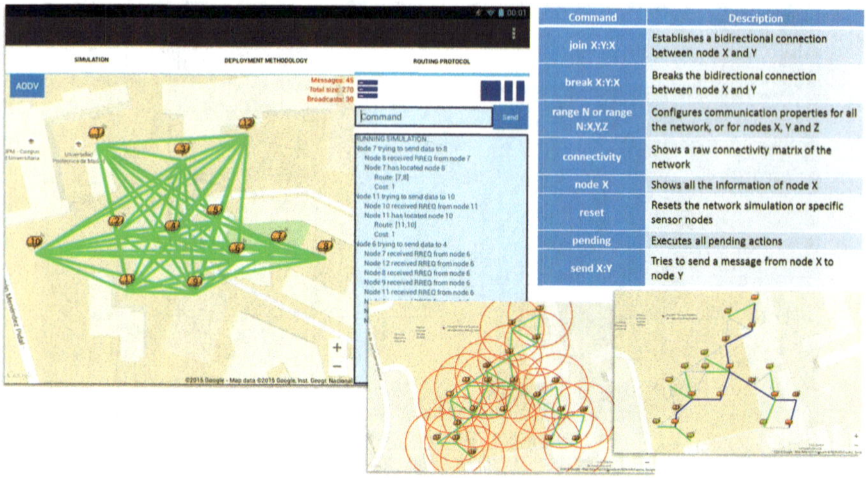


**Fig. 8.4** Routing optimization engine interface and main evaluation action commands

nodes. A comparison among the different routing optimization mechanisms can be performed in situ, particularly targeting those approached in this chapter and integrated into the modular architecture of the sensor platform. First, based on the input management information, neighborhood among the nodes can be highlighted, and parameters related to link metrics and hierarchy status in case of the cluster-based routing strategy is provided to the users. All the updated information that a node handles is displayed when selecting it. Users can access a list of available protocol implementations and select one of them for the performance analysis. Based on this, the routing simulation and evaluation mechanism can be triggered within two different types of executions: A discreet simulation, which involves the communication between two specific remote points of the network according to the selected protocol; and a continuous simulation, in which two random nodes are periodically selected to establish a communication and data dissemination process based on the routing strategy to be adopted, triggering in both cases all protocol routines needed in order to find the destination node and transmit the message. Likewise, as shown in Fig. 8.4 the graphical interface makes use of the Maps API to provide an overall view of the wireless nodes in field, upon which the connectivity and hierarchy lines are highlighted. During the runtime evaluation and when the execution process ends, users can check simulation statistics on the right upper corner of the map and access the updated information gathered by the nodes.

Moreover, the user interface of the routing optimization engine also provides a dynamic console that logs every relevant event or action during the simulation. When the simulation is over, a series of parameters regarding the performance of the selected routing protocol are shown, such as the total number of broadcasts or messages delivery rate, so that a comparison between different protocols can be made. As shown in Fig. 8.4, different actions and events can be triggered by means of using a set of defined commands on the protocol evaluation console. Furthermore, simulations can be run for different types of network topologies and sizes, targeting the analysis and optimization of the system scalability, which also makes it easier to locate the strengths and weaknesses of each protocol implementation. Based on the on-site optimization chain results, a dialog for the execution of the configuration process into the nodes based on the results of the optimization strategy can be shown by selecting the corresponding sensor nodes. During the performance of the deployment methodology, the actual + the modeled link correlation schemes can be overlapped so as to analyze the different wireless network interconnections and metrics from the on-site distributed system.

## 8.7 Experimental Test Cases and Discussion

In order to put the deployment optimization capabilities and the applicability of the support tool chain into different WSN contexts, several groups of test cases are presented, so that the parametrization of the modeling, evaluation and simulation engines can be discussed in addition to providing a comparative analysis of the system optimization results, targeting the deployment/maintenance efficiency enhancement for the wireless distributed sensor networks. The first set of test cases is focused on analyzing different configurations of the deployment optimization algorithms in order to obtain a trade-off between optimal solution and computational time, in addition to evaluate the impact of performing the deployment/maintenance methodology by using the Cp-based optimized sequence.

As shown in Fig. 8.3 (right) the WSN is composed of 52 sensor nodes that will be scaled to verify how the size of the network might affect the corresponding results. Moreover, parameters regarding number of generations, population size, and convergence rate are also tuned so as to define an optimal setup depending on the type of deployment/maintenance to be carried out. In Fig. 8.5 (left) a representation of the impact of the aforementioned parameters is shown in detail. As expected, the cost of the solution improves as the maximum number of generations and the populations get larger, though the computational time also grows. This is the main reason why the execution factor is added, to be able to decide the limits on the generation of an optimal result with respect to an affordable calculation time. It can also be highlighted that the gradient is more accentuated in those regions of the representation where the number of generations increases than in case of the population size. Thus, the convergence rate plays a fundamental role to guarantee that in those situations where the number of generations is oversized, the execution time will not be penalized. Moreover, in Fig. 8.5 (right) it is noticeable the impact of performing the deployment task by using the optimized sequence based on the Cp strategy with respect to a point-to-point two-dimensional configuration sequence. Two main aspects can be remarked in such optimization result. On one hand, the cost of the configuration transitions is enhanced in case of the optimized sequence, which affects positively the overall energy performance of the deployment and

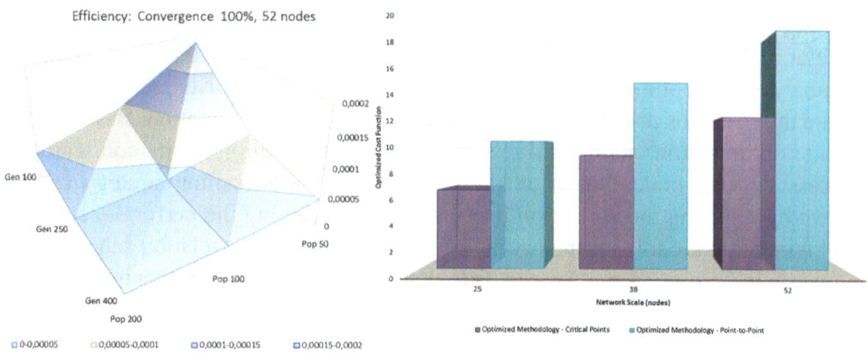


**Fig. 8.5** Deployment optimization algorithm parametrization analysis (*left*) and comparative analysis of deployment methodologies (*right*)

maintenance tasks. On the other hand, as the network gets larger and spread, the redundancy optimization strategy plays a more important role since the associated cost of the dissemination transactions does not shoot up as it can be in the second case. The second set of test cases refers to the application of the deployment support tool chain to the problem of generating an efficient multi-hop mesh network scheme by using the proposed routing optimization mechanisms.

Currently two different strategies are implemented into the system to be evaluated in the Cookies modular platform: the Cluster-based optimization technique as well as an optimized version of the flat-based AODV routing strategy. One of the main goals is to provide users with the ability to compare the performance of both algorithms in order to analyze their suitability and adaptability to the target scenario.

This experimental case involves 30 nodes forming the WSN shown in Fig. 8.6, for which the NDoSq was applied to carry out the node configuration tasks. Based on this, the cluster-based optimization process determines the CHs and the GWs points (in blue and orange, respectively) that correspond to the hierarchy parametrization, in which blue lines refer to the main multi-hop paths while green lines involves secondary routes (the flat scheme is expressed by all the interconnection lines).

Table 8.1 summarizes several route discovery test cases from remote points in order to realize a comparative analysis of both implementations, by using the data processed by the deployment support tool. The main parameters that are taken into account to carry out the experimental comparison are the energy consumption of a routing path, the cost of discovering and maintaining the communication between two remote points, and the transmission delay among the network connections, which are computed as the involved broadcasts of the selected mechanism, the total amount of broadcast spent and the number of hops associated with the discovery paths. Nodes represented in square brackets define those situations where the route reply action is generated by an intermediate point that already had registered a route to the destination point. It can be noticed that in case of the flat-based routing mechanism as the number of data dissemination between remote nodes increases

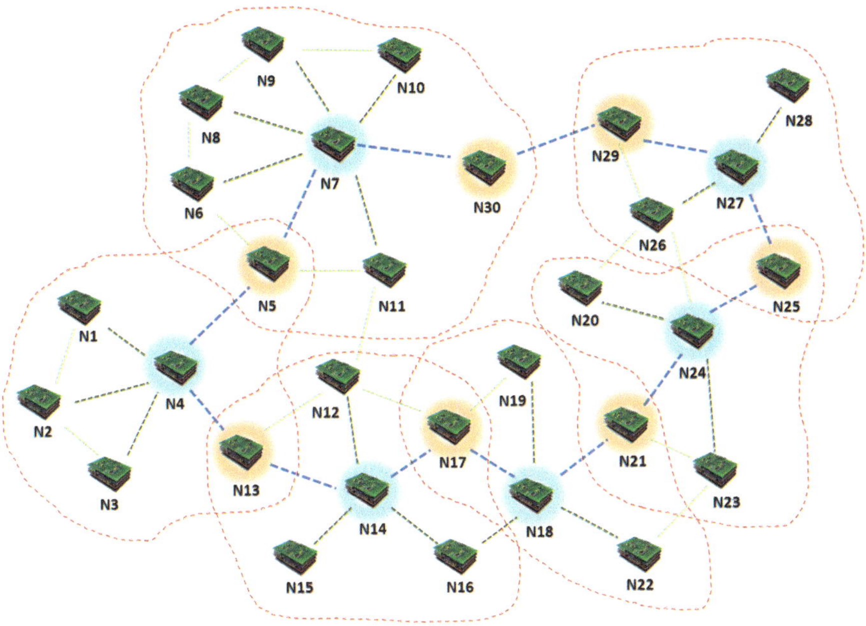


**Fig. 8.6** WSN experimental test case and cluster formation result

as well as the network gets larger, the flooding level tends to increase considerably with every discovery process, whereas in the cluster-based optimization strategy this number increases almost proportional to the scale of the network. In fact, although the generation of the clustering structure needs an initial packet exchange process for the hierarchical configuration purposes, the control frame dissemination and flooding costs of both techniques intersect when the number of pair-points raises 7–8 different requests, which certainly determines the threshold upon which the cluster-based optimization mechanism produces clearer benefits for the mesh networking scenario. Moreover, as shown in the experimental results, such differences are more accentuated when applying the NDoSq based on the Cps parametrization, since the number of dissemination messages is optimized to assure the overall mesh networking connectivity schema with a reduced cost.

Therefore, this produces that the intersection point at the threshold value is obtained more quickly during the operability of the WSN. Another interesting remark is related to the alternative path optimization process performed by the CHs that take part in a request route, which use their available clustering information to generate an adjacent multi-hop interconnection, especially if it provides a more efficient communication path (adjacent routes represented in brackets).

**Table 8.1** Comparative analysis of the routing strategies for the proposed experimental tests

| Request | Route found | | Number of hops | | Broadcasts involved | | | Total Broadcasts | | |
|---|---|---|---|---|---|---|---|---|---|---|
| Cluster formation | - | | - | | 160 | 90 | | 160 | 90 | |
| 3>28 | 3-4-5-7-30-29-[27]-28<br>(3-4-5-7-30-29-27-28) | 3-4-5-7-30-29-27-[28] | 7<br>(7) | 7 | 0 | 0 | 29 | 160 | 90 | 29 |
| 8>22 | 8-7-5-4-13-14-17-[18]-22<br>(8-7-11-12-17-18-22) | 8-7-11-12-17-[18]-22 | 8<br>(6) | 6 | 0 | 0 | 25 | 160 | 90 | 54 |
| 25>15 | 25-24-21-18-17-[14]-15<br>(25-24-21-18-16-14-15) | 25-24-21-18-17-[14]-15 | 6<br>(6) | 6 | 0 | 0 | 27 | 160 | 90 | 81 |
| 24>3 | 24-21-18-17-14-13-[4]-3<br>(24-20-19-17-12-13-4-3) | 24-26-[29]-30-7-5-4-3 | 7<br>(7) | 7 | 0 | 0 | 15 | 160 | 90 | 96 |
| 11>20 | 11-7-30-29-27-25-[24]-20<br>(11-7-30-29-26-20) | 11-12-17-19-20 | 7<br>(5) | 4 | 0 | 0 | 24 | 160 | 90 | 120 |
| 14>7 | 14-13-4-5-[7]<br>(14-12-11-7) | 14-12-[11]-7 | 4<br>(3) | 3 | 0 | 0 | 20 | 160 | 90 | 140 |
| 11>3 | 11-[7]-5-4-3<br>(11-5-4-3) | 11-[5]-4-3 | 4<br>(3) | 3 | 0 | 0 | 15 | 160 | 90 | 155 |
| 20>22 | 20-24-21-[18]-22<br>(20-24-23-22) | 20-24-[23]-22 | 4<br>(3) | 3 | 0 | 0 | 9 | 160 | 90 | 164 |
| 16>29 | 16-14-13-4-5-7-30-[29]<br>(16-18-21-24-26-29) | 16-18-19-20-[26]-29 | 7<br>(5) | 5 | 0 | 0 | 24 | 160 | 90 | 188 |
| 26>10 | 26-27-29-30-[7]-10<br>(26-29-30-7-10) | 26-29-30-[7]-10 | 5<br>(4) | 4 | 0 | 0 | 25 | 160 | 90 | 213 |
| 6>24 | 6-7-30-29-27-25-[24]<br>(6-7-30-29-26-24) | 6-[7]-30-29-26-24 | 6<br>(5) | 5 | 0 | 0 | 4 | 160 | 90 | 217 |
| Local Repair: 12>17 | 12-14-17 | 12-14-17 | 2 | 2 | 0 | 0 | 24 | 160 | 90 | 241 |

Cluster-based Routing performance   Flat-based Routing performance   Deployment Optimization

## 8.8 Conclusions and Future Perspectives

The new era of emerging technologies for WSNs and IoT-based applications poses important challenges regarding efficient methodologies to optimize and evaluate the performance of multi-hop mesh networking capabilities. In this direction, a new perspective for tackling the lack of defined on-site support frameworks to deploy and maintain wireless distributed systems has been approached in this chapter, in which a combination of modeling, optimization, and performance analysis mechanisms are integrated into a unique intelligent tool-chain. This proposed platform tries to cope with the underlying gap between the planning/simulation stages of the development process and the real implementation assessment and enhancement under the actual target scenario, aiming to achieve the long-term operability of large-scale WSN-based applications in an energy/resource-efficient fashion. The system implementation has been conceived in such a modular and heterogeneous way that it can be used in low-rate wireless connectivity environments such as in those distributed systems composed of IEEE 802.15.4-based networking technologies, as well as in WSNs with more demanding communication capabilities where protocols relying on IEEE 802.11 technologies are to be included. This perspective is certainly thought to support novel trends that come out with the use of mesh networking and routing optimization strategies in future smart-city implementations, in which the reliability of the wireless collaborative systems will depend on the quality, efficiency, and adaptability of the dissemination techniques.

## References


1. G. Cardone, P. Bellavista, A. Corradi, L. Foschini, Effective collaborative monitoring in smart cities: Converging MANET and WSN for fast data collection, in *Kaleidoscope 2011: The Fully Networked Human? Innovations for Future Networks and Services*, Dec 2011, pp. 1–8
2. L. Atzori, A. Iera, G. Morabito, The internet of things: a survey. Comput. Netw. **54**(15), 2787–2805 (2010). ISSN 1389-1286, doi:10.1016/j.comnet.2010.05.010
3. Y. Wang, B.M. Kelly, X. Li, On the network connectivity of wireless sensor networks following a random and non-uniform distribution, in *IEEE 9th Conference on Wireless and Mobile Computing, Networking and Communications (WiMob'13)*, 7–9 Oct 2013, pp. 69–74
4. P. Chatterjee, N. Das, Coverage constrained non-uniform node deployment in wireless sensor networks for load balancing, in *Applications and Innovations in Mobile Computing (AIMoC'14)* (2014), pp. 126–132. doi:10.1109/AIMOC.2014.6785530
5. K. Xu, H. Hassanein, G. Takahara, Q. Wang, Relay node deployment strategies in heterogeneous wireless sensor networks. IEEE Trans. Mob. Comput. **9**(2), 145–159 (2010). doi:10.1109/TMC.2009.105
6. S. Misra, S.D. Hong, G. Xue, J. Tang, Constrained relay node placement in wireless sensor networks: formulation and approximations. IEEE/ACM Trans. Networking **18**(2), 434–447 (2010). doi:10.1109/TNET.2009.2033273
7. H.N. Pham, D. Pediaditakis, A. Boulis, From simulation to real deployments in WSN and back, in *IEEE International Symposium on a World of Wireless, Mobile and Multimedia Networks (WoWMoM'07)*, 18–21 June 2007, pp. 1–6. doi:10.1109/WOWMOM.2007.4351800
8. G. Werner-Allen et al., Motelab: a wireless sensor network testbed, *in International Symposium on Information Processing in Sensor Networks (IPSN'05), USA*, Apr 2005, pp. 483–488
9. M. Doddavenkatappa, M.C. Chan, A.L. Ananda, Indriya: a lowcost, 3D wireless sensor network testbed, in *TridentCom*, Apr 2011, pp. 302–316
10. L. Jiao, D. Qin, J. Yang, L. Ge, F. Qin, MRS: a click-based multipath routing simulator, in *7th International ICST Conference on Communications and Networking in China (CHINACOM)*, Aug 2012, pp. 1, 6
11. G.S. Rawat, D. Kumar, S.C. Gupta, An analysis of DYMO, FSR, ZRP and RIP routing protocols using QUALNET, in *Proceedings of the International Conference on Interdisciplinary Advances in Applied Computing (ICONIAAC'14)*, ACM, New York, USA, 2014
12. IEEE Standard for Local and metropolitan area networks—Part 15.4: Low-Rate Wireless Personal Area Networks (LR-WPANs). In IEEE Std 802.15.4-2011 (Revision of IEEE Std 802.15.4-2006), pp. 1–314, 2011
13. J. Portilla, T. Riesgo, A. Abril, A. de Castro, Rapid prototyping for multi-application sensor networking. SPIE Newsroom, 12 November 2007, doi:10.1117/2.1200711.0851
14. Information technology—Telecommunications and information exchange between systems Local and metropolitan area networks–Specific requirements Part 11: Wireless LAN Medium Access Control (MAC) and Physical Layer (PHY) Specifications. In ISO/IEC/IEEE 8802-11:2012(E) (Revison of ISO/IEC/IEEE 8802-11-2005 and Amendments), pp. 1–2798, 2012
15. A.F. Molisch, K. Balakrishnan, C.C. Chong, S. Emami. et al., IEEE 802.15.4a channel Model—Final report. IEEE 802.15 Working Group for WPAN, Technical Document, 2004
16. R. Malhotra, N. Singh, Y. Singh, Genetic algorithms: concepts, design for optimization of process controllers. Comput. Inf. Sci **4**(2), 39 (2011)
17. D.N. Galatchi, R.A. Zoican, Flat routing protocols for ad hoc mobile wireless networks, in *10th International Conference on Telecommunication in Modern Satellite Cable and Broadcasting Services (TELSIKS'11)*, 5–8 Oct 2011, pp. 669–672, doi:10.1109/TELSKS.2011.6143200
18. S. Sharma, S.K. Jena, A survey on secure hierarchical routing protocols in wireless sensor networks, in *Proceedings of the International Conference on Communication, Computing & Security (ICCCS'11)* (2011), ACM, New York, NY, pp. 146–151

19. M. Rezaee, M. Yaghmaee, Cluster based routing protocol for mobile ad hoc networks. INFOCOMP J. Comput. Sci. **8**, 30–36 (2009)
20. C. Perkins et al., Ad hoc on-demand distance vector (AODV) routing. RFC 3561 (2003)
21. J.Y. Yu, P.H.J. Chong, M. Zhang, Performance of efficient CBRP in mobile ad-hoc networks (MANETS), in *Vehicular Technology Conference*, 21–24 Sept 2008, Calgary, BC, pp. 1–7

# Chapter 9
# Wireless Sensor Networks for the Internet of Things: Barriers and Synergies

**Mihai T. Lazarescu**

## 9.1 Introduction

Research and technology advances continuously extend and diversify wireless sensor network (WSN) applicability. As a consequence, WSN designers faced an increasing range of applications and requirements under rising cost and time pressures since the Internet of Things (IoT) paradigm was coined more than 15 years ago [1]. "Typical" requirements for WSN hardware and software are increasingly difficult to define [2] because they continuously adapt to very diverse application requirements and operating conditions at a rate which does not seem slowed down by standardization efforts or proprietary API proposals.

Moreover, although WSN solutions are used for numerous applications, the implementations generally differ under various aspects which significantly reduce the economies of scale. Consequently, both hardware and software of WSN solutions are often application-specific prototypes that carry significant non-recurrent engineering costs and risks (e.g., reliability, optimization, and development time).

Additionally, for various practical reasons WSN deployments are typically developed at lower abstraction levels, which can have two significant undesirable effects. First, this can divert an important development effort from application logic implementation, as shown in Fig. 9.1, which increases development time and cost, and generally decreases reliability. Second, lower abstraction level development often requires competencies that are seldom found among application domain experts, which can lead to higher development cost and more reluctant adoption of WSN-based solutions.

M.T. Lazarescu (✉)
Dip. Elettronica e Telecomunicazioni (DET), Politecnico di Torino, Corso Duca degli Abruzzi 24, Torino, Italy
e-mail: mihai.lazarescu@polito.it



G. Keramidas et al. (eds.), *Components and Services for IoT Platforms*,
DOI 10.1007/978-3-319-42304-3_9

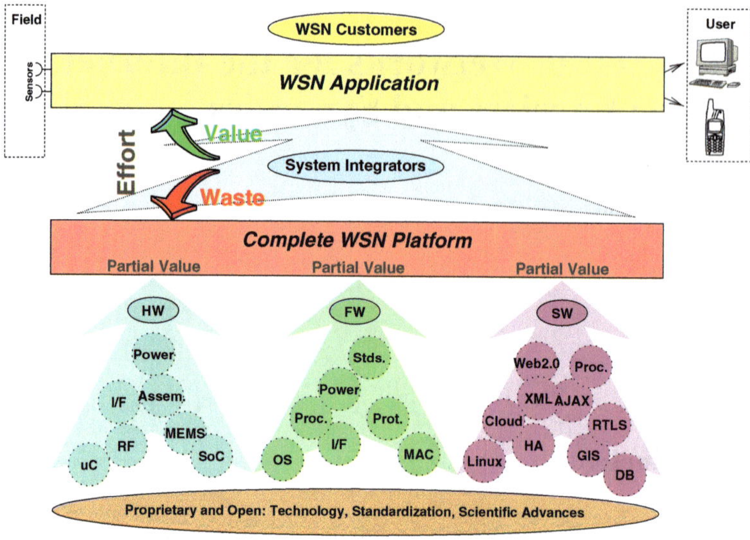


**Fig. 9.1** Value flow for a WSN application and platform

IoT vision to transform and enrich the way in which we perceive and interact with the reality often assumes capillary distributed devices for which WSNs continue to play an important role as one of the key enabling technologies since IoT paradigm inception. They often need to meet stringent requirements such as long maintenance-free lifetime, low cost, small size, ease of deployment and configuration, adequate processing of collected data, privacy and safety, and, not the least, fast and reliable development cycles that evolve on par with the underlying technologies. These are especially important for environmental monitoring applications, both closer to human day-to-day live (buildings, cities) and remote (e.g., open nature and climate monitoring).

As shown in Fig. 9.1, at the top of the WSN *value* chain are WSN application customers, whose needs are addressed by system integrators. The latter leverage on the top of WSN *technology* chain, where we find WSN platforms which can fully cover, end-to-end, WSN applications.

WSN platforms and development frameworks play a central role as the primary interface for accessing the underlying technology, standardization, and research state of the art. A complete platform includes field device hardware and firmware, development framework, and an application server (e.g., for data storage, monitoring, and post-processing). Development framework, IP components, and supported hardware need continuous updates because WSN field evolves fast and these are an important factor in the final value delivered to end users. Also, development framework flexibility, reliability, and ease of use are important factors allowing system integrators and application domain experts to direct most of their effort to WSN application development, where it is valued most.

However important, designing and maintaining a complete, flexible, and reliable WSN development framework with extensive hardware support is costly and requires a broad range of expertises, from advanced web and UI design to low-level embedded software and IDE, and code generation techniques to support high level application specifications. Quality assurance is also very important, since overall unreliability perception is still a limiting factor to wider WSN adoption.

Most hardware vendors provide various degrees of development support for their own WSN devices. They are usually focused on typical uses of the devices and are often hardware-centric instead of application-centric. As such, the toolsets may require significant extension or adaptation to properly cover a broader range of applications. Additionally, they also tend to significantly lag the state of the art, as they are meant to follow the progress of one producer.

Moreover, the up-front integration effort into existing development flows and the proprietary solutions may lock-in to vendor's hardware, often leading to significant hold-up problems that may hamper business potential. All these aspects finally translate into wasting system integrators' effort, missing or lagging market opportunities, and increasing development costs and risks. For new players it may also add to entry and differentiation barriers, effectively limiting the adoption of WSN solutions.

To best meet expectations and optimize the value, WSN development frameworks need to be based on reuse (both code and tools, since no player can efficiently cover all WSN aspects), to be easy to update or upgrade/replace to keep the pace with the fast evolution of the underlying technologies, and to abstract and automate the flow especially for application domain experts. Besides, the flow should favor design portability between different hardware and software solutions to avoid costly and inefficient vendor lock-ins and, last but not least considering the fast evolution pace in the field, to simplify the permeation of promising research to production.

Development frameworks revolve around the concept of a flexible WSN platform, as shown in Fig. 9.1, which is the convergence point of multiple WSN hardware and software components and technologies. Effective development tools should start from top-level application-specific requirement descriptions provided by the developer and automatically find suitable implementations and configurations that support them, based on existing components. At the same time it should provide the developer metrics and tools useful to evaluate solution quality.

This process aims to avoid diverting significant developer effort towards implementation details. This should speed up application development and make the flow more accessible to application domain experts.

At the same time, the flow should simplify the integration and coexistence of tools and technologies from different vendors and projects. Most existing WSN solutions efficiently address specific vertical application domains for various reasons, and not the least because building and maintaining a complete and flexible platform often requires a broad range of competencies and can be very costly. This development flow aims to reduce the effort and cost by:

*Being accessible to application domain experts.* This helps spreading the use of WSN-based solutions to more application domains, while often reducing development cost and time.

*Focusing design effort mostly on application logic.* WSN technology is based on several engineering disciplines. WSN development tools and flows should provide a good separation between the underlying technological details and the application developer.

*Optimizing implementations for cost, power, and reliability.* This is especially important for the quality of service of the WSN application over its lifetime. Moreover, they implicitly reduce the recurrent cost for both node production and their field maintenance.

*Integrating existing or new tools and technologies.* Tool development can be often effort-intensive, hence reusing existing tools, either proprietary or public domain ones, is economical. Besides, the tools may be customized, e.g., for specific hardware or for specific application domains. Effort and cost issues are amplified by the fast evolution of WSN technology.

*Maintainability of the complete development flow.* This is tightly related to tool integration above. The tools should be easy to integrate in the development platform, e.g., in terms of semantics, interfaces, and data formats, in order to simplify the upgrade or replacement of existing tools or the addition of new ones.

*Simplifying the comparison of design results.* The platform should simplify playing *what if* scenarios, in which elements change, e.g., a tool in the development flow, the target node, or the embedded operating system (OS). Since the rest of the platform remains the same, it simplifies the observation of the effects of the change. This should allow a closer reproduction of research results and the comparison between different research tools or approaches. Also, this should allow to select the most effective solution for a given WSN application.

*Facilitating research permeation in commercial applications.* This is tightly related with the above point. The benefits of new research results can be compared with the existing flows by playing adequate *what if* scenarios. Moreover, research tools are already integrated in the platform simplifying their porting to existing production flows based on the platform.

*Building business models.* Last but not least, the purpose of the platform is to be useful for real WSN applications by providing value through vendor- or application-specific customizations of the general purpose flow. Thus, on the one hand, the platform should allow the integration of proprietary tools or protected intellectual property (IP) blocks, e.g., simulation models or functional code. On the other hand, the platform should simplify the contribution of code (custom or general purpose), flows, or other developments made by commercial users.

In Sect. 9.2 we will briefly review some existing WSN development tools and flows in terms of these objectives. In Sect. 9.3 we will review some options for WSN hardware for nodes and server-side software. Then, in Sect. 9.4 we will look into a possible development framework that can fulfill most of the above criteria. Section 9.6 will conclude the work.

## 9.2 WSN Programming Models and Tools

Early WSN implementations were manually coded close to hardware or the embedded operating system [3]. Typically this brings smaller code size and higher execution efficiency at the cost of maintenance, portability and design reuse. It also requires a good understanding of various technologies underlying WSN nodes, which is difficult to find among software programmers and seldom found among application domain experts.

Over time, as WSN cost, time to market and operation reliability increase in importance, higher programming abstractions and design automation techniques get increasing attention. The literature is now rich of WSN design aids, both as languages and their compilers as well as support software, such as middleware and real-time embedded OSs.

In the following we are reviewing some relevant categories of design aids, with an eye on their performance along the lines listed towards the end of Sect. 9.1.

### *9.2.1 Low-Level Programming*

This is a node-centric programming model that ranges from close to hardware and up to some level of abstraction usually provided by an embedded OS or by a virtual machine (VM).

#### 9.2.1.1 Operating System-Level Programming

Among the OSs, we can list TinyOS [4], programmable in nesC [5] which offers a good modularization with well-defined interfaces and functional encapsulation that abstracts implementation details. However, low abstraction level and an event-based programming style without support for blocking operations increase programming complexity.

To reduce the complexity, TinyGALS [6] provides FIFOs for asynchronous message passing on top of the synchronous event-driven model and synchronous method calls. SOS [7] and CoMOS [8] implement priority queues (the former) and preemptive message handling (the latter).

SNACK [9] allows to compose applications by combining services provided by reusable library components. T2 [10] simplifies project reuse and porting on new hardware platforms by combining a horizontal decomposition (to support multiple devices) and vertical decomposition that allows to build the application using hardware-independent functional blocks. OSM [11] extends the TinyOS event-driven model with states and transitions whose actions depend on both events and program state.

Several OSs propose different forms of thread abstraction to simplify programming of event-driven systems, although they are difficult to implement with the limited hardware resources of the target nodes. Fiber [12] provides one blocking context on top of TinyOS. Mantis OS [13] provides preemptive, time-sliced multithreading. TinyThreads [14] provides a stack estimation tool to reduce memory consumption for thread stacks. Protothreads [15] provides a stack-less co-operative multithreading library for Contiki OS [16], another embedded event-based OS. Y-Threads [17] implements efficient preemptive multithreading using a shared stack for the computational parts of the application, while for the control parts are allocated small separate stacks.

#### 9.2.1.2 Virtual Machine or Middleware

These programming models have several important features. One of them is efficient dynamic reprogrammability of small embedded systems. Maté [18] and ASVM [19] are small application-specific VMs built on top of TinyOS. Melete [20] extends Maté to multiple concurrent applications. VMStar [21] allows dynamic update of its own code or the application. Impala [22] is an example of middleware with efficient over-the-air reprogramming and repairing capabilities.

VMs also provide platform-independent execution models for WSN applications. Token machine language [23] models the network as a distributed token machine in which the nodes are exchanging tokens which trigger matching handlers. This model can be used to implement higher-level semantics that distance themselves from the underlying nesC implementation. t-kernel [24] modifies ("naturalize") application code at load time to implement OS protection and virtual memory without hardware support.

### 9.2.2 *High-Level Programming*

High-level programming models typically allow the developers to focus more on application logic, but often at the expense of node performance. For instance, group- or network-level programming can facilitate the collaboration between sensors, which is necessary for a significant part of WSN applications and often challenging to program. However, the designer has few means to check or improve operation efficiency of the nodes within the abstracted units.

In the following we will review a few of the most prominent high-level programming methods and tools.

#### 9.2.2.1 Model-Based Development

These methodologies attempt to offer the developers an application entry interface that can be both more productive and also better suited for application domain experts, while preserving lower-level control over design results, which is necessary to control and optimize WSN performance.

An interesting approach uses several domain-specific modeling languages (DSMLs), one for each of the network-, group- and node-level decomposition of a WSN application [25]. The DSML for data-centric network modeling includes three node types: sources, aggregation/fuse, and sink. The DSML for group modeling allows geographical node grouping, definition of network topology (e.g., tree or mesh), and of the amount of in-network processing (aggregation and fusion). The DSML for node modeling offers several types of tasks: for sensor sampling, data aggregation/fusion, networking, and sink. However, application specification appears to be limited to parametrization of the models. Hence, the approach can be considered a modular way to customize a parameterized application.

REMORA uses an XML-based modeling abstraction [26] for more advanced component-based designs than TinyOS static composition. Models include services (offered and required) and the triggering events, persistent state, and implementation in a C-like language. The event model extends the TinyOS one with attributes, differentiation between application and OS events, configuration and point-to-point, or multicast distribution. The framework has a low overhead over Contiki, but offers much improved encapsulation than simple multithreading.

A framework supporting hardware–software co-design is shown in [27]. It is based on tools widely used in industry like Simulink®[1] and Stateflow®[2] and well-known open projects like OMNeT++/MiXiM [28], TinyOS [4], and Contiki [16]. Graphical high-level application entry is supported by the abstract concurrent models provided by Simulink and Stateflow. Application specification can be simulated at node level and can be automatically translated by the framework in network simulation models, including hardware in the loop for better accuracy, as well as implementation models that can run on top of popular embedded OSs.

#### 9.2.2.2 Group-Level Programming

This programming model provides constructs to handle multiple nodes collectively as a group abstracting its internal operation into a set of external functions. Groups can be logical or neighborhood-based.

[1]Mathworks Simulink http://www.mathworks.com/help/simulink/index.html.

[2]Mathworks Stateflow—Finite State Machine Concepts http://www.mathworks.com/help/toolbox/stateflow/index.html.

Neighborhood-Based Groups

Based on physically neighboring nodes, this type of group is well suited for local collaboration (recurrent in some classes of applications) and the broadcasting nature of WSN communication, improving in-group communication. Abstract Regions [29] and Hood [30] provide programming primitives based on node neighborhood which often fits the application needs to process local data close to its origin. Hood implements only the one-hop neighbor concept, while Abstract Regions implements both topological and geographical neighborhood concepts.

Logical Groups

These groups can be defined based on logical properties, including static properties like node types, or dynamic, such as sensor inputs, which can lead to volatile membership.

EnviroTrack [31] assigns addresses to environmental events and the sensors that received the event are dynamically grouped together using a sophisticated distributed group management protocol.

SPIDEY language [32] represents logical nodes based on the exported static or dynamic attributes of the physical nodes, and provides communication interfaces with the logical neighborhood as well as efficient routing.

State-Centric Groups

These are mostly intended for applications that require collaborative signal and information processing [33]. Around the notion of collaboration group, programmers can specify its scope, define its members, its structure, and the member roles within the group. Pattern-based groups, such as neighborhood- or logical-based, can also be defined. Its flexible implementation allows it to be used as building block for higher-level abstractions.

#### 9.2.2.3 Network-Level Programming (Macroprogramming)

WSN macroprogramming treats the whole network as a single abstract machine which can be programmed without concerns about its low-level inter-node communication.

Database

Database is an intuitive abstraction derived from the main function of the nodes, which is to collect sensing data. Early implementations like Cougar [34] and

TInyDB [35] provide support for declarative SQL-like queries. Both attempt energy optimizations. Cougar processes selection operations on sensor nodes to reduce data transfers. TInyDB optimizes the routing tree for query dissemination and result collection by focusing on where, when, and how often to sample and provide data.

SINA [36] allows to embed sensor querying and tasking language scripts with the SQL queries to perform more complex collaborative tasks than those allowed by SQL semantics.

MiLAN [37] and DSWare [38] provide a quality of service (QoS) extension to queries which can be defined based on the level of certainty of an attribute among those that can be measured by the node with a given accuracy. MiLAN converts a query in an energy-optimized execution plan that includes the source nodes and routing tree. DSWare expresses and evaluates the QoS as a compound event made of atomic events, whose presence/absence define the confidence level.

Although database-like interfaces are simple and easy to use, they are not well suited for applications that require continuous sensing or with significant fine-grained control flows.

### Macroprogramming

WSN macroprogramming provides more flexibility than database models.

#### *Specification of Global Behavior*

Regiment [39] implements a Haskell-like functional language. By preventing the developer to manipulate directly program states it allows the compiler to extract more parallelism.

Kairos [40] is a language-independent programming abstraction that can be implemented as an extension of existing program languages. As such it defines a reduced set of constructs (node abstractions, one-hop neighbors, and remote data access, which includes a weak data consistency model to reduce communication overhead).

#### *Resource Naming*

Spatial programming [41] can reference network resources by physical location and other properties and access them using smart messages. These contain the code, data, and execution state needed to complete the computation once they reach the nodes of interest.

Using SpatialViews [42] the developer can dynamically define a spatial view as a group of nodes that have properties of interest, such as services and location.

Declarative resource naming [43] allows both imperative and declarative resource grouping using Boolean expressions. Resource binding can be static or dynamic and the access to resources can be sequential or parallel.

*Other Metaprogramming Abstractions*

Semantic streams [44] supports declarative queries using semantics associated with sensor data by inference units. Both sensors and inference units are built automatically from a Prolog-based specification.

Software sensor [45] provides a service-oriented architecture in which each hardware sensor is abstracted as a software sensor. The latter can be composed in multiple ways using the SensorJini Java-based middleware in order to define large-scale collaboration within the network.

### 9.2.3 Evaluation of Existing WSN Programming Models and Tools

We will evaluate how the WSN development tools, tool categories, and methodologies reviewed in Sect. 9.2 can be used to reduce overall WSN application cost. We will also evaluate their potential to work in synergy with other tools in a comprehensive development platform that is suitable to cover the widening diversity of WSN applications and keep pace with the rapid technological evolution in the field.

#### 9.2.3.1 Low-Level Programming Evaluation

Low-level programming, be it OS- or VM-based, typically allows a good control over the node and good design optimization, but it often requires in-depth engineering and programming knowledge. This is rarely found among application domain experts and may also divert important design effort from application logic implementation.

Embedded OSs and VMs have typically a layered structure which encapsulates well the hardware-dependent parts in order to facilitate their porting to other hardware nodes, which is important for keeping the pace with technology evolution. They can be maintained with reasonable effort and can be relatively easy integrated in higher-level design flows.

Moreover, embedded OSs usually facilitate the development of library elements that implement specific functions, which can be instantiated in new designs to facilitate design reuse.

The common base offered by the underlying OS or VM can be used as reference to compare the effectiveness of novel solutions. New research results can be included in library elements as well as the OS or VM core design, effectively simplifying their adoption in commercial designs.

Considering the higher complexity of application programming at low levels, commercial services can be offered for, e.g., application development or porting, OS/VM (or library elements) porting to new hardware, and for training and support.

#### 9.2.3.2 High-Level Programming Evaluation

Abstractions at group- or network-level are meant to hide the inner workings of the abstracted entity, including its internal communications.

On the one hand, this is positive because it allows the developer to focus on application logic. On the other hand, the abstractions are often implemented using dedicated or less common (e.g., Haskell-like) programming languages which may be difficult to use by application domain experts.

Also, given that significant parts of the application are handled automatically by the system in order to satisfy higher-level specifications, the developer may not have the means to understand, evaluate, or improve the efficiency with which these are implemented by the system.

The tools implementing such abstract flows are typically developed as a close ecosystem. They are unlikely to share models or IP blocks among them, although they may use common lower-level abstractions (e.g., Regiment [39] uses TinyOS) or may be flexible (and simple enough) to allow its implementation as an extension of other programming languages like Kairos. As such, their development and maintenance can be rather costly. Moreover, the application projects are difficult to port between such frameworks, which limits also the permeation of research to commercial applications.

For these flows, in terms of business we can assume training, support, and design services.

A distinctive note can be made for model-based design (MBD) frameworks (see Sect. 9.2.2.1). The developer can focus most of the effort on application development, for which the tools allow various degrees of liberty. Application entry interface can be suitable for application domain experts, as demonstrated by the wide adoption of Stateflow® interface by experts in various industrial domains. Automated code generation and good integration of the flow with simulation tools (including hardware-in-the-loop) and target OSs simplify design space exploration for optimizations and also allow manual optimization of the generated projects. Integration with existing projects reduces the cost of framework maintenance. Moreover, they provide an observable development environment where the effects of changes to framework features (e.g., code generation) or to its components (e.g., simulators) can be easily compared for existing and new designs. Research advances can be evaluated the same way before being included in commercial design flows. Business models that use these flows can enrich their capabilities using custom IP blocks.

## 9.3 WSN Hardware and Server-Side Support

In the following we will comparatively review some options for WSN node hardware and for server-side software.

### 9.3.1 WSN Hardware

Its main components are the microcontroller and RF device, either stand-alone or combined in a single integrated circuit, RF antenna, energy sources, transducers, printed circuit board (PCB), and package.

#### 9.3.1.1 Microcontroller

There are many families of low and very low power microcontrollers, each covering a broad range of capabilities such as storage size for programs (FLASH), data (RAM and EEPROM), and peripherals. Given the diverse and specialized offering, the design flow should assist the developer in selecting the optimal microcontroller for the application, since it can influence significantly node cost and energy consumption.

Microcontroller component package type and size has both a direct impact on the component cost and an indirect one on PCB cost through its size and number of pins. PCB size can further influence the size and cost of the sensor node package.

Additionally, most microcontrollers can operate using an internal clock source. This has no additional component cost or impact on PCB size or complexity, but its frequency is not very accurate nor stable in time. However, if the WSN application (or the communication protocol) does not require accurate timing, the internal oscillator is worth considering.

Thus, microcontroller selection should mainly consider:

- on-board support for hardware interface with the peripherals (e.g., transducer, RF device), application firmware, and communication protocol;
- the trade-off between package type and its associated direct and indirect costs at sensor node level;
- adequate support for selected RF communication protocol. Its effects at system level should be carefully weighed. For instance, accurate timers and receive capabilities significantly increase sensor node cost, microcontroller requirements, and energy consumption.

#### 9.3.1.2 RF Device

As for the microcontroller, the cost of the RF device significantly depends on its capabilities. For example, a transceiver will typically cost more and require more external components than a transmit-only radio.

Modern RF devices provide application data interfaces at different levels of abstraction. These can range from digital (or analog) signal levels to advanced data packet management. Although the latter may increase RF device cost, it also sensibly reduce microcontroller computation and memory requirements, working with a cheaper and smaller device, and also reduce energy consumption and cost.

Most RF devices use an external crystal quartz for timing and channel frequency and may optionally output a clock for the microcontroller. However, this implies that the system is active only when the RF device is active, which may increase energy consumption.

Commonly used multihop WSN communication protocols require that all nodes have receive capabilities, which increase the cost of the RF device. Moreover, most of them actively synchronize their clocks at least with the neighboring nodes to reduce the energy spent for idle channel listening, which requires an accurate time reference constantly running on-board and more microcontroller resources, all translating to higher device cost and energy consumption. However, if the application does not require bidirectional communication, an asynchronous medium access control in a star topology may sensibly reduce node cost and energy consumption.

Several producers offer general purpose devices that include a microcontroller and a radio in a single package. These can save PCB space and costs, but the integrated microcontrollers may be oversized for some applications, especially those using a smaller communication protocol stacks. The same may apply for the integrated radio devices.

#### 9.3.1.3 RF Antenna

Antenna influences node cost, size, deployment, and performance mainly through its size, cost, and RF characteristics. Most antenna features are correlated with operation frequency, range requirements, and RF properties of the environment.

For instance, star network topologies may require longer RF range to lower the number of repeaters and gateways. In this case, lower RF frequencies (315/433 MHz bands or below) can increase the communication range for a given RF link budget.

WSN application requirements may influence antenna characteristics such as directionality, gain, size, resilience to neighboring conductive objects (tree trunks, metal bodies). Antenna options can range from omnidirectional $\lambda/2$ dipole, or a $\lambda/4$ whip (both costly to ruggedize), to helical antennas radiating in normal mode (NMHA) with a good size-performance trade-off, to PCB trace antennas (better for higher communication frequencies), and ceramic chip antennas with good performance but higher cost.

PCB components may also influence antenna performance.

#### 9.3.1.4 Energy Supply

Along with node energy consumption, energy supply is very important for the overall reliability and exploitation cost of the network.

A first decision concerns the use of energy harvesting or non-regenerative (e.g., a primary battery).

Environmental energy harvesting may suit applications with access to environmental energy sources (e.g., RF, vibration, fluid flow, temperature gradients, and light). Combined energy harvesting solutions can increase the availability [46], although supply reliability is hard to estimate in the general case.

In all other cases, primary batteries should be considered such way to support average energy requirements of the node for the expected lifetime. This cost can be used as the upper bound for the evaluation of energy harvesting-based solutions.

Either way, low node energy consumption is very important for any type of energy source.

#### 9.3.1.5 Transducers

Transducers are used to sense the properties of interest in the environment surrounding the node. Its performance affects the node in several ways. First, the transducer should have low energy requirements and/or allow (very) low duty cycle operation. Its interface with the node should not increase exceedingly microcontroller requirements. Last but not least, the transducer should not require periodic maintenance, which may significantly increase the operation cost of the network.

#### 9.3.1.6 Package

Node package protects node components from direct exposure to environment and defines the node external size, mechanical properties, and physical aspect. Its cost may increase due to special production requirements and its dimensions. Thus, special requests, such as transparent windows for light energy harvesting, should be carefully considered.

It may also provide the means to mount the node in the field, thus the package should be designed to simplify node deployment and maintenance to reduce the overall cost per node.

#### 9.3.1.7 Hardware Nodes

There are many hardware sensor nodes, developed for both research and commercial purposes.[3]

Their characteristics are extremely diverse. For example, in terms of average current consumption they range from a low end of 30 μA of Metronome Systems

[3]Currently there are over 150 node models listed on Wikipedia https://en.wikipedia.org/wiki/List_of_wireless_sensor_nodes.

NeoMote[4] and 100 μA of Indrion Indriya_DP_01A11[5] [47] up to the high end 100 mA of Nokia Technology Platforms NWSP [48] (wearable) and Massachusetts Institute of Technology ubER-Badge [49].

Considering the wide diversity of features and support, node hardware selection can be daunting, especially for application domain experts. Development tools should provide enough flexibility to map the application on different types of nodes and provide the developer adequate guidance to select a suitable target node.

#### 9.3.1.8 Server-Side Support

Servers offer several important functions in a WSN application, such as to receive, store, and provide access to field data. They bridge the low power communication segments, which have important latency-energy trade-offs, with the fast and ubiquitous access to field data needed by humans or applications. Other functions include in-field node configuration and query, as well as software updates.

Global sensor networks (GSN) [50] is a middleware that facilitates WSN deployment, programming, and data processing. It supports integrated sensor data processing towards a vision of a global "sensor Internet." By abstracting the underlying technology, GSN simplifies, among others, platform additions, combination of sensor data, sensor mobility support, and runtime dynamic system configuration adaptation.

Since collaborative aspects can become dominant for IoT applications, they are well supported by projects like Xively[6] (formerly known as Cosm and Pachube) and WikiSensing.[7] These simplify online collaboration over data sets ranging from energy and environmental data to transport services, to generate real-time graphs and widgets for web sites, for historical data analysis, and generation of alarms.

## 9.4 Semi-Automated WSN HW/SW Application Synthesis

We will now discuss a semi-automated hardware–software application synthesis flow to better understand its benefits in terms of the evaluation criteria presented at the end of Sect. 9.1.

The flow can automatically select modules from a previously developed library to perform design composition, both hardware and software, in order to significantly increase the productivity of the developers through design reuse, and to allow fast design space exploration for application implementation optimization.

---

[4] Metronome Systems NeoMote http://metronomesystems.com/.

[5] Indrion Indriya_DP_01A11 http://indrion.co.in/Indriya_DP_01A11.php.

[6] Xively project http://xively.com/.

[7] WikiSensing project http://wikisensing.org/.

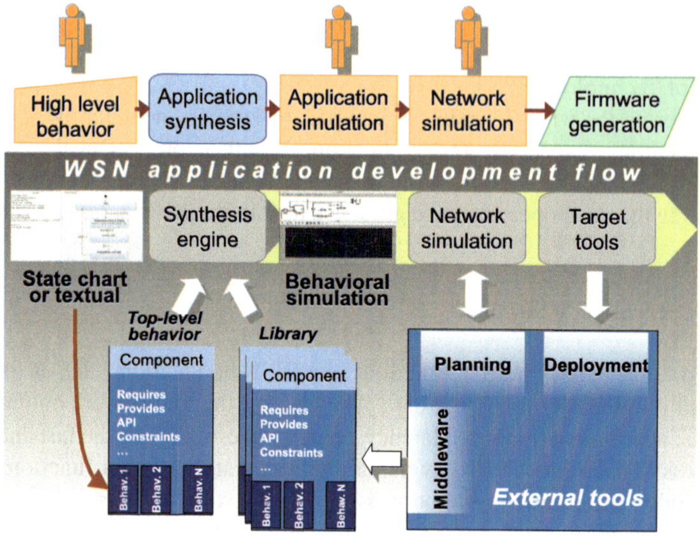


**Fig. 9.2** Main stages of the automated WSN application design flow at node level: input of application behavioral description, automated synthesis of possible solutions (using top-level behavioral description and library components), behavioral simulation, network simulation, node programming code and configuration generation. Developer-assisted phases are tagged using a human body

### *9.4.1 Semi-Automated Development Flow Overview*

Figure 9.2 shows a typical WSN application development flow at node level. The flow receives a high-level node-centric application specification and is well integrated with external tools, each assisting the developer in specific tasks.

Application-specific behavioral input can range from manual source code input to automated code generation from MBD abstractions (such as Stateflow® used in [27] or similar state charts editors [51]) or from UML-based or ad-hoc high-level modeling flows [52, 53]. Either way, the behavioral input is captured in a specific view of the top-level component, which also includes the metadata needed by the subsequent phase of the flow, namely the automated synthesis.

The top-level component and all library components have the same format, which can include more than one view, e.g., behavioral source code or binary, simulation models for various abstraction levels or simulation environments. All views are handled as black boxes by the synthesis engine regardless their format, the synthesis relying only on component metadata. These can be generated manually or automatically by MBD flows [27].

The second phase of the flow shown in Fig. 9.2 is fully automated. A hardware–software system synthesis engine takes the top-level component as input and processes its metadata (such as requires, provides, and conflicts). These properties are the starting point for the synthesis process which iteratively attempts to find all subsets of its library of components that do not have any unsatisfied requirements left and, at the same time, satisfy all constraints of the top-level and other instantiated components. These subsets represent possible solutions for application requirements and can be further examined or modified by the developer.

For each solution, the synthesis tool can create simulation projects, as shown in the next steps of the flow in Fig. 9.2. The simulations are set up to run on external simulators (e.g., OMNeT++ [28]) and can be at various level of abstraction. Basically this is achieved by extracting and configuring the suitable simulation views of the components instantiated in solution into simulation projects.

Besides behavioral models, the components and constraints of the solution can include a bill of materials (e.g., compatible nodes, RF and transducer characteristics, and microcontroller requirements) or software dependencies on specific compilation toolchains or underlying OS features.

Finally, the same mechanism is used to generate the projects that can be compiled using the target tools to create the programs for all WSN nodes. These projects are typically generated in the format expected by the target tools (most often a make-based project).

The solutions generated by the synthesis tool can be used as they are, or the developer can optimize them either by changing the specification and rerunning the synthesis, or by manually editing the generated projects. Either way, the developer can use simulations to validate and evaluate the solutions and their improvements.

As mentioned, the benefits of WSN application automated synthesis are compounded by its integration with external tools, such as simulators and target compilation chains, which can provide inputs or assist the developer in other phases of the flow. For instance, Fig. 9.2 shows some typical interfaces with middleware [54, 55], with WSN planning tools [56] or with deployment and maintenance tools [57].

However, the wide variety of the existing tools and models makes it very difficult to define an exhaustive set of external interfaces. Moreover, any rigidity of tool interfaces or operation models is prone to reduce the value of the tool and hamper its adoption in a context which is characterized by rapid evolution of technology and models, and which does not seem to be slowed down by standardization efforts or proprietary API proposals.

In this context, as we will show later on, an optimal approach for tool integration in the existing and future development flows can be to base its core operation on a model that is expressive enough to encode both high-level abstractions and low-level details. Moreover, it is also important to provide well-defined interfaces and semantics to simplify its maintenance, updates, integration with other tools, and extensions to other application domains.

### *9.4.2 Automated Hardware–Software Synthesis Tool Overview*

The tool covers the following main functions: application input (provide a suitable interface and processing), automated hardware–software synthesis, and code, configuration, and hardware specifications generation.

Application domain experts can benefit most from an interactive user-friendly interface for the description of the WSN application top-level behavior. State charts are well established in this regard for their intuitive use, and they can provide suitable high-level models to facilitate the description of application domain behavior. Alternatively, the synthesis tool can accept application descriptions generated by other tools, such as middleware [58] or metaprogramming [59].

Automated synthesis of hardware–software systems that can support WSN application requirements shields the developer from most time-consuming and error-prone implementation details. At the same time, the synthesis increases the reuse of library components such as software components (e.g., OS, functional blocks, software configurations, and project build setup), hardware components (such as WSN nodes, transducers, radio types or specific devices, and hardware configurations), and specifications (e.g., target compilation toolchain, RF requirements).

Incomplete application specifications are also accepted, because the tool can typically infer default parameters based on values provided by library components and heuristics. This allows the developer to refine application specifications over several design iterations using also the results of previous underspecified synthesis runs. Also, the incomplete systems synthesized from incomplete application specifications still satisfy every requirement, and experienced developers can use these incomplete projects as starting points for manual refinements, to save effort.

Code generation can produce simulation or target compilation projects. Network simulations can be configured using the simulation models of the components instantiated in solutions, their parameters, and the actual configurations. Realistic communication channels defined by a planning tool [56] can be used, if available. In a similar way, the tool uses the implementation code of the components instantiated in solution to generate and configure the project that compiles the code for WSN nodes programming.

Besides this highly automated process, the system synthesized tool allows experienced developers to manually take over the development flow at any stage: design entry, testing and debug, design synthesis, node application simulation, network simulation, and target code generation. Basically, this is achieved by:

- making use of textual data formats that can be edited with general purpose or specialized editors;
- documenting the data formats, their semantics, and processing during each phase of the flow;
- allowing one to run manually the individual tools, even outside the integrated flow, e.g., to explore options and operation modes that are not supported by the integrated flow;
- including well-known tools in the flow with clean and well-documented interfaces to simplify their update or replacement for flow specialization.

### 9.4.3 Automated Synthesis Tool Input Interface

During application specification phase, the developer (or an external tool) provides architectural requirements and top-level behavior as a design component, which becomes the main driver for the subsequent hardware–software system synthesis.

WSN application requirements can be expressed mainly in terms of application-specific behavior and its interfaces, and metadata properties.

As argued above, abstract concurrent state charts are an intuitive and efficient high-level means to simplify top-level application behavior. For design entry within this flow, the state chart tool should also allow to specify the interfaces and metadata for the behavioral part.

Yakindu State Chart Tools[8] is a free source integrated modeling environment based on Eclipse modeling framework (EMF) [60] for the specification and development of reactive, event-driven systems based on the concept of state charts. Its features provide significant support for design entry, especially useful for application domain experts with limited programming experience, such as:

- state chart editing through an intuitive combination of graphical and textual notation. While states, transitions, and state hierarchies are graphical elements, all declarations and actions are specified using a textual notation;
- state chart validation that includes syntax and semantic checks of the full model. Examples of built-in validation rules are the detection of unreachable states, dead ends, and references to unknown events. These validation constraints are checked live during editing;
- state chart simulation models that allow the check of dynamic semantics. Active states are directly highlighted in the state chart editor and a dedicated simulation perspective features access to execution controls, inspection, and setting of variables, as well as raising of events;
- code generation from state charts to Java, C, and C++ languages. The code generators follow a code-only approach. The code is stand-alone and does not rely on any additional runtime libraries. The generated code provides well-defined interfaces and can be easily integrated with other target code.

Yakindu was designed for embedded applications with a meta model based on finite state machines (FSMs), either Mealy or Moore. System behavior is defined by the active state, which is determined by FSM inputs and history. Yakindu meta model is similar to UML state chart meta model except for the following differences which are of particular importance for the flow:

- state charts are self-contained with interfaces defined by events and variables;
- core execution semantics are cycle-driven instead of event-driven, which allows to process concurrent events and to define event-driven behavior on top;

[8]Yakindu SCT project http://www.statecharts.org/.

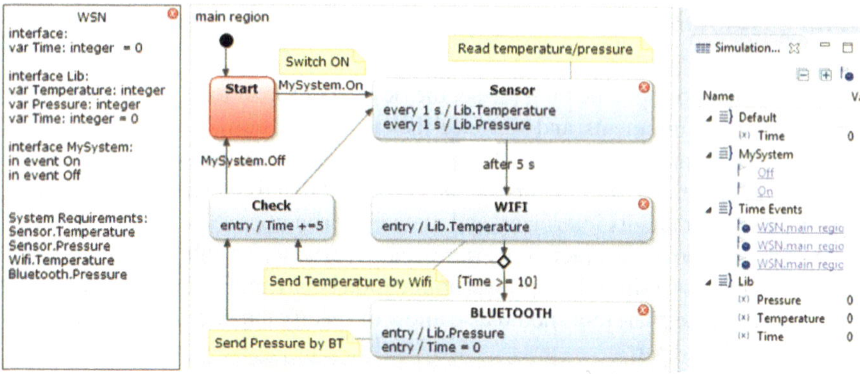


**Fig. 9.3** State chart-based interface for WSN application specification entry and top-level simulation using a customized Yakindu SCT. The *right pane* can display simulation data or a tool pane for state chart editing. The interface of the state chart can be interactively defined on the *left pane*. The active state during simulations is highlighted on the state chart in the *central pane*

- time is an abstract concept for the state charts;
- time control is delegated to environment.

The model interpreter and the code generators adhere to core semantics.

Considering the above, Yakindu can be used and extended in order to provide all functions needed for design entry for the flow. Figure 9.3 shows the main panes of Yakindu interface. On the right side is a tool pane with elements that can be used to edit the state chart in the central pane (such as transition, state, initial state, choice, and synchronization) or simulation data during chart simulation (as shown in the figure). In the left pane is shown the interface of the state chart. It allows to define variables and events that can be used by chart states, and it was extended to accept the metadata necessary for top-level component for the synthesis tool. The editable state chart in the middle pane describes top-level WSN application behavior, which can be interactively simulated or automatically converted into source code along with the interface and the metadata defined in the left pane, such way to be accepted by the synthesis engine.

To further assist application domain experts in using the interface, wherever is required textual input (such as to fill the properties of the state chart components or the interface), the developer is guided by a context-sensitive editor that lists the legal entries for keywords, names, operators, etc. Also, developer input is checked in real-time and errors are highlighted.

All these are captured in the top-level component of the design that is then used to drive the system synthesis engine. Using library components, the engine attempts to automatically compose a hardware and software system that supports the application-specific behavior and provides all its requirements.

For instance, let us consider a WSN application that collects and sends every 5 min the environmental temperature during four intervals of 2 h spread evenly

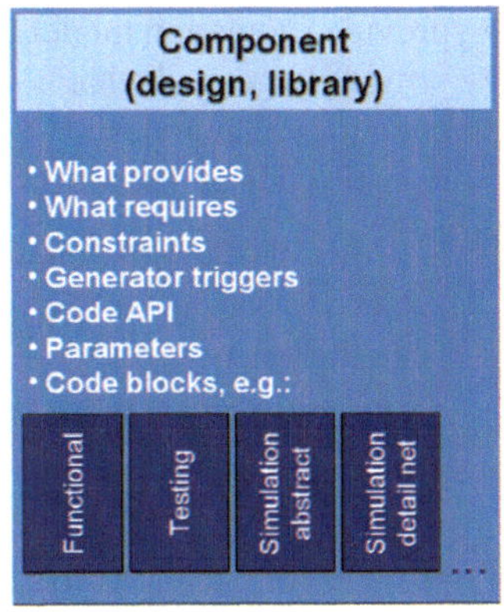


**Fig. 9.4** Top-level application specification component and library components share the same structure: a variable set of views (*shown darker on the bottom*) that are handed as *black boxes* by the system synthesis process, and a set of metadata that express the requirements and the capabilities of the component. The components are encoded in XML (Eclipse EMF XMI)

during the day. The functional description of this application consists of a periodic check if the temperature collection is enabled, if it is enabled then checks if 5 min have passed from previous reading, and if so then it acquires a new reading and sends it to the communication channel. The whole application behavior can be encoded in just a few condition checks and data transfers, plus some configuration requirements to support them (such as timers, a temperature reading channel and a communication channel). The rest of node application and communications are not application-specific, hence the developer should not spend effort developing or interfacing with them. In this flow (see Fig. 9.2), these tasks are automatically handled by the synthesis engine, which attempts to build a system that satisfies all specifications by reusing library components, as will be explained later.

The top-level component can include also several types of metadata properties. For instance, if the 6LoWPAN protocol is a compatibility requirement of the WSN application, a requirement for 6LoWPAN can be added to top-level component, regardless if the top-level component functional code interfaces directly with the field communication protocol. This way, the 6LoWPAN requirement directs the application synthesis to instantiate the functional components from the library that provide this communication protocol. However, the synthesis tool will instantiate only those 6LoWPAN components that satisfy other system requirements that are collected from both the top-level and other instantiated components.

### *9.4.4 Structure of Top-Level and Library Components*

Library components are central to the operation of the system synthesis engine (see Fig. 9.4). They are used for:

- definition by the developer of behavior and requirements of node-level WSN application, modeled as a top-level component;
- definition of library blocks that can be instantiated by the synthesis tool to compose a hardware–software system that satisfies all design specifications;
- interfaces with OS or middleware services, when needed by application behavior;

- provide simulation models, at different levels of abstraction;
- provide target code that is used to build the projects to configure and compile the code for target nodes;
- provide code generators that can be run by the synthesis tool to:
  - either check if the component can be configured to satisfy the requirements derived for the current solution during synthesis, so that it can be instantiated in the solution;
  - or build specialized code stubs, e.g., for API translation and component code configuration that are based on the actual parameters of the solution in which they are instantiated;
- provide specifications for the hardware components that are collected in a bill of materials (BOM);
- provide non-functional requirements, such as for a specific compilation toolchain or special RF requirements.

Library components are designed to be compatible with the concurrency models provided by the OS or the middleware abstractions which they require at runtime. The same stands for the support of inter-component communication infrastructure that is provided by the OS or the middleware services. However, to achieve a consistent system composition all communications need to go through component interfaces in order to be visible to the synthesis engine, so that it can make the proper decisions. In fact, engine selections are also based on require and provide properties, in addition to other metadata.

### *9.4.5 System Synthesis Process*

To exemplify the synthesis process, we show in Fig. 9.5 a simplified representation of just a few of metadata properties both for library components (bottom) and for top-level specification component (top-left).

At the begin of system synthesis process, the synthesis tool is driven by metadata specifications of the top-level component of the design, then is also guided by library components metadata. As system synthesis progresses and library components are instantiated in the partial solution, instantiated components metadata drive tool search alongside the still unsatisfied specifications of the top-level component. During the synthesis process, the top-level component and its metadata are considered mandatory, while library components can be instantiated and removed from solution as necessary to satisfy design requirements.

For example, with the top-level specification and library components shown in Fig. 9.5, the system synthesis engine is able to find several solutions. It starts by loading the top-level component from design entry and the 13 components from the library. Then it explores the possible combinations and reports the following system compositions, each satisfying the specifications and requirements of all instantiated library components:

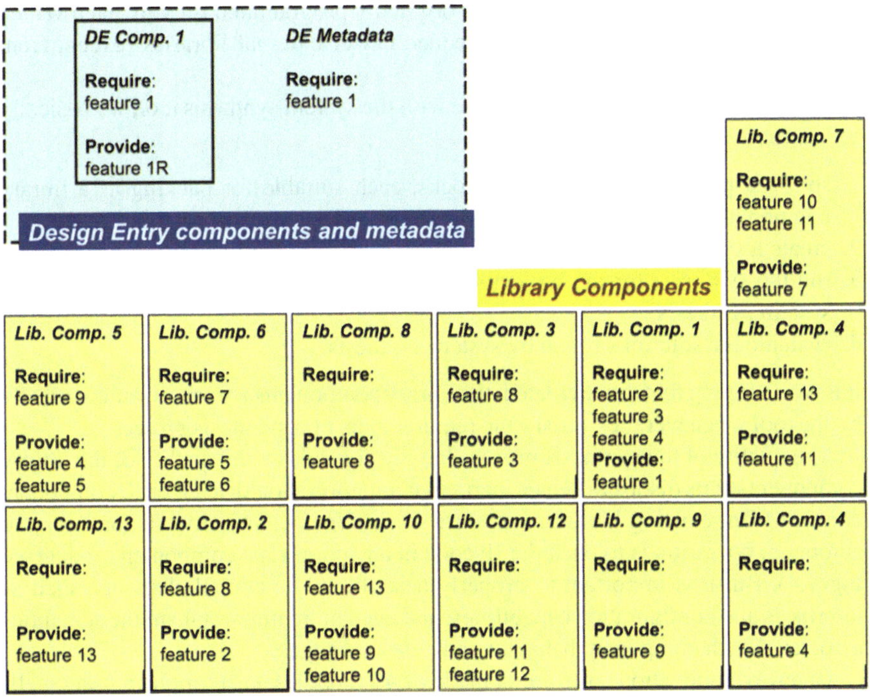


**Fig. 9.5** Simplified example of metadata for design specification component and some library components

1. solution using components 1, 2, 3, 4, 8;
2. solution using components 1, 2, 3, 5, 8, 9;
3. solution using components 1, 2, 3, 5, 8, 10, 13.

## 9.4.6 *Synthesis Use for Legacy Designs*

In the following we will briefly present the use of the synthesis tool for a representative legacy WSN application of practical interest, a self-powered WSN gateway designed for long-term event-based environmental monitoring. Source code is rather complex and optimized. It can handle in real-time messages and state for up to 1000 sensor nodes with an average current consumption of about 1.6 mA. It can also bidirectionally communicate with the application server over the Internet using TCP/IP sockets through an on-board GPRS modem, and receive remote updates. Regardless, gateway hardware requirements are very low, comparable to those of a typical WSN *sensor* node.

To achieve this performance, it was originally programmed fully by handwritten code in C language, without an embedded OS or external libraries (except from some low-level standard C libraries).

To convert this legacy project for use with the system synthesis tool, we basically follow these steps:

1. split the project into functional blocks, each suitable for packing as a library component for the synthesis tool;
2. create a synthesis project by defining its specification top-level component;
3. run the synthesis for the project specification component to perform automatic system synthesis;
4. evaluate the solutions found by system synthesis.

It is worth noting that, once created, the library components are reused automatically by the tool whenever they satisfy the requirements of a synthesis project.

The quality of the synthesis process largely depends on the quality of the library components at its disposal. Hence, particular attention should be given to component creation from existing hardware or software IP blocks. One obvious (and easy to automate) operation is to pack the IP code in an appropriate component model (see Fig. 9.4). But it is important to properly describe its functional elements, such as interfaces and configuration capabilities, and even more important are the semantics associated with component behavior and data exchanges.

Gateway application software is made of 49 modules, each implementing well-defined functions. These can be generic functions, like task scheduler, oscillator calibration or message queue (which are used by most applications), or specialized functions, such as drivers for specific on-board sensors (which are used only by specific applications).

Besides functional blocks for main gateway behavior, the code includes several modules for safety and error recovery, and drivers and processing modules for sensors and auxiliary devices that can be optionally mounted on node, such as:

`adc` Drivers for the ADC peripherals.
The module captures the ADC interrupt and calls the conversion data processing function.

`anemometer` Weather anemometer sensor handling functions.
Driver and controller for the anemometer transducer.

`battery` Utilities for battery reading processing.
The module provides the battery-specific voltage-to-capacity conversion tables and the functions to perform the conversion.

`cc` Field and mesh radio drivers.
The module handles everything related to the field and mesh radio on board the gateway.

`crc` CRC utilities.
Processing utilities (CRC calculation).

`gw` Node status.
Controls the state and configuration of the node.

`hal` Hardware high-level interface.
It processes asynchronous events from the network and on-board switches.

`humidity` Weather humidity sensor handling functions.
Driver and controller for the humidity transducer.

`modem` GPRS modem driver.
Driver for the GPRS modem.

`queue` Message queue.
Storage and processing of the messages queued to be delivered to the server.

`sched` Task scheduler.
Scheduler.

`sensor` Sensor state and data processing.
Maintains the state of the sensors in range based on the contents of their messages (or lack thereof).

`timer` Timer handler.
Provides several timers for use within the node.

`usart` USART drivers.
Drivers for the node USART ports.

`version` Firmware version utilities.
It provides the version of node software.

Figure 9.6 shows the metadata of the library component for a very simple gateway module, *version*. The module stores the version of gateway code and provides methods for its access.

At the top level we can see the main categories *properties*, *views*, *resources*, and *interfaces*. For this simple component, we have just one property that holds the name of the module. Behavioral views include two files with the source code of the module. Resources include one non-functional requirement that tracks component dependency on a toolchain that supports the C language extensions it uses, and a symbolic provided resource which can be used, for instance, to specifically require this component in design specifications or through the requirements of other components. In terms of interfaces, the component provides a behavioral function which retrieves and returns the code version. Additionally, for most metadata properties one can enter a description that can help the developer understand component semantics when it is displayed in a component or solution editor.

Figure 9.7 shows the synthesis result of a minimal gateway system, which requires only the core functions. Moreover, the synthesis tool executed the component configuration helpers to set up their instances in solution with the actual parameters found by the solver (e.g., the scheduler is configured to run only the actual tasks).

Besides the software solution, the synthesis tool collects other requirements of the instantiated components in a list that includes, e.g., hardware node type, radio specifications, and target compilation toolchain.

To find a suitable system composition, the solver reached a recursion depth of 888, matched or wired 230 abstract, 472 functional, and two data requirements in less than 0.8 s on a 1.8 GHz Intel® Core™ i7-2677M CPU.

```
<sgraph:Gss xmi:id="_b2b65395f30689ed09f02e">
    <properties>
        <name>version_component</name><description />
    </properties>
    <views xmi:id="_08f5c2612c510ac5e105e7">
        <behavior>
            <view xmi:id="_5c39ae70c147735f28ad4b" name="version.c"
                type="source" language="C" encoding="base64">
                <description></description>
                <mem>LyoqCiAqIEBmaWxlIHZ [...]</mem>
            </view>
            <view xmi:id="_44e6770ca6e62fc2db54e9" name="version.h"
                type="source" language="C" encoding="base64">
                <description></description>
                <mem>LyoqCiAqIEBmaWxlIHZlc [...]</mem>
            </view>
        </behavior>
    </views>
    <resources>
        <behavior>
            <require><name>avr_libc</name><description /></require>
            <provide><name>version_component</name>
                <description /></provide>
        </behavior>
    </resources>
    <interfaces>
        <behavior>
            <provide>
                <description />
                <function>
                    <name>version_get</name>
                    <return><type>char *</type></return>
                    <port><ord>1</ord><type>char *</type></port>
                </function>
            </provide>
        </behavior>
    </interfaces>
</sgraph:Gss>
```

**Fig. 9.6** Example of a simple library component that includes properties and a code view

## 9.5 Synergies for WSN Development Tools and Platforms

WSN platform development and update is a complex, interdisciplinary, and evolving task. As such, it benefits from allowing all interested parties bring their contribution to its development and extensive use.

WSNs are at the center of a constantly increasing research interest, focused on various functional and technological issues. Research advances the state of the art and most promising results can be made available to many WSN industrial actors through WSN platforms. Reciprocally, the research community would benefit from a common, open, and free WSN platform available for experimentation.

**Fig. 9.7** Result of system synthesis that requires only the main gateway component. It includes 36 out of 49 modules (emphasized), correctly leaving out, e.g., drivers for optional sensors, test suites, and interfaces

| | | | |
|---|---|---|---|
| **adc** | hygrometer | **rccal** | **test_tx** |
| anemometer | **igwc** | rel_mesh | **theft** |
| **battery** | **inst** | **rpc** | **timer** |
| **cc** | **main** | **run_state** | **twi** |
| **crc** | **mesh** | **sched** | **usart** |
| **eeprom** | **modem** | **sensor** | **util** |
| **eeprom_ext** | **msg_filter** | sensor_ppc | **version** |
| fc10 | **obs** | **service** | **wd** |
| **field** | **oc_link** | sio | weather |
| geophone | **power** | **spi** | zlist |
| **gw** | pressure | **sr** | |
| **hal** | **queue** | **sw** | |
| humidity | rain | testing | |

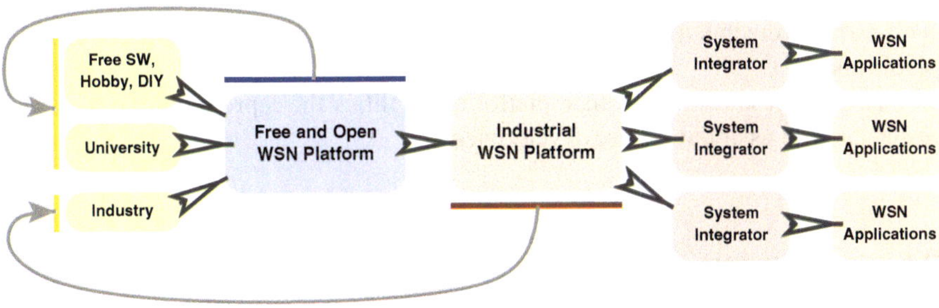


**Fig. 9.8** Ecosystem based on the open WSN platform (for academic research, hobbyists) and its bidirectional synergy with industrialized platform(s) for commercial WSN application development

The same platform can be used also by the wider public, as shown by the growing interest recorded for DIY applications, e.g., in home automation and city environmental monitoring. In exchange, the platform would gain from extensive testing, consolidation, porting to popular hardware, improved development tools, and extensions for innovative applications.

Figure 9.8 suggests a possible ecosystem that includes non-profit and industrial interested parties, which can help both WSN research and spreading the use of WSN technologies to solve real life problems.

The ecosystem revolves around a free and open WSN platform. The platform can include open development tools, like those discussed in Sects. 9.2 and 9.4, server software and open node hardware, either as nodes or node components, as discussed in Sect. 9.3.

Being open, the platform facilitates contributions from several sources, such as academic research, free software community interested in WSN/IoT projects, as well as industrial partners. The latter may become interested because of the business opportunities that can be opened by a platform that helps reducing the effort, cost, and risk of WSN-based solutions.

For this purpose, Fig. 9.8 suggests a productized version of the open platform, which adds the necessary level of reliability, testing, and support for the development of commercial applications. While the open WSN platform reaches its purpose in a continuous state of flux, the industrial product requires stability, ease of use, proprietary IP blocks support, qualified training, and support services.

However, as we argued, developing and maintaining a comprehensive platform is effort-intensive and requires a broad range of engineering skills that are hard to bring together by single entities. But these can be naturally attracted by the open platform, which can serve as foundation to reduce development and maintenance effort for the commercial version(s).

Additionally, the two versions of the platform facilitate the exchanges between academia and industry. Interesting research results can be ported easier to derived commercial platform(s). Conversely is facilitated the contribution of closed source IPs or improvements to platform infrastructure, library, or tools from commercial platforms.

Last but not least, a reference platform simplifies the reproduction of research results as well as collaborative developments.

## 9.6 Conclusion

Although WSNs are object of extensive scientific and technological advances and can effectively achieve distributed data collection for IoT applications, their wide scale use is still limited by a perceived low reliability, limited autonomy, high cost, and reduced accessibility to application domain experts.

Commercial solutions are often effective in addressing vertical application domains, but they may lead to technology lock-ins that limit horizontal composability, and component or design reuse.

We consider WSN platform an essential enabler for effective application design: fast development with low effort and cost, reliable designs based on extensive reuse, flow accessible to application domain experts, and offering maintainable extensive technology coverage.

After examining how existing WSN development tools and flows satisfy these objectives, we propose a node-level hardware–software development flow. It revolves on automated system synthesis starting from a high-level application specification (both behavior and non-functional requirements) and reuses extensively library components.

We argue that this system can foster synergies between academic research, IoT and WSN developer communities, and system integrators developing commercial WSN application, with the distinct possibility of mutual benefit in an ecosystem that merges open and closed source IPs. We consider synergy important, since effective and reliable WSN development requires a wide range of engineering expertise that are difficult to be covered viably by individual players.

## References

1. K. Ashton, That 'Internet of Things' thing. Expert view. RFID J. (2009). http://www.rfidjournal.com/article/view/4986
2. K. Romer, F. Mattern, The design space of wireless sensor networks. IEEE Wireless Comm. **11**(6), 54–61 (2004)
3. R. Szewczyk, A. Mainwaring, J. Polastre, J. Anderson, D. Culler, An analysis of a large scale habitat monitoring application, in *Proceedings of the 2Nd International Conference on Embedded Networked Sensor Systems, SenSys '04* (ACM, New York, 2004), pp. 214–226
4. J. Hill, R. Szewczyk, A. Woo, S. Hollar, D. Culler, K. Pister, System architecture directions for networked sensors. SIGARCH Comput. Archit. News **28**(5), 93–104 (2000)
5. D. Gay, P. Levis, R. von Behren, M. Welsh, E. Brewer, D. Culler, The nesC language: a holistic approach to networked embedded systems. SIGPLAN Not. **38**(5), 1–11 (2003)
6. E. Cheong, J. Liebman, J. Liu, F. Zhao, TinyGALS: a programming model for event-driven embedded systems, in *Proceedings of the 2003 ACM Symposium on Applied Computing, SAC '03* (ACM, New York, 2003), pp. 698–704
7. C.-C. Han, R. Kumar, R. Shea, E. Kohler, M. Srivastava, A dynamic operating system for sensor nodes, in *Proceedings of the 3rd International Conference on Mobile Systems, Applications, and Services, MobiSys '05* (New York, ACM, 2005), pp. 163–176
8. C.-C. Han, M. Goraczko, J. Helander, J. Liu, B. Priyantha, F. Zhao, CoMOS: an operating system for heterogeneous multi-processor sensor devices. Redmond, WA, Microsoft Research Technical Report No MSR-TR-2006-177 (2006)
9. B. Greenstein, E. Kohler, D. Estrin, A sensor network application construction kit (SNACK), in *Proceedings of the 2Nd International Conference on Embedded Networked Sensor Systems, SenSys '04* (ACM, New York, 2004), pp. 69–80
10. P. Levis, D. Gay, V. Handziski, J.-H. Hauer, B. Greenstein, M. Turon, J. Hui, K. Klues, C. Sharp, R. Szewczyk et al., T2: a second generation OS for embedded sensor networks. Telecommunication Networks Group, Technische Universität Berlin, Tech. Rep. TKN-05-007 (2005)
11. O. Kasten, K. Römer, Beyond event handlers: programming wireless sensors with attributed state machines, in *Proceedings of the 4th International Symposium on Information Processing in Sensor Networks, IPSN '05* (IEEE Press, Piscataway, 2005)
12. M. Welsh, G. Mainland, Programming sensor networks using abstract regions, in *Networked Systems Design and Implementation (NSDI)*, vol. 4, pp. 3–3 (2004)
13. H. Abrach, S. Bhatti, J. Carlson, H. Dai, J. Rose, A. Sheth, B. Shucker, J. Deng, R. Han, MANTIS: system support for multimodal networks of in-situ sensors, in *Proceedings of the 2Nd ACM International Conference on Wireless Sensor Networks and Applications, WSNA '03* (ACM, New York, 2003), pp. 50–59
14. W.P. McCartney, N. Sridhar, Abstractions for safe concurrent programming in networked embedded systems, in *Proceedings of the 4th International Conference on Embedded Networked Sensor Systems, SenSys '06* (ACM, New York, 2006), pp. 167–180
15. A. Dunkels, O. Schmidt, T. Voigt, M. Ali, Protothreads: simplifying event-driven programming of memory-constrained embedded systems, in *Proceedings of the 4th International Conference on Embedded Networked Sensor Systems, SenSys '06* (ACM, New York, 2006), pp. 29–42
16. A. Dunkels, B. Gronvall, T. Voigt, Contiki: a lightweight and flexible operating system for tiny networked sensors, in *Proceedings of the 29th Annual IEEE International Conference on Local Computer Networks, LCN '04* (IEEE Computer Society, Washington, DC, 2004), pp. 455–462
17. C. Nitta, R. Pandey, Y. Ramin, Y-threads: supporting concurrency in wireless sensor networks, in *Proceedings of Distributed Computing in Sensor Systems: Second IEEE International Conference, DCOSS 2006*, San Francisco, CA, June 18–20, 2006 (Springer, Berlin/Heidelberg, 2006), pp. 169–184
18. P. Levis, D. Culler, Maté: a tiny virtual machine for sensor networks, in *Proceedings of the 10th International Conference on Architectural Support for Programming Languages and Operating Systems, ASPLOS X* (ACM, New York, 2002), pp. 85–95

19. P. Levis, D. Gay, D. Culler, Active sensor networks, in *Proceedings of the 2nd Conference on Symposium on Networked Systems Design & Implementation - Volume 2, NSDI'05* (USENIX Association, Berkeley, 2005), pp. 343–356
20. Y. Yu, L.J. Rittle, V. Bhandari, J.B. LeBrun, Supporting concurrent applications in wireless sensor networks, in *Proceedings of the 4th International Conference on Embedded Networked Sensor Systems, SenSys '06* (ACM, New York, 2006), pp. 139–152
21. J. Koshy, R. Pandey, VMSTAR: synthesizing scalable runtime environments for sensor networks, in *Proceedings of the 3rd International Conference on Embedded Networked Sensor Systems, SenSys '05* (ACM, New York, 2005), pp. 243–254
22. T. Liu, M. Martonosi, Impala: A Middleware System for Managing Autonomic, Parallel Sensor Systems, in *Proceedings of the Ninth ACM SIGPLAN Symposium on Principles and Practice of Parallel Programming, PPoPP '03* (ACM, New York, 2003), pp. 107–118
23. R. Newton, D. Arvind, M. Welsh, Building up to macroprogramming: an intermediate language for sensor networks, in *Proceedings of the 4th International Symposium on Information Processing in Sensor Networks, IPSN '05* (IEEE Press, Piscataway, NJ, 2005)
24. L. Gu, J.A. Stankovic, t-kernel: providing reliable OS support to wireless sensor networks, in *Proceedings of the 4th International Conference on Embedded Networked Sensor Systems, SenSys '06* (ACM, New York, 2006), pp. 1–14
25. R. Shimizu, K. Tei, Y. Fukazawa, S. Honiden, Model driven development for rapid prototyping and optimization of wireless sensor network applications, in *Proceedings of the 2Nd Workshop on Software Engineering for Sensor Network Applications, SESENA '11* (ACM, New York, 2011), pp. 31–36
26. A. Taherkordi, F. Loiret, A. Abdolrazaghi, R. Rouvoy, Q. Le-Trung, F. Eliassen, Programming sensor networks using REMORA component model, in *Proceedings of the 6th IEEE International Conference on Distributed Computing in Sensor Systems, DCOSS'10* (Springer, Berlin/Heidelberg, 2010), pp. 45–62
27. Z. Song, M.T. Lazarescu, R. Tomasi, L. Lavagno, M.A. Spirito, High-level Internet of things applications development using wireless sensor networks, in *Internet of Things: Challenges and Opportunities* (Springer, Cham, 2014), pp. 75–109
28. A. Varga, R, Hornig, An overview of the OMNeT++ simulation environment, in *Proceedings of the 1st International Conference on Simulation Tools and Techniques for Communications, Networks and Systems and Workshops, Simutools '08* (ICST Institute for Computer Sciences, Social-Informatics and Telecommunications Engineering, Brussels/Belgium, 2008), pp. 60:1–60:10
29. M. Welsh, G. Mainland, Programming sensor networks using abstract regions, in *Networked Systems Design and Implementation (NSDI)*, vol. 4, pp. 3–3 (2004)
30. K. Whitehouse, C. Sharp, E. Brewer, D. Culler, Hood: a neighborhood abstraction for sensor networks, in *Proceedings of the 2Nd International Conference on Mobile Systems, Applications, and Services, MobiSys '04* (ACM, New York, 2004), pp. 99–110
31. T. Abdelzaher, B. Blum, Q. Cao, Y. Chen, D. Evans, J. George, S. George, L. Gu, T. He, S. Krishnamurthy, L. Luo, S. Son, J. Stankovic, R. Stoleru, A. Wood, EnviroTrack: towards an environmental computing paradigm for distributed sensor networks, in *Proceedings of the 24th International Conference on Distributed Computing Systems*, pp. 582–589 (2004)
32. L. Mottola, G. Pietro Picco, Logical neighborhoods: a programming abstraction for wireless sensor networks, in *Proceedings of the Distributed Computing in Sensor Systems: Second IEEE International Conference, DCOSS 2006*, San Francisco, CA, June 18–20, 2006 (Springer, Berlin/Heidelberg, 2006), pp. 150–168
33. J. Liu, M. Chu, J. Liu, J. Reich, F. Zhao, State-centric programming for sensor-actuator network systems. IEEE Pervasive Comput. **2**(4), 50–62 (2003)
34. P. Bonnet, J. Gehrke, P. Seshadri, Querying the physical world. IEEE Pers. Commun. **7**(5), 10–15 (2000)
35. S.R. Madden, M.J. Franklin, J.M. Hellerstein, W. Hong, TinyDB: an acquisitional query processing system for sensor networks. ACM Trans. Database Syst. **30**(1), 122–173 (2005)

36. C. Srisathapornphat, C. Jaikaeo, C.-C. Shen, Sensor information networking architecture, in *Proceedings of the 2000 International Workshops on Parallel Processing*, pp. 23–30 (2000)
37. W.B. Heinzelman, A.L. Murphy, H.S. Carvalho, M.A. Perillo, Middleware to support sensor network applications.IEEE Netw. **18**(1), 6–14 (2004)
38. S. Li, S.H. Son, J.A. Stankovic, Event detection services using data service middleware in distributed sensor networks, in *2003 Proceedings of Information Processing in Sensor Networks: Second International Workshop, IPSN 2003*, Palo Alto, CA, April 22–23, pp. 502–517 (Springer, Berlin/Heidelberg, 2003)
39. R. Newton, G. Morrisett, M. Welsh, The regiment macroprogramming system, in *Proceedings of the 6th International Conference on Information Processing in Sensor Networks, IPSN '07* (ACM, New York, 2007), pp. 489–498
40. R. Gummadi, O. Gnawali, R. Govindan, Macro-programming wireless sensor networks using Kairos, in *Proceedings of the First IEEE International Conference on Distributed Computing in Sensor Systems, DCOSS'05* (Springer, Berlin/Heidelberg, 2005), pp. 126–140
41. C. Borcea, C. Intanagonwiwat, P. Kang, U. Kremer, L. Iftode, Spatial programming using smart messages: design and implementation, in *Proceedings of the 24th International Conference on Distributed Computing Systems* (2004), pp. 690–699
42. Y. Ni, U. Kremer, A. Stere, L. Iftode, Programming ad-hoc networks of mobile and resource-constrained devices, in *Proceedings of the 2005 ACM SIGPLAN Conference on Programming Language Design and Implementation, PLDI '05* (ACM, New York, 2005), pp. 249–260
43. C. Intanagonwiwat, R. Gupta, A. Vahdat, Declarative resource naming for macroprogramming wireless networks of embedded systems, in *Algorithmic Aspects of Wireless Sensor Networks: Second International Workshop, ALGOSENSORS 2006*, Venice, July 15, 2006. Revised Selected Papers (Springer, Berlin/Heidelberg, 2006), pp. 192–199
44. K. Whitehouse, F. Zhao, J. Liu, Semantic streams: a framework for composable semantic interpretation of sensor data, in *Proceedings of the Wireless Sensor Networks: Third European Workshop, EWSN 2006*, Zurich, Feb 13–15, 2006 (Springer, Berlin/Heidelberg, 2006), pp. 5–20
45. E. Wei-Chee Lin, *Software Sensors: Design and Implementation of a Programming Model and Middleware for Sensor Networks* (University of California, San Diego, 2004)
46. S. Bandyopadhyay, A.P. Chandrakasan, Platform architecture for solar, thermal and vibration energy combining with MPPT and single inductor, in *VLSI Circuits, VLSIC*, pp. 238–239 (2011)
47. M. Doddavenkatappa, M.C. Chan, A.L. Ananda, Indriya: a low-cost, 3D wireless sensor network testbed, in *Testbeds and Research Infrastructure. Development of Networks and Communities* (Springer, Berlin/Heidelberg, 2012), pp. 302–316
48. T. Ahola, P. Korpinen, J. Rakkola, T. Ramo, J. Salminen, J. Savolainen, Wearable FPGA based wireless sensor platform, in *29th Annual International Conference of the IEEE Engineering in Medicine and Biology Society, EMBS 2007* (2007), pp. 2288–2291
49. M. Laibowitz, J. Gips, R. Aylward, A. Pentland, J.A. Paradiso, A sensor network for social dynamics, in *Proceedings of the 5th International Conference on Information Processing in Sensor Networks, IPSN '06* (ACM, New York, 2006), pp. 483–491
50. K. Aberer, M. Hauswirth, A. Salehi, Infrastructure for data processing in large-scale interconnected sensor networks, in *Mobile Data Management*, pp. 198–205 (2007)
51. A Mülder, A. Nyßen, TMF meets GMF. Eclipse Mag. **3**, 74–78 (2011). https://svn.codespot.com/a/eclipselabs.org/yakindu/media/slides/TMF_meets_GMF_FINAL.pdf.
52. K. Doddapaneni, E. Ever, O. Gemikonakli, I. Malavolta, L. Mostarda, H. Muccini, A model-driven engineering framework for architecting and analysing wireless sensor networks, in *Proceedings of the Third International Workshop on Software Engineering for Sensor Network Applications, SESENA '12* (IEEE Press, Piscataway, NJ, 2012), pp. 1–7
53. A.R. Paulon, A.A. Fröhlich, L.B. Becker, F.P. Basso, Model-driven development of WSN applications, in *2013 III Brazilian Symposium on Computing Systems Engineering (SBESC)* (2013), pp. 161–166

54. N. Mohamed, J. Al-Jaroodi, A survey on service-oriented middleware for wireless sensor networks. Serv. Oriented Comput. Appl. **5**(2), 71–85 (2011)
55. L. Mottola, G. Pietro Picco, Middleware for wireless sensor networks: an outlook. J. Internet Serv. Appl. **3**(1), 31–39 (2012)
56. A. Ray, Planning and analysis tool for large scale deployment of wireless sensor network. Int. J. Next-Generation Netw. (IJNGN) **1**(1), 29–36 (2009)
57. M.T. Lazarescu, Design of a WSN platform for long-term environmental monitoring for IoT applications. IEEE J. Emerging Sel. Top. Circuits Syst. **3**(1), 45–54 (2013)
58. N. Gámez, J. Cubo, L. Fuentes, E. Pimentel, Configuring a context-aware middleware for wireless sensor networks. Sensors **12**(7), 8544–8570 (2012)
59. L. Mottola, G. Pietro Picco, Programming wireless sensor networks: fundamental concepts and state of the art. ACM Comput. Surv. **43**(3), 19:1–19:51 (2011)
60. D. Steinberg, F. Budinsky, E. Merks, M. Paternostro, *EMF: Eclipse Modeling Framework* (Pearson Education, London, 2008)

# Chapter 10
# Event Identification in Wireless Sensor Networks

**Christos Antonopoulos, Sofia-Maria Dima, and Stavros Koubias**

## 10.1 Introduction

Based on various sources [1–4] an "event" is defined as an important phenomenon that occurs or may have occurred. Consequently, event identification is the procedure through which the respective phenomenon is accurately and reliably identified as well as recorded. Also an event is a specific case that stands out from an otherwise normal situation, exhibiting different data patterns compared to what is expected for a particular scenario. This chapter focuses on such procedures and algorithms specifically designed for or applied to wireless sensor networks (WSNs). Event identification in WSNs is a rapidly evolving research area attracting active interest from both the research and the industrial domains [2, 5–9]. The former is mainly due to its challenges as well as restrictions while the latter can be attributed to the fact that respective implementation is expected to drastically enhance WSNs practical applicability, usefulness, and widespread while at the same time mitigate notorious shortcomings of such networks. This is also indicated by the increasing number of prestigious publications and projects focusing on this objective. In this context the chapter's main objective is to offer a comprehensive survey and classification or different approaches, techniques, and methodologies currently comprising state of the art in event detection targeting WSNs.

Respective functionality is of cornerstone importance in a wide range of applications varying from medical, environmental, mechanical, and virtually any practical scenario [12, 15]. In the context of such scenarios instead of acquiring a complete knowledge and notion of the particular application, through event detection algorithms, the objective is to identify the occurrence or the possible occurrence of a type or set of types of events. The initial, though simplistic, idea of event detection

C. Antonopoulos (✉) • S.-M. Dima • S. Koubias
Department of Electrical and Computer Engineering, University of Patras, Patras, Greece
e-mail: cantonop@ece.upatras.gr



G. Keramidas et al. (eds.), *Components and Services for IoT Platforms*,
DOI 10.1007/978-3-319-42304-3_10

assumes the existence of a specific threshold value, with respect to which and based on the deviation of a particular measurement to that threshold an event can be defined. Temperature is a characteristic example since exceeding a specific level (e.g., 40 °C) can be categorized as an event. However, even in simplistic application scenarios it is identified that such approaches are inadequate to capture complex events depending on multiple inputs. Based on this deficiency a new even detection algorithmic trend is developed utilizing techniques based on pattern recognitions since all events can be represented as specific patterns.

Respective techniques' presentation, analysis, and classification are based on particular set of characteristics which effectively distinguish each approach revealing relative advantages and disadvantages advocating the use of each one to specific application scenarios.

## 10.2 Wireless Sensor Network Characteristics

Design and implementation of event detection algorithms for WSNs have to tackle significant challenges due to the limitations and restrictions posed in such networks. Such limitations mainly stem from scarce resource availability in all aspects of a typical WSN nodes. Without a doubt the most important such resource shortage is energy availability. Aimed to be small, low cost, low complexity, wearable, and effectively expendable a typical node is powered by very small (usual rechargeable) batteries with typical capacity ranging from 450 mAh up to 300 mAh (corresponding to two AA batteries). Thus power conservation comprises probably the most important objective or relative developments.

Consequently, adequate WSN event detection algorithms must be energy efficient and fault tolerant, yet accurate and reliable. Furthermore, an important prerequisite concerns computational and communicational resource conservation and at the same time high configurability and adjustability to wide range of events. In the context of conservative approaches, WSN nodes role was limited to conveying and aggregating raw data to a resource rich central entity (typically referred to as Base Station or Gateway) which was solely responsible for further data analysis and event identification. However, WSN paradigm brings forwards specific characteristics such as multi-hop data transfer, dynamic topologies as well as low bandwidth availability. Such characteristics in combination with centralized approaches can lead to performance shortcomings such as increased event identification delay, unpredictable link breakage, data packets congestion etc. drastically degrading event detection capabilities. On the contrary the contemporary approaches lean towards event identification inside the WSN network through cooperative distributed data processing by adequate subset of WSN nodes. In this way data is being analyzed much faster, events identification delay can be significantly reduced and even more, actuators residing also in the network can react to the identified events much faster, accurately and reliably.

Another critical WSN characteristics influencing event detection algorithm design relates to the data acquisition approach utilized. In that respect three main approaches are typically encountered.

- *Continuous Data Streaming:* In this case data acquired by the sensors are conveyed to the central station periodically without any processing or filtering. Although comprising an easily implemented approach it is considered an ineffective method when large data volumes are aggregated or when data are to be transferred over complex multi-hop paths. Additionally, such data acquisition approach typically entail a centralized data processing architecture often leading to underperformance due to excessive time delay, low data transfer reliability as well as high network congestion scenarios. Last but not least such approach frequently lead to increased energy consumption comprising a critical disadvantage for WSN networks.
- *Query-Drive:* In this case network users effectively insert a query into the network and respective nodes that can actually provide a response transfer required data through the network. On one hand, this approach is much more efficient than continuously streaming data with respect to resource conservation particularly regarding energy and bandwidth consumption. On the other hand, they pose a critical requirement of supporting only a priori known requests which in many cases contradict to the dynamic nature of WSNs.
- *Event driven:* In this case, data are transferred when specific conditions are met (e.g., when predefined thresholds are surpassed) which effectively correspond to an "event," On one hand, such approach is even more efficient in terms of resource consumption while, one the other hand, it allows the creation of complex application scenarios based on fuzzy or complicated data patterns formation.

However, besides optimal data acquisition technique, an important driving force of event detection approach in WSNs relates to the network entities where processing takes place.

- *Base Station:* It can be considered a typical solution but a rather inefficient one for nowadays demanding application scenarios. A Base Station is usually a resource rich WSN node effectively operating as the interface between the WSN network and rest of the work (e.g., an IP network or the internet). Respective hardware offers abundant resources but is also characterized by significant disadvantages such as representing a single point of failure. It can also fail to address strict application requirements in complex network topologies including unreliable links, multi-hop communication paths, and highly mobile WSN nodes.
- *In network:* In this case raw data are being processed by the nodes comprising the WSN network. Therefore, performance characteristics are significantly enhanced since there is no single point in the network where data must flow in order for a decision to be made. Designs following this approach can be further divided into two categories.

  - *Local:* Where a single node can perform the data processing of data acquired.

– *Distributed:* Where a group of nodes (typically schematically correlated) are involved in the data processing. Therefore, data exchange amongst nodes is required in this case.

Concluding, base Station data processing is usually combined with continuous data steam data acquisition approach. Additionally, query-driven and event-driven techniques, although also applicable in Base Station-based algorithms, are increasingly utilized in designs and implementation of algorithms following the in-network data processing paradigm.

## 10.3 Event Detection Challenges in WSNs

Event detection is of paramount importance in WSNs since it allows the efficient management of emergencies and critical citations due to the occurrence of specific event. In most cases the identification of an event is a time constrained functionality. Respective constraints are posed by the specific application and the criticality of the event.

- *Time critical scenarios:* In such cases a specific deadline is defined before which the event must be accurately identified. Omitting specific deadline may lead to degraded performance (usually correlated to soft real time applications) or complete application failure (usually correlated to hard real time applications). The latter case is much more important since respective failure can endanger property or even human lives.
- *Best effort scenarios:* In such cases, it is implied that there is no strict deadline to meet but the event detection algorithms perform to the best of its ability depending on various parameters.

Furthermore, WSNs characteristics also pose critical challenges regarding in-network event detection algorithm design and development. The most important such challenges are as follows:

- *Data unreliability:* It is widely known that WSN data are unreliable and prone to errors. Reasons for such unreliability include (a) environmental conditions effect, since WSN nodes can be exposed to wide range of conditions for extended period of time, (b) effect of unreliable or/and fluctuating power level, (c) inherent error prone wireless communication medium, and (d) low cost, low complexity hardware components usually utilized in WSN platforms. Environmental influence to data acquired is unavoidable since by definition WSN sensors are targeted to be in direct conduct with the modality monitoring while one of the most important advantage of a WSN is supporting deployment in harsh and hazardous environments. Small batteries comprise the main source of energy in WSNs. Therefore, they are subjects to phenomena affecting data accuracy such leakage or/and power variation to specific components with respect to the energy level as the battery is depleted. Wireless communication poor quality is also well known due to signal propagation phenomena as well as ad-hoc communication paradigm typically utilized in WSNs. Finally, in order for WSNs to be widely deployed in vast numbers it is imperative to minimize cost to the level that a single node can be considered expandable. Therefore, respective implementation tends to rely on low cost thus low quality hardware. A common approach able to tackle all aforementioned shortcomings is node redundancy so that specific failure(s) can be compensated from the rest of the nodes.

- *Data Volume:* When event detection takes place in specific central nodes (e.g., the base station) then the problem of excessive data volume comprises a critical challenge since all nodes send their data to the central processing node. On the other hand, when event detection is done in network, the data flow is significantly reduced, however, addressing critical situations demand inter-node collaboration which poses event detection deployment as another challenge to tackle.
- *Complex event patterns:* By definition "events" are patterns of acquired modalities which are different from what is considered a normal pattern. However such patters may vary from very simple and well defined to very complex ones or even unknown. In such cases data modeling, pattern recognition, and categorization techniques are required so as to extract patterns which are quite challenging to be developed and even efficiently executed in a computationally scarce environment such the one typically encountered in WSNs deployments.
- *Network Heterogeneity and Dynamicity:* WSNs are characterized by high degree of heterogeneity since they are comprised by nodes of diverse capabilities, resources, and communication techniques and data acquisition characteristics. Furthermore, decentralized operation, unpredictable node mobility, energy depletion, and varying communication conditions result in an extremely dynamic topology and volatile network characteristics. Consequently, event detection techniques must be able to efficiently handle and adjust to such conditions.
- *Adjustability:* Due to versatile functionality of a WSNs, respective event detection algorithms must also be characterized by a significant degree of adjustability. In that respect it is expected that a specific event detection technique can be utilized so as to identify more than one event. Adjustability implies that the technique must take into consideration the setup and the deployment of the network. A respective prominent approach is to adopt algorithm that can be easily adjusted by, e.g., appropriate weight selection.

## 10.4 Data Fusion Categorization

Probably the most basic and fundamental differentiation concerns single modality and multi-modality events [16]. The former concern events the identification of which is based only on a single characteristic, or type of measurement which is referred to as modality. In the latter category the event identification is based on concurrently, combining multiple inputs (e.g., sensors for WSN networks). To effectively utilize a plethora of different acquired type of signals (i.e., modalities) new sophisticated algorithms are required. Although the concept of multimodal data processing was initially exploited for military and robotic applications it is now utilized in all types of WSN applications. It is worth noting that in the context of this research domain terms like data fusion, sensor fusion, multimodal data fusion, multimodal sensor fusion, information fusion, etc. tend to be used interchangeably with little actual difference. The term, multimodal data fusion, refers to the exploitation of multiple, different, and diverse wireless sensors through a sophisticated process combining elementary features like association, correlation, and combination of data so as to derive a refined event detection decision.

The fundamental goal of multimodal data fusion is the detailed description of the phenomenon through the acquired data and the efficient utilization of this representation so as to increase the event accuracy. Here lays the main point of superiority over the single modal approaches. Single modality techniques offer only partial and incomplete system representations. Therefore, respective developments lead to uncertain conclusions which omit critical insight of the specific system leading to increased probability of erroneous indications. Furthermore, relying only on a single modality, respective solutions inherently suffer from single point failure which can be caused by the sensor itself, the communication channel or any random and unpredictable event.

### *10.4.1 Data Fusion Algorithmic Approaches*

Regarding the core algorithms' characteristics state-of-the-art approaches can be classified as follows [17]:

*Competitive approaches* [18, 19]: According to this approach the algorithm combines the same modality measurements so as to minimize respective deviations due to hardware, communication, and other sources of failure. Through such approaches it is possible to identify potential problems and estimate which measurement is more accurate in each particular case.

*Complementary approaches* [20, 21]: In this case, data fusion is exploited to combine partial data in order to indirectly formulate a complete and accurate model of a system or phenomenon under investigation. In this case the sensors do not directly depend on each other, but can be combined in order to give a more complete representation of the phenomenon under observation. This resolves the incompleteness of sensor data. An example for a complementary configuration is the employment of multiple cameras each observing disjunct parts of a room. Generally, fusing complementary data is easy, since the data from independent sensors can be appended to each other.

*Cooperative approaches* [16, 22]: In this case data fusion combines multiple sensor data amongst which a direct correlation exist so as to derive a higher level conclusion, decision, or indication. An example for a cooperative sensor configuration is stereoscopic vision—by combining two-dimensional images from two cameras at slightly different viewpoints a three-dimensional image of the observed scene is derived.

Summarizing competitive data fusion increases data reliability and data confidence, whereas complementary and cooperative data fusion techniques tend to lead to higher level abstraction measurement and thus more reliable conclusions. In any case combination of more than one approaches can also be considered.

### *10.4.2 Levels of Data Fusion*

Depending on the level of data that is actually fused three different categorizes can be identified [23]. It is noted that these categorizes effectively represent abstraction layers typically utilized at application level. Therefore, more layers can be identified with respect to specific application scenarios while combination of different categories can also be envisioned and exploited.

*Direct level fusion:* This category includes algorithms and techniques that fuse together raw data so as to derive to a decision.

*Feature level fusion:* In this case features extracted from acquired data (usually in the form of vectors) comprise the entities that are actually fused and combined so as to offer higher abstraction knowledge and relative decisions.

*Decision level fusion:* Finally, different and diverse decisions extracted from initial algorithms can be fed into a data fusion algorithm leading to even more abstract indications and conclusions.

## 10.5 Classification of Event Detection Approaches

We choose to classify state-of-the-art approaches with respect to their underlying core techniques. Following an extensive relative literature analysis, authors followed an elicitation process to deduce the most suitable approaches for WSNs mainly based on [2, 5, 24, 25]. As indicated in Fig. 10.1 current approaches can be initially categorized as follows:

- Pattern Matching
- Model Based
- Artificial Intelligence and Machine Learning Based

Furthermore, pattern matching approaches can be further classified as signature matching and prototype matching.

Model-based approaches, on the other hand, are analyzed into Arithmetic Models, Statistical and Probabilistic Methods and Map-Based. AI-based approaches,

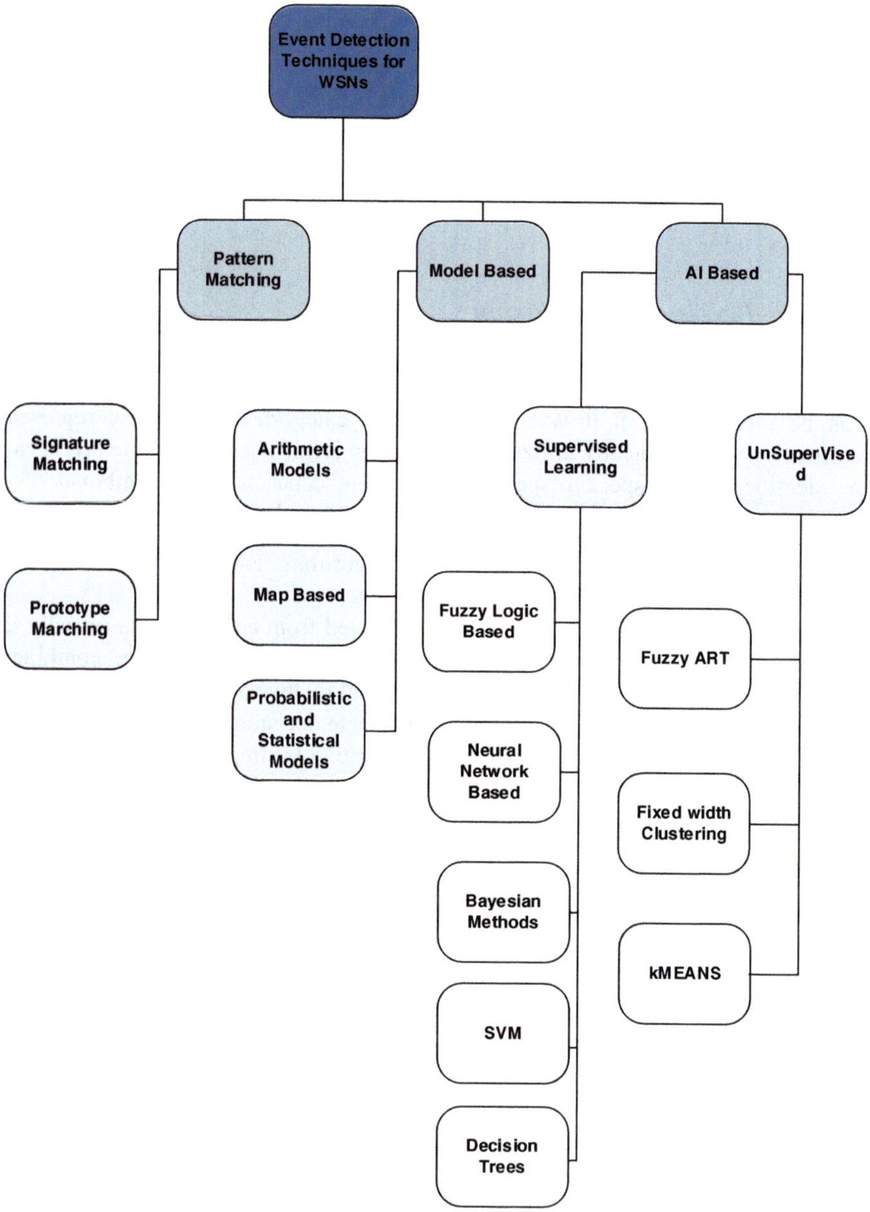


**Fig. 10.1** Taxonomy of event detection techniques

however, comprising probably the most active category, can be further divided into two subcategories, i.e., supervised and unsupervised learning. Each one of these subcategories leads to specific approaches as indicated in Fig. 10.1 comprising the most sophisticated and prominent of the state-of-the-art approaches.

### *10.5.1 Model-Based Approaches*

In the context of this category respective techniques aim to model an event in a specific form such as mathematical formulas or maps. Respective implementations reveal the following properties:

1. They are able to handle complex systems since they represent a, mainly, non-linear even model
2. They are typically dependent upon the specific application and thus suffer from lack of flexibility so as to be applied in other domains
3. An expert is required to accurately configure the model parameters
4. They tend to be computational intensive due to the complexity of the respective models.

#### 10.5.1.1 Arithmetic Model-Based Approaches

Arithmetic model-based techniques utilize discrete or continuous mathematical models in order to model events and identify them according to the degree, acquired data, and converge to predefined models. Bager et al. [26] propose a voting graph neuron (VGN) algorithm for event detection in distributed wide scale WSNs. VGN algorithm is based on a distributed cooperative idea of problem solving according to which the problem of interest is effectively segmented into smaller parts. In this context the event patterns are stored in a distributed graph in the network. Therefore, events are detected by matching data of each WSN node with a subset of the graph. The particular proposal has been evaluated in the context of Matlab-based simulation effort. Zhang et al. [27], on the other hand, propose two new arithmetic model-based techniques aiming to locally identify events in a WSN node. These techniques are based on if-then-else arithmetic rules which are able to correlate real data against models so as to quantify respective convergence. Validation is done in the context of the widely used WSN Castalia simulator trying to detect overheating events and compare respective performance against threshold-based analogous implementations.

#### 10.5.1.2 Map-Based Approaches

The main idea upon which respective techniques are based is that maps comprise a way to accurately represent the physical world and physical phenomena in space and time. Thus a map uniquely corresponds to an event and adopting an adequate process it can actually assist in identifying events that occur or have occurred in a specific environment.

Specifically Khelil et al. [28] present a distributed event detection algorithm based on Map-based world model (MWM) in which each sensor node forms a map of its immediate surrounding and sends the resulting model to the sink node targeting environmental event detection. WSNs, on the one hand, are inherently embedded in the real world, the goal being to detect specific spatial and temporal physical world's ambient characteristics, such as temperature, air pressure, gas presence, and oxygen density. On the other hand, maps present a powerful/efficient tool to model the spatial and temporal behavior of the physical world being an intuitive aggregated view of it. As stated before, the main objective behind the deployment of WSNs is to create an appropriate model for the physical world. Therefore, without loss of generality, the authors model the world as a stack of maps presenting the spatial and temporal distributions of the sensed attributes of interest in the physical world. Authors argue that the specific techniques are indeed able to identify event rapidly and accurately. Li et al. [29] propose a distributed event detection approach based on the creation of a 3D map and an aggregation method. Authors argue that nodes reside in a three-dimensional environment and therefore are able to model this environment as a cubic map cell. Such cubic cell maps can be aggregated in a cluster head or a sink node so as to form an extended cubic cell map containing all environments monitored. In final stage of the event detection process, the map of the entire environment represents the event as well as the event location.

#### 10.5.1.3 Probabilistic/Statistical Model-Based Approaches

These approaches analyze the distribution of data or other statistical metrics so as to form the models and then identify the events. Statistics is the traditional field that deals with the quantification, collection, analysis, interpretation, and drawing conclusions from data. It is concerned with probabilistic models, and specifically inference on these models using data. Statistical techniques are driven by the data and are used to discover patterns and build predictive models. Statistical approaches are generally characterized by having an explicit underlying probability model, which provides a probability of being in each class rather than simply a classification. In addition, it is usually assumed that the techniques will be used by statisticians, and hence some human intervention is assumed with regard to variable selection and transformation, and overall structuring of the problem [35–37].

Static threshold event detection is by far the simplest and most computationally straightforward method of statistical event detection. Event detections are reported when the monitored parameter exceeds a predetermined threshold value, and the detection condition persists as long as the parameter value exceeds the threshold set point. Once the parameter falls within acceptable bounds, the detection condition is met. Threshold values may be determined based upon historical parameter values, for example, to similar sensors and systems, engineering estimates, or parametric analysis. The static threshold method exhibits a "memoryless" property from one observation to the next, as the current observation and detection condition is independent of all prior observations. However, observed values are (usually) dependent upon prior observed values, and one would not reasonably expect the observed values to radically change in the short period of time between successive observations. Many references describe the benefits and utility of static threshold event detection methods, and Kerman et al. [30] baseline the results of the static threshold method against a composite event detection method.

Techniques falling into this category are probably the most straightforward based on simple if-then-else rules aiming to control whether acquired real time measurements deviate and to what degree with respect to predefined thresholds levels. Respective developments tend to exhibit the following characteristics: (1) it is very easy to implement since they effectively comprise by a set of if-then-else rules, (2) they usually require specialized and in-depth knowledge of the system of phenomenon to adequately configure the threshold values, and (3) they tend to be inaccurate in complex scenarios since they model events as linear functions. Vu et al. [31] propose a complex event detection threshold-based technique, which aims to decompose complex events to a set of sub-events of lower complexity. Therefore an event is identified if all sub-events are identified (occurred either concurrently or sequentially). For example, an explosion event is decomposed into two sub-events: (1) the occurrence of a loud noise and (2) the increase of heat. Consequently the event of an explosion is identified if both aforementioned events occur concurrently. The particular study offers the possibility of distributed processing when applied in a wide scale networks and is evaluated in the context of simulation environment. A distributed event detection threshold-based scheme for heterogeneous WSNs is presented in [32]. In these cases the authors propose a two layer clustering approach. Specifically one layer acts as parent-nodes and the final node acts as the sink node. The study proposes a compete framework for both data collection and event identification in the context of the COMis middleware. Another relative event detection approach characterized as consensus-based techniques is presented in [33], aiming to assist in volcano monitoring and respective event identification. The authors suggest a complete framework enabling accurate volcano activity monitoring applied on the well-known (http://www.willow.co.uk/html/telosb_mote_platform.php) WSN platform.

In [34], a biomedical application of wireless sensor networks is presented under the assumption that a small wireless node with an accelerometer is attached to a human wrist, like a wristwatch. The accelerometer provides instantaneous measurement of acceleration (caused by a person's movements) that is currently acting on the device. In this configuration, the accelerometer with accompanying algorithms can be used to classify the subject's movement into one of a few categories. This paper focuses on two threshold-based algorithms, which attempt to identify movements that are potentially harmful or indicative of immediate danger to a patient. The first algorithm seeks to identify rapid shaking movements that usually accompany myoclonic, clonic, and tonic-clonic seizures. Automated, quick seizure detection has the capability of alerting medical personnel when the patient is not physically capable of requesting assistance. The second algorithm is designed to generate an alarm when a patient has sustained an extended period of inactivity, potentially indicative of coma onset or loss of consciousness triggered by an acute brain injury. Like shaking movements, detecting inactive periods also has the potential to alert medical personnel to a problem more expediently than other means. Upon detecting an abnormal event, both algorithms trigger an auditory alarm from the wrist-device and transmit an alarm message (with necessary patient identification) through a ZigBee multi-hop wireless network to a patient monitoring station controlled by medical personnel.

NED (Noise-Tolerant Event and Event Boundary Detection) [38] represents a scheme able to identify events as well as the thresholds of events focusing on WSN networks. It is based on a moving average algorithm so as to minimize noise and a statistical method for event and event levels identification. A very interesting in-network event detection algorithm based on statistical metrics is presented in [39]. According to this approach the results of the algorithm are conveyed to the sink node (or cluster head node) so that the final evaluation is performed. Authors argue that their proposal is accurate, fault tolerant, and energy efficient through simulation-based evaluation.

A probabilistic model uses probability theory to model the uncertainty in acquired data. A probabilistic model describes a set of possible probability distributions for a set of observed data, and the goal is to use the observed data to convert the distribution (usually associated with parameters) in the probabilistic model that can best describe the current data.

Probabilistic event detection methods consist of those methods in which the probability of event occurrence and other related probabilities and parameters are computed and assessed based on specific preexisting assumptions, rather than based on computing and testing statistics derived from sample data sets. Ihler et al. [11], for instance, develop a probabilistic framework for unsupervised event detection and learning based on a time-varying Poisson process model that can also account for anomalous events. Their experimental results indicate that the proposed time-varying Poisson model provides a robust and accurate framework to adaptively separate unusual event plumes from normal activity. This model also performs significantly better than a non-probabilistic, threshold-based event detection technique.

In the context of this approach, probability theory is utilized to model and describe events able to be identified and thus accurately indicate the occurrence of the event. In that context spatio-temporal event detection (STED) [40] comprises a real time in-network event detection scheme able to detect events using a belief propagation technique. Resulting implementation has been evaluated both in the context of real WSN network (utilizing TmoteSky in a small scale network) and in the context of a simulation environment (able to configure a large scale network).

### *10.5.2 Pattern Matching-Based Approaches*

The main idea of this class is that events form a specific data pattern. Therefore, an event can be identified if the pattern of the real time acquired data match the event pattern. To achieve their objective respective techniques depend upon specific equations which are able to evaluate patterns or tendencies of data and thus decide on the occurrence or not of a specific event. These techniques share the following attributes: (1) are able to handle complex events while searching for data pattern formation using non-linear equations, (2) they are easily configurable and adaptable to a wide range of applications, (3) demand specialized knowledge for the correct configuration of the techniques' parameters, and (4) they usually lead to computational intense implementation due to their complexity.

#### 10.5.2.1 Prototype Matching Techniques

Prototype matching is a method of pattern recognition that describes the process by which a sensory unit registers a new stimulus and compares it to the prototype, or standard model, of said stimulus. Unlike template matching and feature analysis, an exact match is not expected for prototype matching, allowing for a more flexible model. An object is recognized by the sensory unit when a similar prototype match is found. A prototype is usually a vector of numbers derived from the sensors characteristics. Consequently, data acquired during the occurrence of an actual event are different from data acquired during a non-event time period. Therefore, comparing such prototypes with continuously acquired data can indicate the occurrence of specific event or events. In [1] the authors propose a distributed technique for event detection based on the construction and comparison of predefined such prototypes. In the context of this study prototypes are constructed during a training phase.Then following the integration of isolated decision by each particular node, events are identified or not at a wider scale. Respective implementation has been evaluated utilizing the ScatterWeb MSB-430 WSN platform in the context of a fence monitoring application scenario.

#### 10.5.2.2 Signature Matching Techniques

Signatures are created by converting data into feature domain. Having created signatures for specific events, acquired data corresponding to event occurrences can be reliably distinguished from the rest of the aggregated data. The difference between prototype and signature matching approaches concerns mainly the methodology utilized when data are transformed from one domain to another such as frequency or symbolic domain.

In [10] authors suggest a local event detection scheme utilizing principal component analysis (PCA) in order to extract the event signature. They also use a threshold to differentiate event data from no-event data. Respective implementation is evaluated on MicaZ nodes. Following another perspective Zoumboulakis et al. [41] define complex events as set of data points effectively defining a pattern and then events are identified utilizing symbolic aggregate approximation (SAX). The idea is to transform data into a symbolic domain and then based on minimum distance estimation between already acquired data patters and real time acquired data try to estimate whether an event has occurred. The implementation concerns local application and it is evaluated using a Matlab simulation environment. Another interesting approach is presented by Martincic et al. [42] by segmenting the whole network into cells. Then each cell by its own attempts to identify event occurrences by comparing cell acquired signatures to prerecorded event signatures. Signatures in this case are actually two-dimensional matrixes containing specific values characterizing the event. Consequently event identification is basically the process of comparing these types of matrixes to matrixes created from acquired data. In the context of WSNs this approach is evaluated considering the simulation environment of TinyOS assuming $2 \times 2$ and $5 \times 5$ sensor networks.

### *10.5.3 Artificial Intelligence and Machine Learning-Based Approaches*

As the title indicates, these techniques use AI methodologies (also known as machine learning) as the core foundation of event detection. Respective proposals can be further classified into supervised and non-supervised categories. The former category requires annotated data during a training phase while the latter does not require or assume any a priori knowledge. It is also worth noting that AI-based event detection algorithms share a similarity with pattern matching-based approaches in their objective of identifying data patterns of tendencies using non-linear equations. An advantage of paramount importance of AI-based techniques is that they don't require specialized knowledge into order to configure the approach's parameters. Instead such a scheme is able to configure and calibrate its own parameters through the knowledge extracted from acquired data. Additionally, with respect to WSNs, it must be highlighted that respective implementations tend to be less computational intensive compared to other approaches like the statistical model-based ones. Since events tend to follow a specific pattern, learning-based techniques comprise a very promising category in WSNs event detection algorithms because they can continuously increase their knowledge without human intervention. Therefore new trends in the context of decision support through accurate event detection increasingly exploit artificial intelligence-based algorithms.

#### **10.5.3.1 Supervised Learning**

As previously mentioned, schemes belonging in this category require annotated data and a well-defined training phase.

##### Fuzzy Logic Techniques

Event-oriented data aggregation (EDA) is a distributed approach for event detection based on fuzzy logic engines, applied on TelosB WSN platform and in the context of an ocean surveillance application scenario. Another fuzzy logic-based distributed scheme is proposed by Xuan et al. in [43]. This approach segments the network in clusters and then each cluster defines a confidence value. This confidence value represents the probability of an event occurring in each cluster. A local event detection scheme also based on fuzzy logic is presented in [13, 14]. As an application scenario, authors aim to evaluate the applicability of fuzzy logic in fire event detection focusing on home residents. A main objective of this approach is to be able to run independently on each WSN nodes without inter-node communication requirement.

#### Neural Network-Based Techniques

Neural networks consist of layers of interconnected nodes, each node producing a non-linear function of its input. The input to a node may come from other nodes or directly from the input data. Also, some nodes are identified with the output of the network. The complete network therefore represents a very complex set of interdependencies which may incorporate any degree of nonlinearity, allowing very general functions to be modeled. In the simplest networks, the output from one node is fed into another node in such a way as to propagate "messages" through layers of interconnecting nodes. More complex behavior may be modeled by networks in which the final output nodes are connected with earlier nodes, and then the system has the characteristics of a highly non-linear system with feedback. It has been argued that neural networks mirror to a certain extent the behavior of networks of neurons in the brain.

Neural networking approaches combine the complexity of some of the statistical techniques with the machine learning objective of imitating human intelligence; however, this is done at a more "unconscious" level and hence there is no accompanying ability to make learned concepts transparent to the user. A distributed event detection approach based on neural network is presented in [44]. This research effort aims towards forest fire event detection through WSN networks while fire event detection occurs both on the sensor and on the cluster head. Specifically on each sensor a threshold-based event detection algorithm approximation is executed, while the cluster head concentrates all indications and produces a final decision for the whole forest based on neural networks. The respective evaluation is simulation based and considers both small and large scale networks.

#### Bayesian Techniques

In a Bayesian network, the graph represents the conditional dependencies of different variables in the model. Each node represents a variable, and each directed edge represents a conditional relationship. Essentially, the graphical model is a visualization of the chain rule.

Bayesian networks are used from inference to prediction to modeling, whereas neural networks are used exclusively to predict. Compared to Bayesian, in a neural network, each node is a simulated "neuron." The neuron is essentially "on" or "off," and its activation is determined by a linear combination of the values of each output in the preceding "layer" of the network. A critical advantage of Bayesian networks compared to artificial neural networks (ANNs) with respect to WSNs concerns the surprisingly well performance assuming very small amounts of training data. On the other hand, ANNs are characterized by significantly higher complexity (in many cases not adequate for WSNs) allowing them to benefit from huge amounts of data compared to Bayesian-based methods which tend to exhibit a maximum performance over a threshold.

Such a technique based on a distributed version of the Bayes algorithm is presented in [45]. Authors indicate scenarios where faulty WSN nodes may exhibit specific patterns. The identification of such a pattern may be correlated to a specific event. It is mainly designed for large scale networks and is evaluated through simulation-based scenarios. A different approach is presented in [46] where an event detection method is presented based on merging utilizing the Bayesian approach where a Kalman Evaluator is utilized to evaluate missing data. Effectively, authors propose an outlier detection scheme based on Bayesian belief networks, which captures the conditional dependencies among the observations of the attributes to detect the outliers in the sensor streamed data. Data are then identified distributed following a Bayesian scheme. This approach is also evaluated in the context of simulation environment.

Finally, in [56] the authors propose an outlier detection scheme based on Bayesian belief networks, which captures the conditional dependencies among the observations of the attributes to detect the outliers in the sensor streamed data.

### Support Vector Machines

SVM classification method consists of two main components: a kernel function and a set of support vectors. The support vectors are obtained via the training based upon specific training data. New data are classified using a simple computation involving the kernel function and support vectors only. In [47], the authors solve the localization problem with the following modest requirements: (1) existence of beacon nodes, (2) a sensor may not directly communicate with a beacon node, and (3) only connectivity information may be used for location estimation (pairwise distance measurement not required). Requirement (1) is for improved localization accuracy. Requirement (2) relaxes the strong requirement on communication range of beacon nodes. Requirement (3) avoids the expensive ranging process pertaining to specialized sensorial equipment. All these requirements are reasonable for large networks where sensor nodes are of limited resources. The authors propose LSVM—a novel solution that satisfies the requirements above, offering fast localization, and alleviating the border problem significantly. LSVM is also effective in networks with the existence of coverage holes or obstacles. LSVM maps the network using the learning concept of support vector machines (SVM). With respect to the localization problem, a set of geographical regions is defined in the sensor field and each sensor node is classified into these regions. Then its location can be estimated inside the intersection of the containing regions. The training data is the set of beacons, and the kernel function is defined based on hop counts only.

Decision Trees

According to this classification method input data is traversing all possible branches of a learning tree [7, 48]. The goal of this process is to compare input data features in relation to different conditions until a specific category is reached. DT-based approach is particularly important for WSNs since they can effectively address respective challenges. Characteristics features applied in WSNs include loss rate, corruption rate, mean time to failure (MTTF), and mean time to restore (MTTR). Finally a critical requirement posed by such solutions concern the necessity for linearly separable data while the process of building optimal learning trees is NP-complete [49].

#### 10.5.3.2 Unsupervised Learning

Contrary to supervised learning approaches, unsupervised algorithms offer the critical advantage of not requiring annotated data, or training phase of any kind.

Fuzzy Adaptive Resonance Theory

The specific theory comprises a neural network category performing the work of clustering. Specifically it concerns the integration of fuzzy logic approach to an adaptive resonance theory algorithm [50] thus increasing the applicability of that algorithm. In [51] such an algorithm is presented in order to detect abnormal events, which is not the common case. In other words, abnormal events can be considered as outliers. Hence, the main idea is to cluster data and the cluster with the minimum population are considered events.

#### 10.5.3.3 Fixed Width Clustering

In [52] authors propose a constant amplitude clustering technique able to detect intrusion events. The idea of constant amplitude clustering aims towards data clustering among a dynamic number of groups during a training phase. Then based on the population of each group a decision is made about which group corresponds to an intrusion event. Following a training phase, data being closer to the intrusion group are identified as intrusion event. Another technique of this group is presented in [53]. Authors in this case propose that isolated nodes aggregate data and send them to a sink (or cluster) node where these groups are merged together in order to detect anomalies. Respective implementation is based on C++ programming language and evaluated on data from the Great Duck Island (sensor acquiring temperature, humidity, and atmospheric pressure) for detecting any kind of anomalies.

K-Means

K-Means is one of the simplest unsupervised learning algorithms and most widely used to solve the well-known clustering problem. K-Means effectively is a numerical, non-deterministic, and iterative method. K-Means clustering requires assigning the number of clusters K beforehand. Additionally, it partitions the data set into K separate groups and every group (cluster) is determined by the distances between vectors.

The main advantage of K-Means clustering is its pace, as in practice it requires only a few iterations to converge. Thus, it is a prominent candidate to run in the sensor node in real time, and is robust and relatively computationally efficient as long as the K value is adequately low. However, the disadvantages include a fixed number of clusters which can make it difficult to predict a K value which is sensitive to the presence of noise data and outliers because they influence the calculation of the clusters' centers. Randomly choosing an initial cluster center can result in different final clusters and as a consequence to an unsatisfactory result, due to different runs occurring for the same input data. Additionally, this method cannot handle the highly overlapping data because the clusters do not overlap. However, it is possible overcome most of these limitations in the preprocessing stage. K-Means is highly sensitive to the initial placement of the cluster centers. Because of initial cluster centers are chosen randomly, the K-Means algorithm cannot guarantee a unique clustering result. Additionally, choosing an ill-fitting initial placement of the cluster centers can lead to slow converge, empty clusters, and a higher chance of standing still in bad local minima [54].

In [55] authors use K-Means for detecting leak and burst events relying on offline techniques which collect loggers' data from multiple locations. We believe that detecting such events in real time, with smart sensors nodes, could improve monitoring operations and save operational costs.

## 10.6 Performance and Behavioral Characteristics of Event Detection Techniques

In order to select the adequate event detection techniques in WSN networks, various performance and behavioral characteristics must be taken into consideration. In this context aspects like the location where the algorithm is executed (i.e., distributed or centralized), time constrained requirements, scalability, and sensor characteristics (e.g., homogenous or heterogeneous data) are of paramount importance in a relative elicitation effort. Existing approaches vary greatly on addressing such issues, therefore they represent objective metrics enabling useful and practical comparison.

### *10.6.1 Processing Model of Event Detection*

Relative model describes how data are handled and the location(s) where the event detection actually takes place. Historically the approach followed in WSNs assumed that all sensor data are collected in a central network entity (sink node, base station, and cluster head) where data were being processed offline. The increasing demand for time constrained (or even real time) performance in contemporary WSNs has effectively render such centralized approaches inadequate. Therefore, nowadays two different models are attracting most of the research interest, specifically (1) local processing mode and (2) distributed processing model.

A local processing model assumes that all processing occurs in each isolated node based on each node own capabilities. Therefore, in such implementation no communication (or limited) is actually required. On the contrary in the context of the distributed processing model events are actually detected following a cooperative approach entailing specific communication collaboration amongst nodes. Thus the fundamental idea of distributed processing model is a collaboration of multiple nodes towards accurate and reliable event detection.

### *10.6.2 Technique's Scalability*

The scale of an actual WSN can drastically vary from few tens to many hundreds or even thousands of nodes depending on the specific application scenario. Consequently selected approach must be considering this parameter or be adaptive to changes of this parameter. Therefore, when a very small scale network is considered then local data processing could be preferred over a distributed approach posing some overhead (communication and computation) without actually enhancing performance. However, when the scale of the network, as well as complexity increases, distributed approaches are clearly more preferable offering better adaptability and enhanced accuracy and reliability.

### *10.6.3 Sensor Data Types*

The degree of data homogeneity or heterogeneity can play a critical role on event detection techniques selection. Event detection technique design able to efficiently handle heterogeneous data is quite challenging since the increased sensor space availability analogously augments the dimension of data to be handled thus requiring sophisticated data processing approaches.

### 10.6.4 Time Constrained Performance Demands

In demanding application scenarios the delay between an event occurrence and event detection can be of cornerstone importance for performance as well as safety (of either equipment, material, or even personnel). Therefore, respective time constrained requirements (implying that a deadline miss leads to performance degradation) and real time requirements (where deadline miss could lead to a system failure) are increasingly required in domains like industry, health, hazardous environments monitoring, etc. Therefore a critical metric of real implementation concerns the capability of an approach to actually meet such demands.

### 10.6.5 Density

A parameter that can drastically affect event detection performance has to do with the expected network node density. It is worth noting that despite its importance and degree of influence it is usually omitted when evaluating respective solutions. On the contrast authors usually focus solely on scalability which, however, only partially cover this aspect.

### 10.6.6 Evaluation Approach

A critical aspect when characterizing a technique is the framework, infrastructure, and methodology used to actually evaluate it since these pertain to the techniques usefulness, objectivity, and most importantly the applicability with respect to the particular characteristics of a WSN network. Therefore, in many cases respective studies emphasize the development of the proposed algorithm in the context of real WSN development platforms in order to validate the proposal's feasibility. It is also important to offer details on the computation cost, the communication cost, and the space complexity of the implementation. Such aspects are critical since they pertain to the resource expenditure of the solution, required to efficiently execute required algorithms. The majority, however, of existing proposal and studies focus on simulation-based evaluation efforts. Such efforts, however, fall short with respect to realistic measurements, feasibility validation, and taking into consideration the dynamic and stochastic behavior of a real network environment especially in wireless sensor network deployments.

## 10.7 Conclusions

Having discussed the most prominent approaches and highlighted all respective main characteristics of each event detection class, this section aims to summarize our survey and extract useful conclusions. We also hope that this effort can serve as a roadmap for future research efforts in this research domain. Undoubtedly event detection comprises a prominent research domain in WSNs, since by definition in most realistic commercial WSN application scenarios, the main objective is to recognize specific situations, as opposed to, for example, continuous streaming of raw data. However, for respective solutions to be effective careful consideration and adequate attention must be paid to WSN specific characteristics and even more scares resource limitation drastically differentiating them from any other, resource rich, wireless technology. Despite, the criticality of this requirement it is found that many proposal omit to take it into consideration failing to offer sufficient performance and behavioral characteristics. Probably the most compelling characteristic stemming from the aforementioned requirements has to do with optimal task allocation amongst nodes clearly advocating distributed strategies especially in wide scale WSN deployments. Furthermore, from the elicitation effort devoted in this chapter it has been extracted that applicability of artificial intelligence comprises one of the most prominent approaches, future event detection algorithms should pursue offering enhanced capabilities with respect to flexibility and adaptability to a wide range of real life application scenarios. Additionally, existing implementation exhibits advanced performance in terms of accuracy and reliability compared to the rest of classification categorizes. Finally, an aspect often overlooked but drastically influencing the added value of any relative proposal pertains to the validation/evaluation environment selected. In that respect the majority of the efforts rely on simulation-based environments which although offering specific advantages in early stages of design and development or preliminary performance indications in wide scale networks suffer from low objectivity, accuracy as well as myopic consideration of the dynamic nature of real WSNS. Therefore, in conjunction to the wide availability of powerful WSN platform nowadays, we believe that real life experimentation of proposed solutions should be an indispensable part of any relative research or/and development effort.

## References

1. G. Wittenburg, N. Dziengel, C. Wartenburger, J. Schiller, A system for distributed event detection in wireless sensor networks, in *Proceedings of the 9th ACM/IEEE International Conference on Information Processing in Sensor Networks* (ACM, New York, 2010), pp. 94–104
2. M. Bahrepour, *Artificial Intelligence Based Event Detection in Wireless Sensor Network* (University of Twente, Enschede, 2013)

3. M.C. Kerman, W. Jiang, A.F. Blumberg, S.E. Buttrey, *Event Detection Challenges, Methods, and Applications in Natural and Artificial Systems*, 14th International Command and Control Research and Technology Symposium (ICCRTS), Jun 15–17, 2009, Washington, DC
4. D. Li, K.D. Wong, Y.H. Hu, A.M. Sayeed, Detection, classification, and tracking of targets. IEEE Signal Process. Mag. **19**(2), 17–29 (2002). ISSN 1053-5888. doi:10.1109/79.985674
5. M. Zoumboulakis, Pattern matching and detection in extremely resource constrained wireless sensor networks. Ph.D. Thesis, 2011
6. N.D. Phung, M.M. Gaber, U. Rhm, Resource-aware online data mining in wireless sensor networks, *in Proceedings of the 2007 IEEE Symposium on Computational Intelligence and Data Mining (CIDM'2007)*
7. M.A. Alsheikh, S. Lin, D. Niyato, H.-P. Tan, Machine learning in wireless sensor networks: algorithms, strategies, and applications. IEEE Commun. Surv. Tutorials **16**(4), 1996–2018 (2014)
8. A. Saoji, P. Lambhate, Survey paper on event detection techniques in wireless sensor network for disaster recovery. Int. J. Eng. Res. Technol. **2**(12), 120–124 (2013)
9. M. Bahrepour, N. Meratnia, P.J.M. Havinga, Automatic fire detection: a survey from wireless sensor network perspective. http://doc.utwente.nl/65223/, Dec 2008
10. J. Gupchup, A. Terzis, R.C. Burns, A.S. Szalay, Model-based event detection in wireless sensor networks, in *Proceedings of the Workshop on Data Sharing and Interoperability on the World-Wide Sensor Web (DSI)* (2007)
11. A. Ihler, J. Hutchins, P. Smyth, Adaptive event detection with time-varying Poisson processes, in *The Twelfth International Conference on Knowledge Discovery and Data Mining (Association for Computing Machinery)* (2006). Retrieved from 28 Feb 2008
12. T.B. Trafalis, H. Ince, M.B. Richman, Tornado detection with support vector machines. Lect. Notes Comput. Sci. **2660**, 708 (2003)
13. K. Kapitanova, S.H. Son, K.-D. Kang, Using fuzzy logic for robust event detection in wireless sensor networks. Ad Hoc Netw. **10**(4), 709–722 (2012)
14. K. Kapitanova, S.H. Son, K.-D. Kang. Event detection in wireless sensor networks—can fuzzy values be accurate? in *Ad Hoc Networks* (Springer, Berlin/Heidelberg, 2010)
15. A. Tavakoli, J. Zhang, S. Son, Group-based event detection in undersea sensor networks, in *The Second International Workshop on Networked Sensing Systems* (2005)
16. H. Medjahed, D. Istrate, J. Boudy, B. Dorizzi, Human activities of daily living recognition using fuzzy logic for elderly home monitoring, in *IEEE International Conference on Fuzzy Systems, 2009. FUZZ-IEEE 2009,* Jeju Island, 20–24 Aug 2009
17. E.F. Nakamura, A.A.F. Loureiro, A.C. Frery, Information fusion for wireless sensor networks: methods, models, and classifications. ACM Comput. Surv. **39**(3) (2007). doi:10.1145/1267070.1267073
18. J. Žurauskienė et al., A graph theoretical approach to data fusion. bioRxiv (2015). doi:10.1101/025262
19. J.I.-Z. Chen, Y.-N. Chung, A data fusion methodology for wireless sensor systems. Int. J. Comput. Commun. Control **VII**(1), 39–52 (2012). ISSN 1841-9836, E-ISSN 1841-9844
20. N. Ahmed, Y. Dong, T. Bokareva, S. Kanhere, S. Jha, T. Bessell, M. Rutten, B. Ristic, N. Gordon, Detection and tracking using wireless sensor networks, in *Proceedings of the 5th International Conference on Embedded Networked Sensor Systems* (2007). doi:10.1145/1322263.1322328
21. P. Zou, Y. Liu, An efficient data fusion approach for event detection in heterogeneous wireless sensor networks. Appl. Math. Inf. Sci. (2015) doi:10.12785/amis/090160
22. P. Manjunatha, A.K. Verma, A. Srividya, Multi-sensor data fusion in cluster based wireless sensor networks using fuzzy logic method, in *IEEE Region 10 and the Third international Conference on Industrial and Information Systems, 2008 (ICIIS 2008)*, 8–10 Dec 2008, pp. 1–6. doi:10.1109/ICIINFS.2008.4798453
23. F. Castanedo, A review of data fusion techniques. Sci. World J. **2013**, 19 pp. (2013), Article ID 704504. doi:10.1155/2013/704504

24. S.B. Kotsiantis, Supervised machine learning: a review of classification techniques, in *Proceedings of the 2007 Conference on Emerging Artificial Intelligence Applications in Computer Engineering*
25. M. Donald, D.J. Spiegelhalter, C.C. Taylor, J. Campbell (eds.), *Machine Learning, Neural and Statistical Classification* (Ellis Horwood, Upper Saddle River, NJ, 1995)
26. M. Baqer, A.I. Khan, Event detection in wireless sensor networks using a decentralised pattern matching algorithm. White Paper (2008)
27. C. Zhang, C. Wang, D. Li, X. Zhou, C. Gao, Unspecific event detection in wireless sensor networks, in *International Conference on Communication Software and Networks, 2009 (ICCSN '09)* (2009)
28. A. Khelil, F.K. Shaikh, B. Ayari, N. Suri, Mwm: a map-based world model for wireless sensor networks, in *Proceedings of the 2nd International Conference on Autonomic Computing and Communication Systems, Autonomics '08*, ICST (Institute for Computer Sciences, Social-Informatics and Telecommunications Engineering) (2008)
29. M. Li, Y. Liu, L. Chen, Non threshold-based event detection for 3D environment monitoring in sensor networks. IEEE Trans. Knowl. Data Eng. **20**(12), 1699–1711 (2008)
30. M.C. Kerman, W. Jiang, A.F. Blumberg, S.E. Buttrey, The application of a quantile regression metamodel for salinity event detection confirmation within New York harbor oceanographic data. J Oper Oceanogr **2**(1), 49–70 (2009)
31. C.T. Vu, R.A. Beyah, Y. Li, Composite event detection in wireless sensor networks, in *Performance, Computing, and Communications Conference, 2007 (IPCCC 2007)* (2007), IEEE International, pp. 264–271
32. A.V.U. Phani Kumar, V. Adi Mallikarjuna Reddy, D. Janakiram, Distributed collaboration for event detection in wireless sensor networks, in *Proceedings of the 3rd International Workshop on Middleware for Pervasive and Ad-hoc Computing*, ACM (2005)
33. G. Werner-Allen, K. Lorincz, J. Johnson, J. Lees, M. Welsh, Fidelity and yield in a volcano monitoring sensor network, in *Proceedings of the 7th Symposium on Operating Systems Design and Implementation*. USENIX Association (2006)
34. T.R. Burchfield, S. Venkatesan, Accelerometer—based human abnormal movement detection in wireless sensor networks, in *Proceedings of the 1st ACM SIGMOBILE International Workshop on Systems and Networking Support for Healthcare and Assisted Living Environments* (2007), ACM, New York, pp. 67–69
35. A.O. Sykes, An introduction to regression analysis. Chicago Working Paper in Law & Economics (1993), University of Chicago Law School. Retrieved 4 Oct 2008
36. J. Sauvageon, A.M. Agogino, A.F. Mehr, I.Y. Tumer, Comparison of event detection methods for centralized sensor networks, in *Proceedings of the 2006 IEEE Sensors Applications Symposium, 2006*
37. G. Welch, G. Bishop, *An Introduction to the Kalman Filter* (University of North Carolina at Chapel Hill, Chapel Hill, 2006). TR 95-041
38. G. Jin, S. Nittel, Ned: an efficient noise-tolerant event and event boundary detection algorithm in wireless sensor networks. in *7th International Conference on Mobile Data Management, 2006 (MDM 2006)* (2006)
39. C.V. Easwaran, An efficient in-network event detection algorithm for wireless sensor nodes, in *Novel Algorithms and Techniques in Telecommunications, Automation and Industrial Electro Electronics* (Springer, The Netherlands, 2008). ISBN 978-1-4020-8736-3
40. J. Yin, D.H. Hu, Q. Yang. 2009. Spatio-temporal event detection using dynamic conditional random fields. In Proceedings of the 21st international jont conference on Artifical intelligence (IJCAI'09), ed. by H. Kitano. Morgan Kaufmann Publishers Inc., San Francisco, CA, pp. 1321–1326
41. M. Zoumboulakis, G. Roussos, Escalation: complex event detection in wireless sensor networks, in *Smart Sensing and Context* (Springer, Berlin/Heidelberg, 2007). ISBN 978-3-540-75695-8
42. F. Martincic, L. Schwiebert, Distributed event detection in sensor networks. *International Conference on Systems and Networks Communication* (2006)

43. K.-X. Thuc, K. Insoo, A collaborative event detection scheme using fuzzy logic in clustered wireless sensor networks. AEU Int. J. Electron. Commun. **65**(5), 485–488 (2011)
44. L. Yu, N. Wang, X. Meng, Real-time forest fire detection with wireless sensor networks, in *Proceedings. 2005 International Conference on Wireless Communications, Networking and Mobile Computing* (2005)
45. B. Krishnamachari, S. Iyengar, Distributed Bayesian algorithms for fault-tolerant event region detection in wireless sensor networks. *IEEE Transactions on Computers* (2004)
46. M. Moradi, J. Ghaisari, J. Askari, M. Gorji, A new method for detection of a distributed event in wireless sensor networks, in *2011 19th Iranian Conference on Electrical Engineering (ICEE)* (2011)
47. D.A. Tran, T. Nguyen, Localization in wireless sensor networks based on support vector machines. IEEE Trans. Parallel Distrib. Syst. **19**(7), 981–994 (2008)
48. T.O. Ayodele, Types of machine learning algorithms, in *New Advances in Machine Learning* (InTech, Rijeka, 2010)
49. S.R. Safavian, D. Landgrebe, A survey of decision tree classifier methodology. IEEE Trans. Syst. Man Cybern. **21**(3), 660–674 (1991)
50. C.P. Smith, Fuzzy adaptive resonance theory: applications and extensions. Masters Theses, 2015
51. A. Kulakov, D. Davcev, Tracking of unusual events in wireless sensor networks based on artificial neural-networks algorithms, in *International Conference on Information Technology: Coding and Computing (ITCC'05)—Volume II* (2005)
52. C.E. Loo, M.Y. Ng, C. Leckie, M. Palaniswami, Intrusion detection for routing attacks in sensor networks. Int. J. Distrib. Sens. Netw. **2**(4), 313–332 (2006)
53. S. Rajasegarar, C. Leckie, M. Palaniswami, J.C. Bezdek, Distributed anomaly detection in wireless sensor networks, in *2006 10th IEEE Singapore International Conference on Communication Systems (ICCS 2006)* (2006)
54. M.E. Celebi, Improving the performance of K-means for color quantization. Image Vis. Comput. **29**, 260–271 (2011)
55. M.M. Mohamed Ibrahim, W. WU, Lightweight unsupervised event detection approach in wireless smart sensors for monitoring water distribution system, in *E-proceedings of the 36th IAHR World Congress 28 June–3 July, 2015*
56. D. Janakiram, V. Adi Mallikarjuna Reddy; A.V.U. Phani Kumar, Outlier detection in wireless sensor networks using Bayesian belief networks, in *2006 First International Conference on Communication System Software and Middleware (Comsware 2006)* (2006), pp. 1–6. doi:10.1109/COMSWA.2006.1665221

# Part IV
# Efficient Data Management and Decision Making for IoT

# Chapter 11
# Integrating IoT and Fog Computing for Healthcare Service Delivery

**Foteini Andriopoulou, Tasos Dagiuklas, and Theofanis Orphanoudakis**

## 11.1 Introduction

Internet of Things (IoT) is a network paradigm consisted of a huge variety of network connected devices named *Things* such as wearable sensors, wireless sensors, and mobile devices. IoT enables things to be connected, generate data and interact with each other remotely as well as with people [6, 10]. Due to the increasing number of the medical equipment, sensor, and mobile devices that are interconnected through the Internet, IoT has gained the attention of the healthcare domain. IoT is considered a promising solution for the healthcare industry since it can move the treatment process away from hospitals and provide to patients the ability to access care remotely, self-manage their own disease, and receive assistance in emergency cases through a mobile (and potentially ubiquitous) infrastructure [2, 10, 24]. However, the sensor and smart devices used for monitoring patients' current healthcare status have limited computation and storage capabilities, therefore they are not able to deal with the large amount of generated data. IoT poses many challenges in terms of data exchange, interoperability, and availability of resources as well as security and privacy [6, 18, 24].

The cloud computing approach was adopted to address the aforementioned challenges posed by IoT [9, 18]. The cloud computing paradigm provides the functionality in order to store, process, and manage the increasing amount of data generated by IoT autonomous systems through convenient, on-demand access to a

F. Andriopoulou (✉) • T. Orphanoudakis
Hellenic Open University, Patras 26335, Greece
e-mail: fandriopoulou@eap.gr; fanis@eap.gr

T. Dagiuklas
Division of Computer Science and Informatics London South Bank University,
103 Borough Road, London, SE1 0AA
e-mail: tdagiuklas@lsbu.ac.uk



G. Keramidas et al. (eds.), *Components and Services for IoT Platforms*,
DOI 10.1007/978-3-319-42304-3_11

shared pool of resources [23]. Cloud computing offers many advantages from the user point of view as it provides ubiquitous access to the content and resources (hardware, data, infrastructure, applications, and service development) without any need to keep large storage files and maintain powerful computing devices [9]. Through the cloud computing approach, everything is available as a service over the Internet. The patient can use his/her smartphone as an interface to connect and have access to the remote medical data or store new content to cloud data centers [12, 27]. Even though the centralized cloud approach enables healthcare professionals to have ubiquitous access to patients' medical records away from hospitals and share large amounts of content even through their smartphones, the centralized nature of the cloud restricts itself from providing low latency, location awareness, and geo-distribution that are crucial for the IoT applications [41].

As every device and appliance can potentially be connected to the Internet, it is predicted that the number of mobile-connected devices will exceed the number of people on earth and by the end of 2018 there will be 1.4 mobile devices per person [11] while in 2020, 24 billion devices are expected to be connected in Internet [18]. Though the cloud computing paradigm provides high scalability and can address the exponential growth of the IoT devices, the need of numerous devices to be connected to the Internet, communicate with each other, transmit continuously data and retrieve medical records, stresses further the network and cloud infrastructure. The cloud computing approach requires constant high speed reliable Internet connectivity with sufficient bandwidth and low latency [16]. In case that either the bandwidth is insufficient or any other network failure occurs, cloud can lead to increased delays and become a single point of failure that the healthcare service provisioning cannot tolerate. Moreover, the continuous communication of the IoT devices with the cloud increases energy consumption. Therefore, cloud computing cannot satisfy the requirements of IoT applications for real-time information processing since it demands continuous and reliable interactions among the IoT technologies and healthcare services with low latency. While healthcare service provisioning becomes more dependent on the network connectivity, network failures may interrupt or delay the healthcare service delivery with adverse effects in patients' quality of life even leading to mortality.

This chapter deals with the aforementioned challenges posed by the cloud computing through the integration of the IoT technologies and fog computing paradigm in order to provide reliable healthcare service provisioning. The concept of transferring the computing intelligence closer to the user, to the edge network, seems to be the most promising solution for reliable healthcare service provisioning [35]. Fog computing approach operates on network edge and acts as a bridge between the end-user devices and the cloud infrastructure of the healthcare provider. An architectural model is presented that shows how integration is achieved through the IoT, fog, and cloud technologies cooperation. The main component of the system model is the fog server that is a virtualized platform based on cloud computing approach with limited operations, deployed closer to the terminal equipment in order to enable protocol conversion, data storage, processing, and evaluation. As fog computing is deployed at the edge network, it can provide enhanced features including user mobility support, location awareness, dense geographical distribution, low latency, and delays that could not be supported inherently by the cloud computing approach. These characteristics are significant for provisioning delay-sensitive services such as healthcare and emergency services. In this context, the successful integration of the IoT-enabled technology with the fog computing approach can provide a number of benefits such as scalability, less demands for bandwidth, location awareness, reduction of data traffic, improved quality of service (QoS) in terms of low response time, and greater user experience. As a consequence, the integration of IoT and fog computing for healthcare purposes can provide numerous advantages such as faster and accurate treatment delivery, reduction of medical costs, improvement of doctor–patient relationships, and the delivery of personalized treatment oriented to users' needs and preferences. A number of use cases illustrate the opportunities and the benefits that the integration of IoT, fog, and cloud computing approach offers for cost effective, efficient, timely, and high-quality ubiquitous healthcare service delivery. These use cases include daily monitoring and healthcare service provisioning as well as extended eCall service delivery.

The rest of the chapter is organized as follows. Section 11.2 describes briefly the state of the art and presents the need for integration among various technologies in order to provide value-added healthcare services to the end users. Section 11.3 presents an overview of the fog computing. A definition and the key characteristics of the fog computing approach are presented as well as the benefits of the integration with the IoT-enabled technologies. In Sect. 11.4, it is presented an overview of the integration of IoT–fog and cloud computing. The system model and the fog server architecture are described. Section 11.5 illustrates a set of use cases that illustrate the benefit of the IoT–fog and cloud integration for provisioning healthcare services in real-time conditions. Finally, Sect. 11.6 concludes the chapter.

## 11.2 Related Work

Hospitals were the first organizations that incorporated the computing systems for delivering healthcare applications based on the Hospital Information System (HIS) architecture [43]. Since the evolution of the ICT and cloud computing has dramatically changed the way the healthcare services are delivered. The development of ambient intelligent systems empowered patients to self-manage their own disease and provide well-being application solutions away from hospitals. Many research works have been published and a variety of applications have been implemented ([13, 17, 25, 30, 42, 44], etc.) that enabled patients use medical sensors, wearable devices, and actuators so as to monitor vital bio-signals such as blood pressure and body temperature. Moreover, sensors are also used in the home environment in order to monitor patients' activities (i.e., internet connected cameras, accelerometers, gyroscopes, etc.) and environmental conditions (i.e., temperature, humidity, luminosity, etc.). The information from sensory devices from both medical and environmental is transmitted to remote medical centers (in the cloud) where the information is processed and stored in order to be accessible by the healthcare professionals. Mobile applications and commercial products have also been developed that enable wearable sensors such as smartphones and watches to gather bio-signals or control the movement tracking of user's daily activities (i.e., calories burned, heartbeat, steps walked, etc.), providing suggestions, and prevent aggravation of user's healthcare conditions (iFall [32], iWander [33], Philips Hospital to Home [26], Vitaphone [39], and Bosch Telemedicine [8]).

The evolution of the IoT technologies expanded further the opportunities of the healthcare domain [10]. The RFID technology, key component of the IoT technology, allowed any object in terms of sensor devices, patients, or appliances in patent's environment to be tracked and monitored automatically using an RFID tag. RFID technology enables the identification of anything supporting location awareness and geo-distribution [6]. Lu and Liu [22] proposed the insertion of the IoT technology for healthcare purposes in China. They provided the structure and the functionality of the IoT in order to provide hospital care and remote real-time ECG monitoring. Using the RFID technology the healthcare professionals are able to retrieve medical data from the bedside easily for each patient and provide immediately assistance whenever an emergency episode occurs. Jara et al. [20] proposed an IoT based personal diabetes management system in order to monitor patient's blood glucose concentration and provide blood glucose management and insulin therapy. The proposed architecture offers a set of mobile assistance services for healthcare professionals in order to configure and reporting patient's health status based on RFID. It aims to prevent and anticipate the hyperglycemia and/or hypoglycemia episodes and evaluate the risks of new medications for insulin therapy. In [29], the authors proposed an IoT approach for supporting rural healthcare applications. A patient wears an active RFID sensor and whenever the sensor detects an abnormal value, the doctor of the medical center is informed. Doctor can have access to patients profile using the RFID that identified the user. This approach could be adopted for continuous healthcare monitoring since it eliminates in half the energy consumption. Lee and Ouyang [21] proposed an IoT based intelligent application for risk factor assessment. The risk factor is calculated through the communication and cooperation of various IoT devices with correlated data. NightCare system [3] is an ambient intelligent platform based on the RFID tags for monitoring the state of a person during the night. RFID sensors are placed into the bed, underneath the bed, the carpet, and the user's clothes or integrated with other RFID sensors (i.e., humidity, fever, etc.) in order to detect and report abnormal events such as the presence or absence of the user in the bed and demand assistance.

Although the wide acceptance of the RFID technology and the variety of healthcare applications have been developed, there are limited efforts towards realizing remote healthcare service delivery since it requires the integration of the Wireless Sensor Networks (WSNs) with the RFID sensors [2]. This integration faces the interoperability and translating challenges among the heterogeneous devices and the multiple network protocol, respectively. In [45], the authors proposed an IoT gateway (i.e., home-gateway, mobile phone, set-of-box, etc.) in order to deal with the interoperability and integration of the heterogeneous IoT devices. These gateways provide a set of functionalities such as interoperability and translating between the protocols used in various IoT devices. iHome system [40] is an intelligent home based platform for providing remote monitoring of elderly people. The iHome consists of the iMedBox, the iMedPack, and the Bio-Patch. The iMedPack and Bio-Patch are responsible for reminding and monitoring services while the iMedBox acts as a gateway that provides interoperability and IoT network connectivity. Cubo et al. [12] proposed an IoT platform based on the cloud approach for Ambient Assisted Living (AAL) applications that enables remote monitoring and access of the data generated by sensor and android devices. The data generated by the IoT devices are aggregated to a gateway that is responsible to transfer the data to the cloud based platform for processing and remote access. To face the heterogeneity issues that appeared from the different devices, the gateway incorporated the service oriented approach in order to orchestrate the devices according to their behavior and determine the order of the exchanged messages. Tabish et al. [37] proposed an indoor and outdoor remote ECG monitoring system that supports patients for normal conditions and emergency cases. In emergency cases, the system provides live streaming remote monitoring and reading of ECG signals even if it consumes a lot of energy while in daily conditions the data are stored locally to a smart gateway and then are transmitted to the cloud server so as to be easily reached by the healthcare professionals.

As the number of the IoT devices has grown rapidly, the communication among the IoT devices and the huge amount of data that are transmitted to the cloud via the internet have increased stretching the network and cloud infrastructure that cannot satisfy all the requirements of QoS and resource allocation. Fog computing [16, 41] was introduced in order to reduce the data transferring to the cloud and the unnecessary communication of all the things that can connect to the internet and generate data. Aazam et al. [1] proposed the data preprocessing and trimming before sending to the cloud. A smart gateway is co-located with a smart network, the fog that performs temporary data storage, data filtering, and processing with a secure way and according to predefined rules. Fog connects to the cloud in order to upload the data and then erase the temporary data. In this context, Shi et al. [31] discussed the characteristics and the features of the fog computing and its benefits for the healthcare domain. Rahmani et al. [28] proposed an intelligent gateway that enhances the functionality of the smart gateway and offers local storage, real-time local processing, and data mining services. The smart e-Health gateway acts as a bridging point among the medical sensors, the smart appliances in the home of the user or the hospital, and the cloud platform and copes with interoperability, reliability, and scalability issues. The smart e-Health gateway can be considered as a fog computing approach and provides both network interoperability and local data processing before the data are transferred to the cloud server. Stantchev et al. [34] proposed a three-layer architecture for supporting elderly people at home. It integrates IoT, cloud, and fog computing approach enabling any kind of sensors to connect to the infrastructure. Fog devices provide tasks such as data computing, data storage, local management of the sensors, and handle mobility. Then, fog device transmits the data to the cloud for further evaluation from the doctors. In this approach, the fog device acts as a backup point in case the link to the cloud is faulty.

## 11.3 Fog Computing

Fog computing is a new term coined by *Cisco* [7] in order to describe the act of bringing cloud computing closer to the end user. Deploying services closer to the user's proximity can provide various advantages including reduction in congestion and latency, elimination of data transmission in the network infrastructure, improvement of scalability, flexibility, and reassure the security of the encrypted data. Consequently, the fog computing paradigm can provide efficient delay-sensitive and healthcare related services that require reliable data transfer with low latency [16].

### *11.3.1 Definition*

The rapidly growing number of IoT devices that generate an increasing amount of data and consume cloud services have forced many ICT operators to introduce different approaches in order to provide powerful computing models located closer to the end user and confront the network strains of data transmission to the clouds.

Cisco defines fog computing as a virtualized platform, built at the edge of the network, providing computing, storage, and networking services between the end devices and the cloud servers. Fog computing can be considered an extension of the cloud computing approach as it incorporates various characteristics and features of cloud computing paradigm [7].

Vaquero and Roberto-Merino [38] defined fog computing as "a scenario where a huge number of heterogeneous (wireless and sometimes autonomous) ubiquitous and decentralized devices communicate and potentially cooperate among them and with the network to perform storage and processing tasks without the intervention of third parties. These tasks can be for supporting basic network functions or new services and applications that run in a sandboxed environment. Users leasing part of their devices to host these services get incentives for doing so."

Until now, there is not any standardized definition and architecture of fog computing despite the fact that many ICT operators are developing architectures that aim to transfer the data aggregation to an edge server. Most of these architectures use edge servers as dumb intermediate devices to the cloud. However, fog computing has more capabilities than an intermediate device since it is an extension of the cloud approach. Therefore, it should be defined appropriately in order to avoid confusion with other technologies.

### 11.3.2 Characteristics

Fog computing is a cloud computing approach that extends the capabilities of the cloud. Similar to the cloud, it provides features for on-demand provisioning of storage, computing, and network resources. But it is differentiated from cloud in three main characteristics: (1) its proximity to the end user, (2) its dense geographical distribution, and (3) support of user's mobility. These features could not be supported inherently by the cloud computing approach due to its centralized structure and its physical distance from the end user. The key characteristics of fog computing are analyzed as follows:

- Proximity to the user: Fog computing is located to the edge network, thereby, it is physically closer to the end devices that generate data. This means that the IoT devices and the fog are in the same Local Area Network (LAN) that enables them to exchange information with low delays. This feature enables fog to reduce the latency, delay, and jitter which are critical requirements for provisioning reliable delay-sensitive applications such as healthcare service delivery and emergency services.
- Dense geographical distribution: The wide geographical distribution of fog approach provides many advantages compared to the centralized cloud deployment. Data is located, stored, and processed at the most appropriate edge server in the network. The closest the data are stored and processed to the server, the shortest routing and less transmissions through routers and switches are taking place decreasing the probability to be under attack. Hence, the distribution of the data and the data storage and processing closer to the edge of the network increases the security of the sensitive medical data and improves flexibility and agility. Such agility and flexibility are necessary for IoT applications that require low latency.
- Mobility support: Fog computing can support user's mobility and provide location awareness. This is achieved by means of geographical distribution and its location at the edge network. The edge location of the fog can provide various network and context information aggregated by network traffic and analytics as well as from numerous IoT devices that can communicate with each other. Location awareness is a key feature for the healthcare service provisioning in order to support user's mobility and provide a wide range of mobile personalized applications.

These key features provide the main advantages of the fog computing contrary to the cloud computing approach. Table 11.1 shows the main differences between the fog and cloud computing and their impact in provisioning healthcare services. Due to the geographical distribution and the proximity closer to the end devices, fog can support user's mobility, location awareness, the reduction of latency, delay and jitter, elimination of data transmission in the network infrastructure, improvement of scalability, flexibility, and reassure the security of the encrypted data. However, fog has limited computational power, therefore, it cannot replace cloud computing.

**Table 11.1** Cloud computing vs fog computing [16]

| Requirements | Fog computing | Cloud computing |
|---|---|---|
| Mobility | High | Limited |
| Geographical distribution | Distributed | Centralized |
| Location of server | At the edge of the network | Within the Internet |
| Distance between the client and server | In the physical proximity (one hop) | Faraway (multiple hops) |
| Geographic coverage | Wide | Global |
| Location awareness | Yes | No |
| Latency | Low | High |
| Delay Jitter | Low | High |
| Bandwidth | Low | High |
| Response Time | Seconds to minutes | Minutes, days, or weeks |
| Hardware | Limited storage/compute resources | Scalable storage/compute resources |
| Data storage | Temporary | Permanent |
| Security | Can be defined | Undefined |
| Flexibility | High | Limited |
| Agility | High | Limited |
| Denial of Service (DoS) attack | Low probability | High probability |
| Type of last mile connectivity | Wireless | Leased line |

The integration and cooperation between them is necessary in order to enable big data analysis and support the creation, deployment, and execution of new healthcare services in order to provide personalized healthcare services oriented to users' needs and preferences based on their current health status.

### 11.3.3 Benefits of Integration with the IoT Technology for Healthcare Service Provisioning

As discussed previously, the fog computing approach is an extension of cloud computing that offers on-demand storage, compute, and network resources between the end devices and the cloud. Since fog has the ability to provide storage and computing functionality closer to the end devices, it is enabled to aggregate, process, and store a huge amount of information enabling real-time analysis. Due to the fact that medical sensors generate data with high frequency, the performance of the real-time analysis can be improved providing intelligent data analysis and decision making according to local policies and the network resources available to the end users.

The proximity to the end user can enable fog to handle data immediately without network constraints. The location of the fog closer to the end user reduces the latency, delay, and jitter. Thereby, it can improve the quality of the healthcare service and can provide timely response actions and reliable healthcare services without the need to transfer data to the cloud.

Mobility is a significant characteristic for providing ubiquitous healthcare services since the majority of the end users are mobile users that demand to have access and receive healthcare services wherever they are located. The dense geographical distribution of fog computing can provide mobility to the end users and can support properly the wide distribution of the medical data and the IoT devices.

Therefore, the integration of IoT and fog computing for provisioning healthcare services represents a promising solution that can significantly reduce the data forwarding and routing across the network infrastructure, resulting in reduced latency and reliable healthcare service provisioning with enhanced quality.

## 11.4 Integration of IoT–Fog and Cloud

Ubiquitous healthcare applications generate numerous sensor data that ought to be managed properly by means of timely processing and analysis. As mentioned in Sect. 11.3.2, fog computing can provide low latency and delay that lead to timely service provisioning since it is located to the edge network, closer to the end user. This section presents the system model of the integration between IoT–fog and cloud computing as well as the architecture of the fog server, the main component of the system that provides the basic functionality of protocol conversion, data storage, processing, and analysis.

### *11.4.1 System Model*

The main components of the architecture are

(a) Fog node: Fog nodes support multiple communication protocols in order to aggregate data from various heterogeneous IoT devices (i.e., medical and environmental sensors, smartphones, sport watches, etc.).
(b) Fog server: Fog server is a lightweight cloud server that is responsible to collect, store, process, and analyze the information from IoT devices and provides predefined applications and services on-demand. Moreover, the fog server has to decide what information must be sent to the cloud, with which data format and when. As fog server can be considered any computational entity with high capabilities that is deployed at the edge of the network such as gateways, set-top-boxes and low power single-board computers located at home, shopping center, bus terminals, train stations, and parks.
(c) Cloud: Cloud provides the data warehouse for permanent storage, performs big data analysis and other back-end applications.

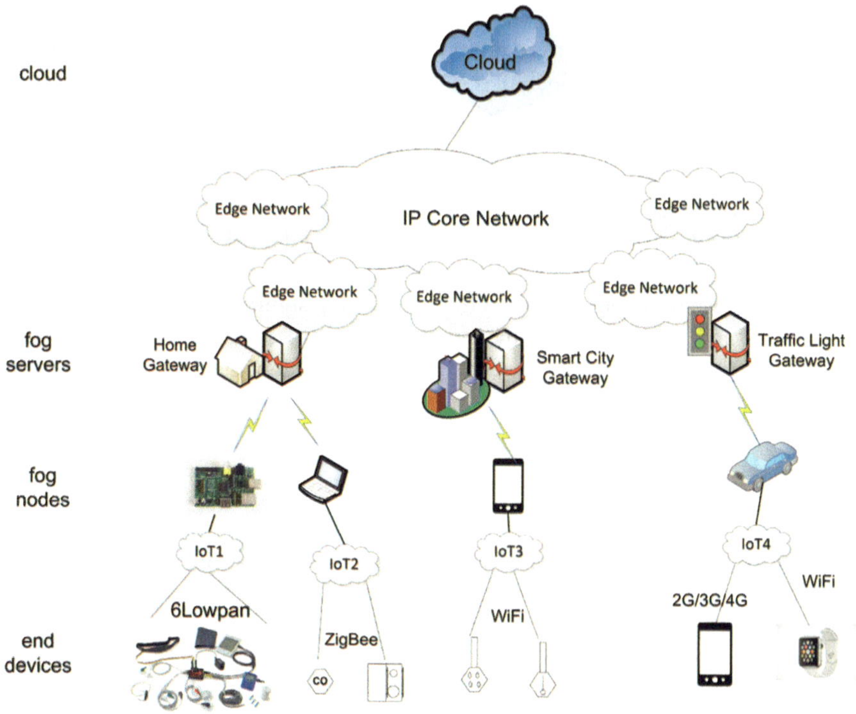


**Fig. 11.1** The architecture of the integration between IoT–fog and cloud computing

The architecture of the integrated IoT based and fog computing approach for healthcare service provisioning is presented in Fig. 11.1. It should be mentioned that fog nodes, fog servers, and IoT devices are distributed entities according to their geographic location while the cloud is centralized and located at the healthcare provider's premises.

In this architectural model, the end users are equipped with a variety of IoT devices such as medical sensors wearable or implanted, devices that monitor presence, motion, mobile devices, and environmental sensors. IoT devices interact among each other with different communication technologies such as 3G, LTE, WiFi, WiMAX, 6Lowpan, Ethernet, or ZigBee. These devices are connected directly or via wired and wireless communication protocols to a sink node such as smartphone or computer, known as *fog node*. The fog node acts as an intermediate point between the IoT devices and the edge network. Fog nodes support different communication protocols in order to enable data collection from different IoT technologies and handle user's mobility and provide local management of the IoT and sensory devices. The fog node performs limited computational operators and provides pushing services for receiving and uploading data. Then, the information from IoT devices is transmitted to the *fog server*. Fog server performs protocol conversion, data storage, processing, filtering, and analysis. According to the data evaluation, fog is able to make decisions and provide services to the end users based on predefined policies and rules without the necessity to interact with the cloud server due to the fact that the fog server is actually a lightweight cloud server. However, the fog server has limited hardware capabilities and cannot fully support the creation of new services and applications. In this context, the fog server has to interact with the cloud in order to provide prediction models, enable big data analysis as well as new service execution and orchestration. Therefore, fog cannot replace the cloud but they have to cooperate in order to provide timely value-added services to the end user with enhanced QoS. Moreover, the fog server can operate as a backup mechanism in case the link to the cloud is faulty.

The data which has been processed, analyzed, and temporarily stored in the fog server can be transmitted to the cloud for permanent storage and further analysis or if there is not any necessity to be stored, they are removed from the fog server.

### *11.4.2 Fog Server Architecture*

The fog server is based on a lightweight approach of cloud computing that is enabled to provide the basic operations of protocol conversion, data storage, processing, filtering, analysis as well as application and service provisioning. In order to be able to provide all the aforementioned functionality, it is composed of the following key modules, as presented in Fig. 11.2:

- Listener: Listener is permanently "alive" and ready to accept or send messages for control or download. As soon as it receives a message from fog node, it checks the validity of the message and the users' capabilities [4]. The listener forwards the messages into a queue and wait for authentication by the security and identity manager and validation of the message by the message broker. If the user is authenticated and the messages are validated, then, they are forwarded to the message database. Otherwise, the sender of the message is properly informed about the cause of failure.

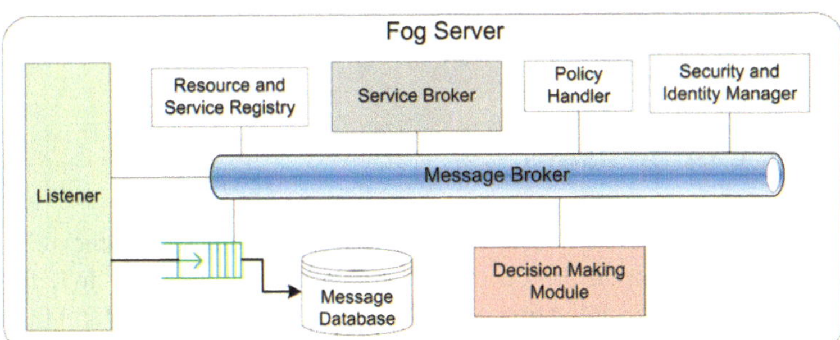


**Fig. 11.2** The architecture of the fog server

- Security and Identity Manager: It verifies both cloud service providers and consumers in order to prevent malicious and unauthorized access. It blocks malicious traffic and manages the overall security of the fog server. Furthermore, firewalls are used for the detection of intrusions focusing on protecting data, resources and services from malware, viruses, worms, or trojans.
- Message Database: It is a database where all the validated messages are stored for recovery and security purposes.
- Decision-Making Module: The decision-making module processes, analyzes, and evaluates the data aggregated by the IoT devices. According to the evaluation results, the decision-making module makes decisions about the most suitable services that should be triggered in order to provide healthcare services and treat the user [5].
- Message Broker: Message broker is a message bus that is responsible to provide protocol translation, message analysis, and processing in order to enable message communication from and to the IoT devices. It is the appropriate messaging environment for efficient and consistent exchange of information among the IoT devices as well as the communication establishment with the services provided by the 3rd party providers [4].
- Service Broker: Service broker interacts with the service registry and policy handler in order to find, bind, invoke, formulate, execute, and interpret the required service according to negotiation algorithms and mathematical models so as to be optimized for each end user and provide the appropriate [4]. Service broker triggers the message broker so as to provide the required services to the end user in a transparent and interoperable way.
- Resource and Service Registry: Resource and service registry is a publish/subscribe database where each cloud provider has published their own catalogue with applications and services for being easily accessed by the service broker. Moreover, resource and service registry stores all the IoT-enabled physical and virtual devices that are managed by the same fog server so as to be easily accessed by the other IoT devices.
- Policy Handler: It contains a repository with all the service level agreements (SLAs) published by cloud providers. Whenever an end user (consumer) requires services, the service broker negotiates and matches the consumer's SLA with the services each cloud provider offers and finds the most appropriate matching.

## 11.5 Use Case Scenarios

This section introduces two use case scenarios for remote healthcare service provisioning that illustrate the benefits of the integration between IoT, fog, and cloud computing in the application layer. It is considered that the patient (actuator) is equipped with an RFID tag medical card that identifies him/her so as to be easily discovered in a building with multiple users or outdoors. In the actuator's smart environment there are various appliances (i.e., IP video cameras, laptops, tablets,

PCs, TV sets, microphones, and speakers) and environmental sensors (i.e., humidity, temperature, smoke detector, motion, and alarm). Moreover, a variety of wearable medical sensors (i.e., pulse oxygen in blood (SPO2), airflow (breathing), body temperature, electrocardiogram (ECG), glucometer, galvanic skin response (GSR—sweating), blood pressure (sphygmomanometer), patient position (accelerometer), and muscle/electromyography (EMG) sensor) are used for healthcare monitoring. The patient's position is also captured through accelerometers, IP video cameras, and motion recognition sensors.

### *11.5.1 Daily Monitoring and Healthcare Service Provisioning*

Healthcare service provisioning can be deployed for AAL. Typical examples are elderly people who usually forget to take their medications, do the predefined exercises from the doctors, report their healthcare condition to the physicians or forget the gas open. Sometimes the forgotten event is critical and user's life may be at risk. In this use case, we illustrate the benefits of the integration of IoT and fog computing in order to monitor user remotely and provide notification services such as reminders for medications and/or warnings or healthcare services for emergency cases such as fall accidents or an aggravation of user's health condition. Figure 11.3 illustrates the use case of the daily monitoring and healthcare service delivery.

The environmental and medical data from the sensors may be connected to a low power single-board computer using IEEE 802.15.4 standard that acts as a sink node (fog node) (step 1a and 1b). Environmental and medical data from a fog node are sent to the home-gateway (fog server) for processing and storage (step 2). The home-gateway (i.e., laptop, PC, etc.) has more computational capabilities from the sensor devices and can analyze all the aggregated information, perform complex algorithms, and correlate data in order to detect abnormal events. The term abnormal is referred to any measurement above the defined normal thresholds or any odd sound such as an accident inside a home, fall, fire, or medical episode. The Listener receivers the information, forwards it to a queue and when the validation and identification process is completed successfully, the data is transferred to the Message Database. The collected data from diverse IoT devices is processed and evaluated by a decision framework located to the home-gateway (Decision-Making module of fog server) using rule-based and machine learning techniques and the abnormal measurements trigger an alarm system (step 3). In case there are no abnormal measurements, the data remains locally stored and is transferred to the cloud whenever it is permitted by the network conditions such as available bandwidth and network traffic. On the other hand, the alarm system indicates the source(s) of the alarm(s), the severity of the event and triggers the Service Broker in order to provide applications and services to the end user. If there is not any predefined service, then, it interacts with the healthcare cloud provider for supporting the patient (step 4). In this case, the data is transmitted to the cloud for permanent storage and further processing based on machine learning algorithms

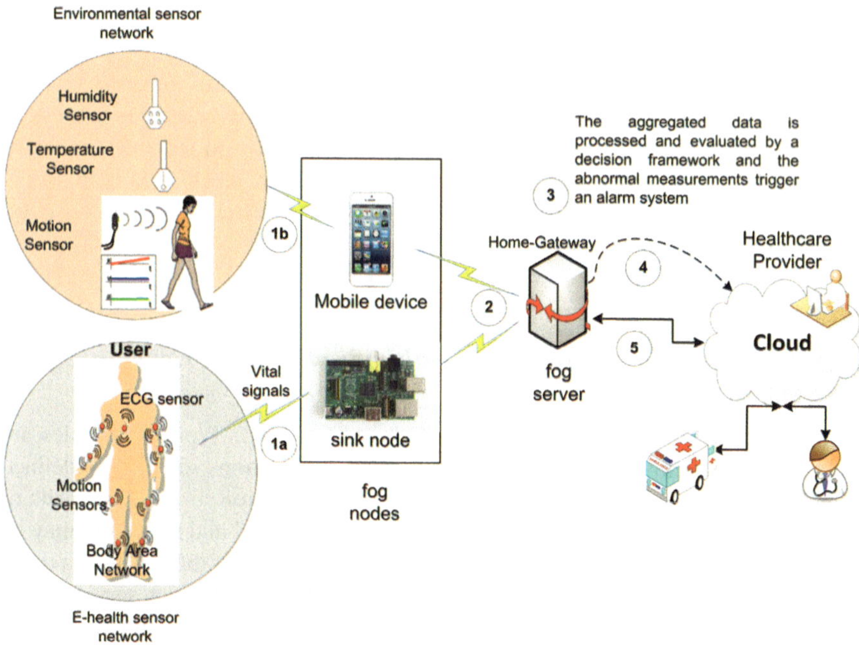


**Fig. 11.3** The use case of daily monitoring for provisioning timely healthcare services with low latency

in order to provide predictions and aggravations in patient's current health status (step 5). The healthcare professionals can have access to the data stored in the cloud, check the medical data and the actions performed, give advice, prescript medicines, or communicate with the patient.

In this approach, it should be mentioned that the doctor's intervention is not necessary in order to provide notification services. Therefore, a reminder application is required to inform the patient to receive his/her medication, follow the predefined diet and exercises subscribed from the physicians, or turn off the gas. Through the integration of the IoT devices and fog computing, the appropriate application service can be provided through the fog server without the necessity to transfer data to the cloud, reducing network traffic, and releasing storage space in the cloud for significant data.

As an example, when a fall is detected through the accelerometer (patient position sensor) and microphones, the blood pressure sensor and ECG sensor are triggered. The IP camera is turned on and starts to send video streaming to the home-gateway (fog server). Home-gateway performs image and position recognition algorithms in order to detect if a fall occurs, if the patient is still fallen, the duration, the location, and the severity of the fall. Home-gateway receives also the data from the medical sensors (blood pressure sensor and ECG), processes, and evaluates them so as to determine if a heart failure or a hypoglycemic sock was the cause of the fall. If the patient is unconscious, injured, or the fall has occurred due to a medical condition, the home-gateway creates an alarm and triggers the healthcare provider reporting the severity of the event, the location of the patient and transfers the collected data. The healthcare provider can further process the data and take decisions for the treatment plan. Otherwise, the data remains in the local database of the home-gateway and a report with the event and the medical data are sent to the cloud and are merged with the patient's profile. Once the data is uploaded on the cloud, there is no any other necessity to be stored locally at the fog. Therefore, they are removed from the fog storage releasing storage space.

### *11.5.2 Extended eCall Service Delivery*

Numerous efforts have been undertaken in order to increase road safety, reduce road accidents and road deaths. eCall system [14] is an initiative to provide immediate assistance to people involved in a collision wherever they are located in the European Union [19]. eCall deploys an in-vehicle system that initiates emergency calls automatically in case of a serious road accident. In order to detect a collision, the vehicle is equipped with vehicle sensors such as crash sensing sensors that provide information about the vehicle's description and its location. Then, through the cellular network, the Public Safety Answering Point (PSAP) is informed in order to provide assistance [19]. However, accidents still occur. The transmission of the data must be reliable and fast, around 14–17 s, in order to provide accurate service delivery and save lives [15]. In this use case, we illustrate an extended eCall system that integrates a variety of IoT devices and fog computing in order to provide quicker response for emergency services and decrease the number of dead or injured passengers (Fig. 11.4).

Except of the vehicle's sensors that monitor the vehicle's current position and status, a variety of IoT devices are used. Some of these devices include: video cameras located in traffic lights that calculate the traffic, microphones, speakers, environmental sensors, and medical wearable sensors that the driver or passengers use. Vehicle sensors and IoT devices interact with a fog node that collects the aggregated information from sensors and mobile devices through multiple public or private access points (i.e., WiFi,3G, LTE) (step 1). The fog node in this case is the traffic light station that acts as a sink node. The traffic light transmits the aggregated information from sensors to the fog server. The fog server can be any dedicated gateway located in public or private buildings (i.e., train terminals, bus stations, bus stops, etc.). The listener receivers the data from the IoT-enabled devices, forwards them to a queue, and performs validation processes. If the messages are validated, then the information is stored temporally to the Message Database and the Decision-Making Module performs a set of processing, filtering, and analyzing actions (step 2). In case an abnormal event is detected (e.g., a collision or heart attack), the fog server may either provide services to the end user or dictate the fog nodes to periodically collect data with a different frequency and from various end devices

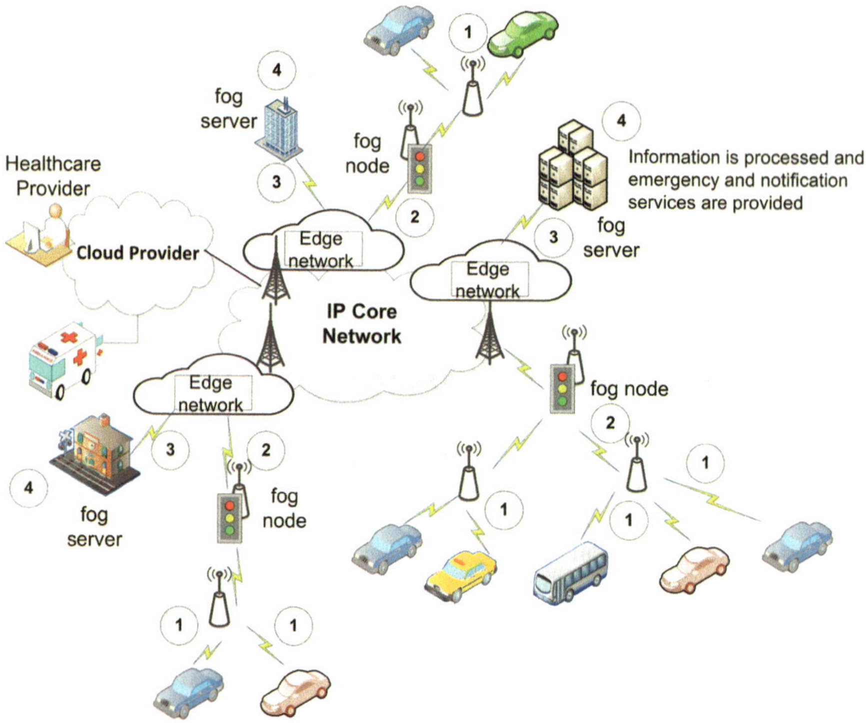


**Fig. 11.4** The use case of extended eCall services increasing road safety and providing timely response for road accidents

in order to have a global view of the emergency event (step 3). Moreover, the information is processed using movement and acceleration patterns that can prevent other road accidents or traffic problems that may occur as a consequence of the present accident as well as can send warning messages to the vehicle driver in order to stop the vehicle or other vehicles and passengers in the nearby area, notify volunteers to assist the patient or injured passengers, and provide an ambulance service (step 4). The data remains in the fog server where the volunteers and first aid responders can have access to the stored data or trigger IoT devices such as video cameras and microphones to have a look to the scene of the emergency event. In this context, it may need information from the neighboring fog nodes in order to map the traffic, detect different vehicles on the same road, detect the ambulances, and open lanes to pass through the traffic [36]. Once the critical-mission event has been confronted, the data can be transmitted to the cloud for permanent storage or if there is not any need to be stored, they are removed from the fog server.

As seen in this use case, integration among IoT technologies, fog, and cloud computing provides a variety of beneficial features such as location awareness, low latency and improves the quality of the provisioning services. The cloud computing approach itself has not the ability to provide reliable data processing for delay-sensitive services. Through the fog computing the decision-making process is transmitted to the edge network, near the end user that increases the response time for provisioning notifications and emergency services such as ambulances, first aid responders, and rescue teams.

## 11.6 Conclusion

The rapidly growing number of IoT devices that generate an increasing amount of data and consume cloud services have overstretched the network and cloud infrastructure. In order to confront the network strains of data transmission to the cloud, many ICT operators have introduced different approaches for providing powerful computing models located closer to the end user. Cisco [11] coined the approach of fog computing in order to optimize the cloud approach and deliver timely delay-sensitive services to the end user. Since fog computing is deployed at the edge network, it can provide low latency, geographical distribution, timely processing of data streaming, and mobility support. Based on these features, fog computing seems to be the most promising solution for reliable healthcare applications.

This chapter presents the benefits of the integration of the IoT technologies, cloud and fog computing paradigm in order to provide reliable novel healthcare services. A system model is presented that shows how integration and cooperation among the IoT, fog, and cloud technologies is achieved. The main component of the system model is the fog server that exploits the cloud computing in order to enable protocol conversion, data storage, processing, and evaluation. Moreover, the fog server provides enhanced features including user mobility support, location awareness, dense geographical distribution, low latency, and delay. These characteristics are significant for provisioning delay-sensitive services such as healthcare and emergency services. In this context, the integration of IoT and fog computing for healthcare purposes can provide numerous advantages such as faster and accurate treatment delivery, reduction of medical costs, improvement of doctor–patient relationships, and the delivery of treatment. A number of use cases illustrate the opportunities and the benefits that the integration of IoT, fog, and cloud computing approach offers for cost effective, efficient, timely, and high-quality ubiquitous healthcare service delivery. These use cases include daily monitoring and healthcare service provisioning as well as extended eCall service delivery.

As a conclusion, it should be mentioned that fog computing operates better than cloud computing in provisioning healthcare applications and services with low latency, delays, jitter and supports user's mobility. However, it cannot replace cloud computing and the cooperation between them is necessary in order to enable big data analysis, support prevention as well as create and execute new distributed healthcare services in order to provide personalized treatment oriented to users' needs and preferences based on their current health status. Fog has limited computational power, therefore, it cannot provide the creation and execution environment for the deployment of new services. Moreover, upgrading the fog server so as to provide the functionality of cloud would increase the operational costs.

**Acknowledgements** The present work was undertaken in the context of the "nExt generation eMergencY commuNicatiOnS - (EMYNOS) project" with contract number 653762. The project has received research funding from the H2020 European Framework Programme.

## References

1. M. Aazam, E.-N. Huh, Fog computing and smart gateway based communication for cloud of things, in *2014 International Conference on Future Internet of Things and Cloud (FiCloud)* (IEEE, New York, 2014), pp. 464–470
2. H. Alemdar, C. Ersoy, Wireless sensor networks for healthcare: a survey. Comput. Netw. **54**(15), 2688–2710 (2010)
3. S. Amendola, R. Lodato, S. Manzari, C. Occhiuzzi, G. Marrocco, Rfid technology for IoT-based personal healthcare in smart spaces. IEEE Internet Things J. **1**(2), 144–152 (2014)
4. F.G. Andriopoulou, L.T. Kolovou, D.K. Lymberopoulos, An integrated broker platform for open eHealth domain, in *Wireless Mobile Communication and Healthcare* (Springer, Berlin, 2012), pp. 234–246
5. F.G. Andriopoulou, K.D. Birkos, D.K. Lymberopoulos, u-MCHC: a predictive framework for ubiquitous management of exacerbations in chronic diseases, in *2013 35th Annual International Conference of the IEEE Engineering in Medicine and Biology Society (EMBC)* (IEEE, New York, 2013), pp. 6976–6979
6. L. Atzori, A. Iera, G. Morabito, The internet of things: a survey. Comput. Netw. **54**(15), 2787–2805 (2010)
7. F. Bonomi, R. Milito, J. Zhu, S. Addepalli, Fog computing and its role in the internet of things, in *Proceedings of the First Edition of the MCC Workshop on Mobile Cloud Computing* (ACM, New York, 2012), pp. 13–16
8. Bosch Healthcare: Health Buddy System (2016). Accessed 30 Jan 2016
9. A. Botta, W. de Donato, V. Persico, A. Pescapé, Integration of cloud computing and internet of things: a survey. Futur. Gener. Comput. Syst. **56**, 684–700 (2016)
10. N. Bui, M. Zorzi, Health care applications: a solution based on the internet of things, in *Proceedings of the 4th International Symposium on Applied Sciences in Biomedical and Communication Technologies* (ACM, New York, 2011), p. 131
11. Cisco Visual Networking Index: Global mobile data traffic forecast update, 2015–2020 (2016). Accessed 24 Jan 2016
12. J. Cubo, A. Nieto, E. Pimentel, A cloud-based internet of things platform for ambient assisted living. Sensors **14**(8), 14070–14105 (2014)
13. C. Doukas, T. Pliakas, I. Maglogiannis, Mobile healthcare information management utilizing cloud computing and android OS, in *2010 Annual International Conference of the IEEE Engineering in Medicine and Biology Society (EMBC)* (IEEE, New York, 2010), pp. 1037–1040
14. eCall in all new cars from April 2018 (2016). Accessed 28 Jan 2016
15. eCall: Time saved = lives saved (2016). Accessed 28 Jan 2016
16. M. Firdhous, O. Ghazali, S. Hassan, Fog computing: will it be the future of cloud computing? in *Proceedings of the 3rd International Conference on Informatics & Applications*, Kuala Terengganu, pp. 8–15 (2014)
17. G. Fortino, M. Pathan, G.D. Fatta, Bodycloud: integration of cloud computing and body sensor networks, in *2012 IEEE 4th International Conference on Cloud Computing Technology and Science (CloudCom)* (IEEE, New York, 2012), pp. 851–856

18. J. Gubbi, R. Buyya, S. Marusic, M. Palaniswami, Internet of things (IoT): a vision, architectural elements, and future directions. Futur. Gener. Comput. Syst. **29**(7), 1645–1660 (2013)
19. Hero: Harmonised eCall European Pilot (2016). Accessed 28 Jan 2016
20. A.J. Jara, M.A. Zamora, A.F. Skarmeta, An internet of things—based personal device for diabetes therapy management in ambient assisted living (AAL). Pers. Ubiquit. Comput. **15**(4), 431–440 (2011)
21. B.M. Lee, J. Ouyang, Intelligent healthcare service by using collaborations between IOT personal health devices. Blood Press. **10**, 11 (2014)
22. D. Lu, T. Liu, The application of IOT in medical system, in *2011 International Symposium on IT in Medicine and Education (ITME)*, vol. 1 (IEEE, New York, 2011), pp. 272–275
23. P. Mell, T. Grance, The NIST definition of cloud computing. Commun. ACM **53**(6), 50 (2010)
24. D. Miorandi, S. Sicari, F. De Pellegrini, I. Chlamtac, Internet of things: vision, applications and research challenges. Ad Hoc Netw. **10**(7), 1497–1516 (2012)
25. S.K. Mouleeswaran, A. Rangaswamy, H.A. Rauf, Harnessing and securing cloud in patient health monitoring, in *2012 International Conference on Computer Communication and Informatics (ICCCI)* (IEEE, New York, 2012), pp. 1–5
26. Philips Hospital to Home: Redefining healthcare through innovation in telehealth (2016). Accessed 31 Jan 2016
27. R. Piyare, S.R. Lee, Towards internet of things (IoTs): integration of wireless sensor network to cloud services for data collection and sharing. arXiv preprint (2013). arXiv:1310.2095
28. A.-M. Rahmani, N.K. Thanigaivelan, T.N. Gia, J. Granados, B. Negash, P. Liljeberg, H. Tenhunen, Smart e-health gateway: bringing intelligence to internet-of-things based ubiquitous healthcare systems, in *2015 12th Annual IEEE Consumer Communications and Networking Conference (CCNC)* (IEEE, New York, 2015), pp. 826–834
29. V.M. Rohokale, N.R. Prasad, R. Prasad, A cooperative Internet of Things (IoT) for rural healthcare monitoring and control, in *2011 2nd International Conference on Wireless Communication, Vehicular Technology, Information Theory and Aerospace & Electronic Systems Technology (Wireless VITAE)* (IEEE, New York, 2011), pp. 1–6
30. C.O. Rolim, F.L. Koch, C.B. Westphall, J. Werner, A. Fracalossi, G.S. Salvador, A cloud computing solution for patient's data collection in health care institutions, in *Second International Conference on eHealth, Telemedicine, and Social Medicine, 2010. ETELEMED'10* (IEEE, New York, 2010), pp. 95–99
31. Y. Shi, G. Ding, H. Wang, H.E. Roman, S. Lu, The fog computing service for healthcare, in *2015 2nd International Symposium on Future Information and Communication Technologies for Ubiquitous HealthCare (Ubi-HealthTech)* (IEEE, New York, 2015), pp. 1–5
32. F. Sposaro, G. Tyson, ifall: an android application for fall monitoring and response, in *Annual International Conference of the IEEE Engineering in Medicine and Biology Society, EMBC 2009*, pp. 6119–6122 (IEEE, New York, 2009)
33. F. Sposaro, J. Danielson, G. Tyson, iWander: an android application for dementia patients, in *2010 Annual International Conference of the IEEE Engineering in Medicine and Biology Society (EMBC)* (IEEE, New York, 2010), pp. 3875–3878
34. V. Stantchev, A. Barnawi, S. Ghulam, J. Schubert, G. Tamm, Smart items, fog and cloud computing as enablers of servitization in healthcare. Sensors Transducers **185**(2), 121 (2015).
35. I. Stojmenovic, Fog computing: a cloud to the ground support for smart things and machine-to-machine networks, in *2014 Australasian Telecommunication Networks and Applications Conference (ATNAC)* (IEEE, New York, 2014), pp. 117–122
36. I. Stojmenovic, S. Wen, The fog computing paradigm: scenarios and security issues, in *2014 Federated Conference on Computer Science and Information Systems (FedCSIS)* (IEEE, New York, 2014), pp. 1–8
37. R. Tabish, A.M. Ghaleb, R. Hussein, F. Touati, A.B. Mnaouer, L. Khriji, M.F.A. Rasid, A 3g/wifi-enabled 6lowpan-based u-healthcare system for ubiquitous real-time monitoring and data logging, in *2014 Middle East Conference on Biomedical Engineering (MECBME)* (IEEE, New York, 2014), pp. 277–280

38. L.M. Vaquero, L. Rodero-Merino, Finding your way in the fog: towards a comprehensive definition of fog computing. ACM SIGCOMM Comput. Commun. Rev. **44**(5), 27–32 (2014)
39. Vitaphone Telemedicine (2016). Accessed 30 Jan 2016
40. G. Yang, L. Xie, M. Mantysalo, X. Zhou, Z. Pang, L.D. Xu, S. Kao-Walter, Q. Chen, L.-R. Zheng, A health-IOT platform based on the integration of intelligent packaging, unobtrusive bio-sensor, and intelligent medicine box. IEEE Trans. Ind. Inf. **10**(4), 2180–2191 (2014)
41. S. Yi, C. Li, Q. Li, A survey of fog computing: concepts, applications and issues, in *Proceedings of the 2015 Workshop on Mobile Big Data* (ACM, New York, 2015), pp. 37–42
42. M. Yuriyama, T. Kushida, Sensor-cloud infrastructure-physical sensor management with virtualized sensors on cloud computing, in *2010 13th International Conference on Network-Based Information Systems (NBiS)* (IEEE, New York, 2010), pp. 1–8
43. D. Zhang, Z. Yu, C.-Y. Chin, Context-aware infrastructure for personalized healthcare. Stud. Health Technol. Inform. **117**, 154–163 (2005)
44. J. Zhou, T. Leppanen, E. Harjula, M. Ylianttila, T. Ojala, C. Yu, H. Jin, L.T. Yang, Cloudthings: a common architecture for integrating the Internet of things with cloud computing, in *2013 IEEE 17th International Conference on Computer Supported Cooperative Work in Design (CSCWD)* (IEEE, New York, 2013), pp. 651–657
45. Q. Zhu, R. Wang, Q. Chen, Y. Liu, W. Qin, IOT gateway: bridging wireless sensor networks into Internet of things, in *2010 IEEE/IFIP 8th International Conference on Embedded and Ubiquitous Computing (EUC)* (IEEE, New York, 2010), pp. 347–352

# Chapter 12
# Supporting Decision Making for Large-Scale IoTs: Trading Accuracy with Computational Complexity

**Kostas Siozios, Panayiotis Danassis, Nikolaos Zompakis, Christos Korkas, Elias Kosmatopoulos, and Dimitrios Soudris**

## 12.1 Introduction

During the last years, there are tremendous improvements in the domain of embedded devices. To begin with, the new process technologies enable the underline hardware to become smaller, cheaper, and more powerful. This trend in conjunction to the improvements at networking infrastructure enables the majority of the devices to be extended with communication capabilities. Hence, embedded devices nowadays are able to connect, interact, and cooperate with their surrounding environment. This new platform paradigm, also known as "Internet of Things" (IoT), is as a network of objects (or things) capable of detecting and communicating information between each other.

Defining things and recognizing what a particular thing is and what it represents in the context of IoT requires a careful analysis of what philosophers, such as the Aristotle and Philoponus have to say and how their philosophical thoughts can transcend into the near future. Specifically, in the work "The Categories" Aristotle gives strikingly general and exhaustive account of things that are beings. According to this opinion, beings include substance, quantity, quality, as well as relation among others. Hence, from the philosophical point of view, the word "things" is

K. Siozios (✉)
School of Physics, Aristotle University of Thessaloniki, Thessaloniki, Greece
e-mail: ksiop@microlab.ntua.gr

P. Danassis • N. Zompakis • D. Soudris
School of ECE, National Technical University of Athens, Athens, Greece
e-mail: panosd@microlab.ntua.gr; nzompakis@microlab.ntua.gr; dsoudris@microlab.ntua.gr

C. Korkas • E. Kosmatopoulos
Department of ECE, Democritus University of Thrace, Xanthi, Greece
e-mail: chriskorkas@iti.gr; kosmatop@iti.gr



G. Keramidas et al. (eds.), *Components and Services for IoT Platforms*,
DOI 10.1007/978-3-319-42304-3_12

not restricted to material things but can apply also to virtual things and the events that are connected to "things." In the context of IoT, a "thing" might be defined as a real (physical) or digital (virtual) entity, which is capable of being uniquely identified, and that exists and moves in space and time.

The challenge of integrating embedded computing and physical processes with feedback loops, where physical processes affect computations and vice versa has been recognized for some time, allowing applications with enormous societal impact and economic benefit to be developed by harnessing these capabilities across both space and time domains. Such a pooling of system's resources and capabilities together to create a new, more complex system which offers more functionality and performance than simply the sum of the constituent sub-systems. In the IoT paradigm, numerous objects that surround us will be on the network in one form or another. This trend is inline to the projection that in the near future it is expected that computing and communication capabilities will be embedded in all types of objects and structures in the physical environment [1]. To do so, flexible yet efficient protocols, architectures, and technologies are absolutely necessary, in which information and communication systems are transparently embedded in the enrolment around us.

While technology and platform capabilities ought to be equally important in order to design next-generation of smart systems, it is also crucial to manage the system's complexity under the constraint posed by the large amounts of data that these systems will produce. According to a study from Cisco Systems, there will be as many as 50 billion embedded systems and other portable devices connected to the internet by 2020. Because these systems are so inexpensive and the networks so pervasive, they could provide a wealth of data that industry could use to monitor and improve operations. For instance, by 2020, the digital universe will reach 44 zettabytes—a tenfold increase from 2013 [2]. On top of this, analysts will need deep knowledge of the specific target application domains to ensure they incorporate the right data to generate useful insights. Although new methodologies, services, and design technologies are absolutely necessary to address this challenge, the trend nowadays is to enable smart things and their services to be fully integrated in the developed systems by reusing and adapting technologies and patterns commonly used for traditional local-/wide-area networks. Specifically, the IoT is developing over time by a way of co-evolution, with technology, applications, and stakeholders' understanding of the implications driving each other.

The integration of physical processes and computing devices is not new. Embedded systems have been in place for a long time and these systems often combine physical processes (e.g., through digital/analog sensors) with computing. However, the core differentiator between an IoT and either a typical control system, or an embedded system, is the communication feature among system's components, which adds (re-)configurability and scalability, allowing instrumenting the physical world with pervasive networks of sensor-rich embedded computation [3]. The goal of an IoT architecture is to get maximum value out of a distributed large system by understanding how each of the individual (sub-)systems work, interface and are used. This trend is also supported by the continuation of Moore's law, which imposes that the cost of a single embedded computer equipped with sensing, processing, and communication capabilities drops towards zero [4]. Thus, it will be economically feasible to densely deploy networks with very large quantities of sensor readings from the physical world, compute quantities, and take decisions out of them. Such a very dense networks offer a better resolution of the physical world and therefore a better capability of detecting the occurrence of an event; this is of paramount importance for a number of foreseeable applications.

Apart from the technology-oriented parameters that affect the efficiency and/or the flexibility of an IoT-based system, the supporting tools are also crucial for deriving an optimum solution. The current solutions targeting to support the development of these platforms rely mainly on stand-alone tools that tackle distinct aspects of the system's development. Consequently, the constraint propagation among the employed tool-sets is the current viable way to enable the design of more complex platforms. Although this concept seems straightforward and promising, it relies on the fundamental premise that models are freely interchangeable amongst tool vendors and have interoperability amongst them. In other words, this imposes that models can be written, or obtained from other vendors, while it is known a priori that they will be accepted by any vendor tool for performing different steps of system's prototyping (e.g., architecture/topology analysis, simulation, implementation, etc.). On contrast to this "ideal" approach, the existing flows rarely support either model interoperability or independence between model and software tools. Also, due to the problem's complexity, the existing "constraint propagation" design technique will not be a viable approach for designing large-scale IoT platforms. Towards this direction, and as research and industry pushes for efficient designs, novel frameworks that tackle the entire system's complexity are of high importance [5].

In accordance with this trend, throughout this chapter we introduce a low-complexity decision-making mechanism targeting to IoT-based systems. For evaluation purposes, the efficiency of introduced solution will be demonstrated with the usage of a smart thermostat usecase that supports the building's cooling/heating control.

## 12.2 The Smart Thermostat Usecase

The studied usecase concerns the climate control of buildings equipped with their own renewable energy generation elements (e.g., photovoltaic arrays, wind turbines, geothermal energy, etc.) along with a collection of automatic control elements for affecting the building thermal characteristics e.g., automatically control the heating, ventilation, and air conditioning (HVAC) system set-points. Such a solution is part of home automation—a field that refers to the use of computer and information technology to control home appliances and features (such as heating and cooling). Depending on the building's size, the developed systems can range from simple remote control to complex computer/micro-controller-based networks with varying degrees of intelligence and automation, similar to the case depicted in Fig. 12.1. The popularity of these solutions has been increasing greatly in recent years due to much

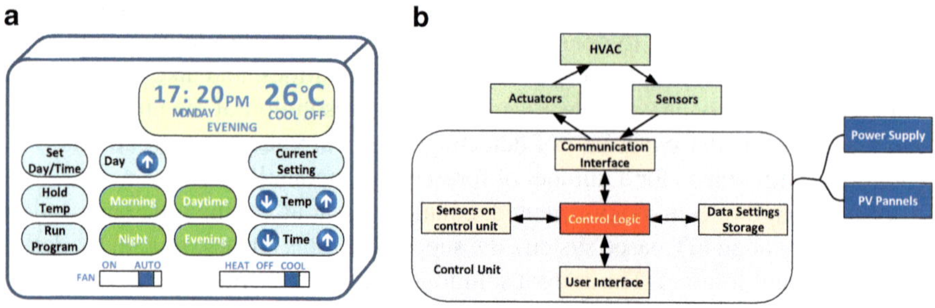


**Fig. 12.1** The employed smart thermostat (**a**) component's view and (**b**) the functional blocks

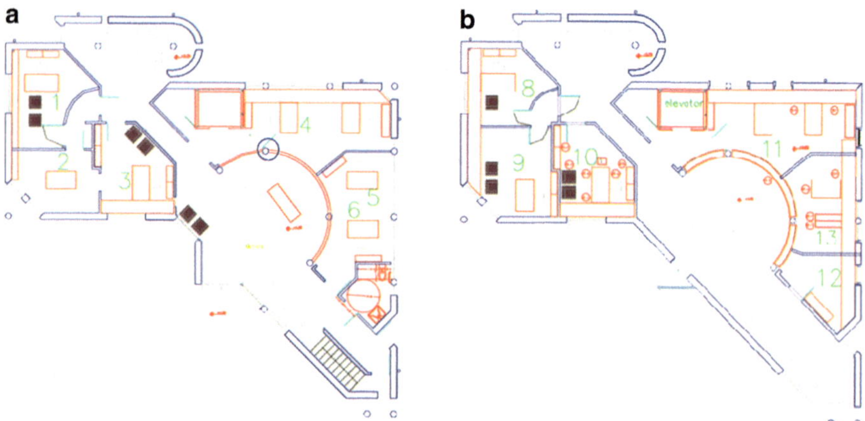

**Fig. 12.2** Floorplans for the (**a**) ground and (**b**) first floor of our smart building

higher affordability and simplicity through smartphone and tablet connectivity, while the concept of the IoT has tied in closely with the popularization of home automation.

To test and evaluate the introduced decision-making framework, a building block located in Chania, Greece, will serve as the employed usecase. The floorplan of this building is depicted schematically in Fig. 12.2. The main reason for choosing this building is the fact that the building is equipped literally with multiple renewable energy generation elements and energy/thermal influencing elements that may found in a real-life building. Moreover, the overall building control infrastructure comprises a quite complex hierarchical system with different decision-making elements affecting the building thermal and energy-consuming performance at different levels of abstraction and in a quite complex manner. Furthermore, it has to be emphasized the fact that the building contains a large number of rooms and offices with totally different characteristics and purposes (laboratories, office rooms, conference spaces, data centre rooms, etc.). Finally, by studying the efficiency of this building with real weather data affecting a whole year (as they were provided by National Renewable Energy Laboratory [6]), depicts that our solution is subject to severe and abrupt weather as well as occupant behavior changes.

In order to guide the aggressiveness of introduced framework, the employed building uses a number of sensors for monitoring temperature values (both inside and outside the building), as well as the variation of sunshine and humidity. For our experimentation we assume (without affecting the generality of introduced solution) that these values are acquired once per 10 min, as the weather data is not expected to modified considerably within this time period. The analysis discussed at the rest of this manuscript considers the problem of operating air-conditioners during the summer period (June, July, and August), in order to cool-climate the rooms. The objective is to define a solution that takes into consideration both the energy efficiency and the user comfort satisfying level with the minimal computational complexity, as compared to existing state-of-the-art control solvers for similar problem. Furthermore, instead of relevant approaches which mainly rely on statistical data, the proposed solution does not consider weather forecasts; thus, the efficient addressing of decision-making problem becomes far more challenging.

### 12.2.1 System Modeling

In order to describe in more thoroughly the system's architecture, Fig. 12.3 presents in UML form the main functionalities performed by the smart thermostats regarding the general form of the employed usecase. Specifically, starting by collecting a number of weather-related data (e.g., temperature, humidity, and radiation), as they were acquired by the weather station, it is possible to analyze the efficiency of different thermostat configurations. The results are fed as input to the scenario analysis in order to compute the optimum scenario set in a season basis (winter, spring, summer, and autumn). Then, the employed controller implements the desired policies in order to maximize user's comfort with the minimum energy cost.

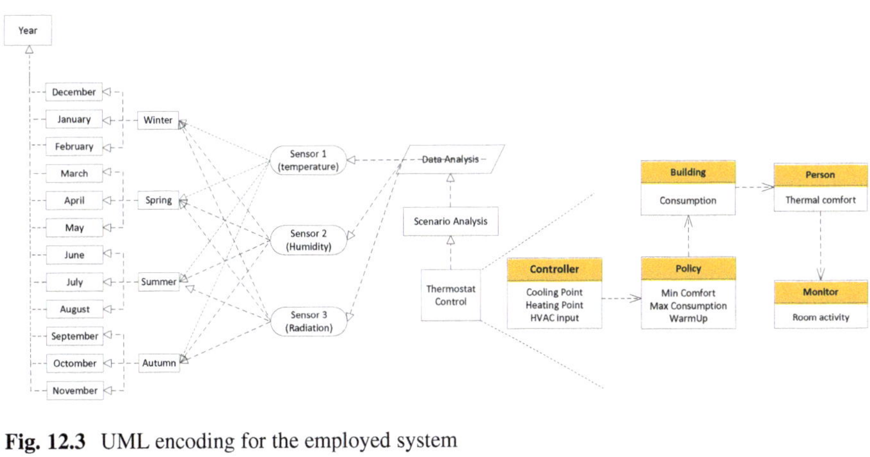


**Fig. 12.3** UML encoding for the employed system

During this phase, alternative techniques for further energy savings (e.g., building warm-up phase, switch-off heating/cooling in case there is no motion detection) are also applicable. However, these techniques are beyond the scopes of the analysis discussed throughout this chapter, as we focus solely at the decision making.

## 12.3 Challenges and Motivation

This chapter describes in detail the proposed efficient control system that is easy- and inexpensive-to-deploy in everyday buildings in order to support the task of decision making for HVAC systems: no expensive infrastructure or modeling tools and effort are required. Instead of the typical adopted procedure for developing decision-making systems for building climate control, which relies on modeling initially the building dynamics using one of the existing building modeling tools (e.g., EnergyPlus [7], Modelica [8], etc.) and then developing a model-based control using the model for the building dynamics, our approach relies on a different strategy. The objective of our solution is to provide thermal comfort and acceptable indoor air quality with the minimum possible energy cost. More thoroughly, the problem at hand is a quite challenging problem where the control system attempts to exploit "as much as it can" the renewable energy so as to reduce the demand for non-renewable energy (coming from the grid) or during time-slots of low-cost tariffs, while maintaining user comfort (i.e., making sure that the building occupants are *satisfied* with the in-building temperature and other thermal conditions). This will be done without requiring the deployment of an "expensive," elaborate, and complete sensor infrastructure, a prerequisite for the deployment of state-of-the-art building climate control systems—each of the constituent systems is provided only with information about its own state and energy costs.

Despite the significant progress made in optimal nonlinear control theory [9, 10] the existing methods are not, in general, applicable to large-scale systems because of the computational difficulties associated with the solution of the Hamilton–Jacobi partial differential equations. Similarly, model predictive control for nonlinear systems, a control approach which has been extensively analyzed and successfully applied in industrial plants during the latest decades [11, 12], faces also dimensionality issues: in most cases, predictive control computations for nonlinear systems amount to numerically solving on line a non-convex high-dimensional mathematical programming problem, whose solution may require a quite formidable computational burden if on line solutions are required. Another family of approaches employ optimization-based schemes to calculate the controller parameters [13]. These approaches require analytical calculation of Jacobian and Hessian matrices, which in large-scale systems is a very time-consuming process. Existing simulation-based approaches are not able to efficiently handle systems of large-scale nature as this requires solving a complex optimization problem with hundreds or thousands of states and parameters [14, 15].

The previously mentioned problem imposed that novel techniques able to support efficient decision making with almost negligible computational and/or storage complexity are of high importance. These lightweight solutions would be ideal to support tasks related to control of IoT-based systems, as these systems usually rely on battery-based (low-performance) embedded processors. Additionally, the feature of hierarchical decision-making is of similar importance because such an approach further reduced the problem's complexity. Note that the absence of developing lightweight solutions (able to be executed onto embedded platforms) for supporting large-scale system's decision making is not due to neglect, but rather due to its difficulty. Also, as we have already highlighted, this problem becomes far more challenging in case the decision making has to be made under run-time (or real-time) constraints. In such a case, usually a compromise between the desired accuracy and the processing overhead is performed.

## 12.4 Support Decision Making with the Fuzzy Inference Systems

The task of determining the temperature set-points for each thermal zone can be accomplished by a mechanism that relies on two competing fuzzy inference systems (FIS). The first one (*availability FIS*) assesses the "availability" of energy ($A^E(t)$), while the second one (*demand FIS*) determines the predicted "demand" for energy consumption ($D^E(t)$).

The "availability" of energy ($A^E(t)$) is derived based on the weather data, the available funds for purchasing energy ($AF$) and the current market's energy trading rate ($P(t)$). Intuitively it is a metric that determines how affordable is to purchase energy from the grid. On the other hand, the "demand" for energy consumption ($D^E(t)$) is determined according to the occupants' thermal comfort. Intuitively it reflects the occupants' need to spent more energy depending on their satisfaction. The aforementioned FIS compete with each other until they reach an approximate equilibrium $A^E(t) \simeq D^E(t)$, i.e., being able to afford to spend the required energy (through the appropriate modification at the thermostat's set-point) while taking into account restrictions posed by the occupants' thermal comfort violation. To that end we start by determining an initial temperature set-point[1] and use it to compute the predicted occupants' thermal comfort based on a model for general thermal satisfaction called predicted mean vote (PMV).[2] By modifying the thermostat's set-point we change the PMV per room and consequently the demand for energy consumption ($D^E(t)$). Normally, improving the PMV comes

[1]The initial set-point can be a static value for the whole year (e.g., 23 °C), a plethora of predetermined values depending on the season, or a set-point derived by another engine (e.g., by using an artificial neural network).

[2]The PMV model is an index that provides the average thermal sensation according to the ASHRAE thermal sensation scale [−3, +3]. It was developed by Fanger in the 1970s and stands among the most recognized thermal comfort models.

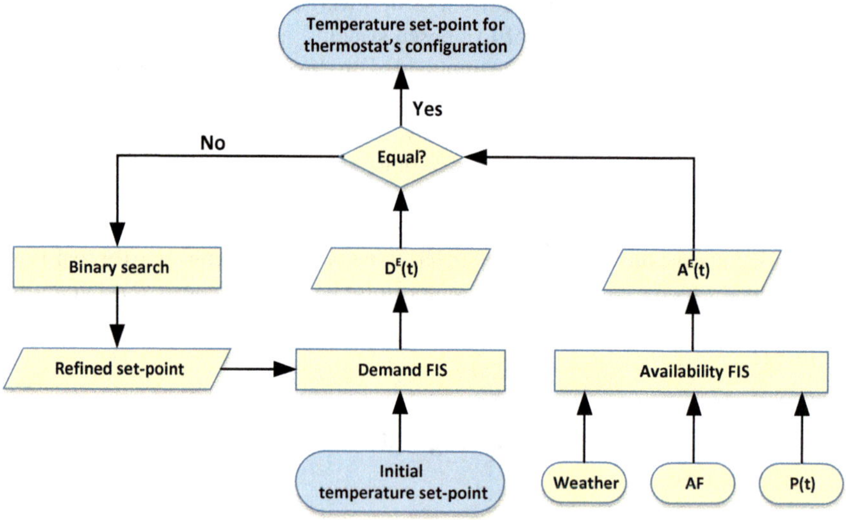


**Fig. 12.4** Proposed methodology regarding the competing FIS

with a higher cost in energy consumption and that trade-off is the basis at which our two FIS compete on. For this purpose, the proposed mechanism performs a binary search[3] within the accepted temperature range [18–28 °C] and appropriately configures the thermostat's operational set-point to the temperature that results to the closest equilibrium. The aforementioned process is depicted in Fig. 12.4. Note that the "availability" of energy $A^E(t)$ remains constant during a specific time-step, since this value is independent of the examined set-point.

In the subsequent section we provide additional details on the architecture of the employed FIS.

### 12.4.1 *Fuzzy Inference System's Architecture*

An FIS uses *fuzzy* logic and sets to formulate a mapping between an input space to an output space. The term *fuzzy* refers to the fact that the logic involved extends the classical boolean logic in order to handle the concept of partial truth. Classical logic assumes the "Principle of Bivalence" which states that every sentence can be either true or false. In contrast, humans think using "degrees of truth" and linguistic terms

[3]Note that the binary search algorithm takes a monotonic function as input. In our case, we ensure that our function is monotonic by limiting our search space to a specific range [18–28 °C]. That way we guarantee that increasing the temperature during the winter and decreasing it during the summer result in an improvement in the occupants' thermal comfort.

such as "a little" or "too much" instead of absolute values. In order to deal with these "issues of vagueness," fuzzy logic assigns any real number between 0 and 1 as the truth value of a statement. In essence, in fuzzy logic the truth of any statement becomes a matter of degree, making the FIS a convenient way to create a reasonable model for a complex system that is tolerant to imprecise data.

An FIS consists of a rule (or knowledge) base and a reasoning mechanism called fuzzy inference engine. The rule base is a collection of *if-then* rules provided by experts that use fuzzy sets in the hypothesis and conclusion part. The inference engine combines these rules using fuzzy reasoning and produces a mapping from the input to the output space. Additionally, there is a fuzzification mechanism that transforms the input space into fuzzy sets and a defuzzification mechanism that performs the reverse procedure, i.e., transforms the fuzzy set obtained by the inference engine into crisp values on the range of the output space. The aforementioned process is depicted in Fig. 12.5.

A fuzzy set is defined by a function that assigns to each entity in its domain a value between 0 and 1, representing the entity's degree of membership in the set. Such a function is called membership function (MF). The membership functions associated with the PMV input variable for the *demand FIS* are depicted in Fig. 12.6. More thoroughly, if, for example, the computed PMV is $-0.15$, then the temperature is mostly excellent but there will be some people that will feel a bit cold. In essence these functions represent the vagueness which is the norm rather than the exception in real life and constitute the basis for converting crisp values to fuzzy sets.

The principal component of each FIS, though, is its rule base, which is used to formulate the expert's knowledge. Evaluating a fuzzy rule involves the fuzzification of the premise, the application of any fuzzy operators and finally the implementation of the implication operator that gives the consequent of the rule. The fundamental difference compared to classical boolean rules is that fuzzy rules can be partially applied, depending on the degree of membership of each premise of the rule. Aggregating the consequent part of all the rules creates the final output fuzzy

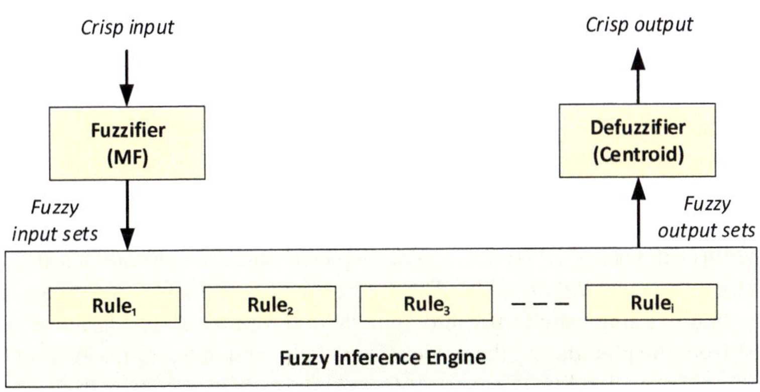


**Fig. 12.5** Architectural organization of the employed FIS

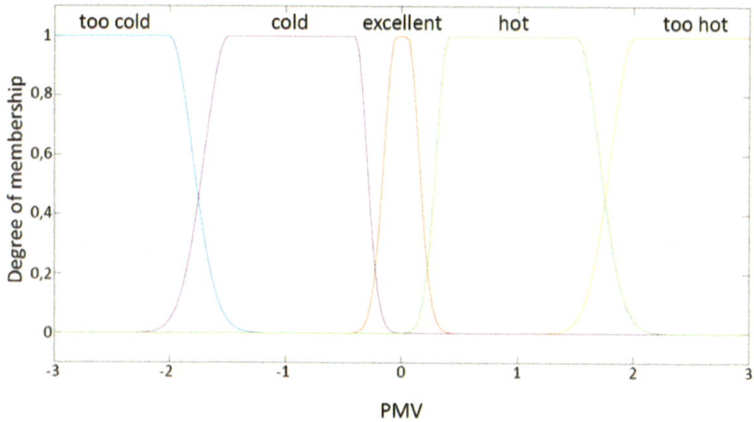


**Fig. 12.6** Membership functions regarding the PMV input of the *demand FIS*

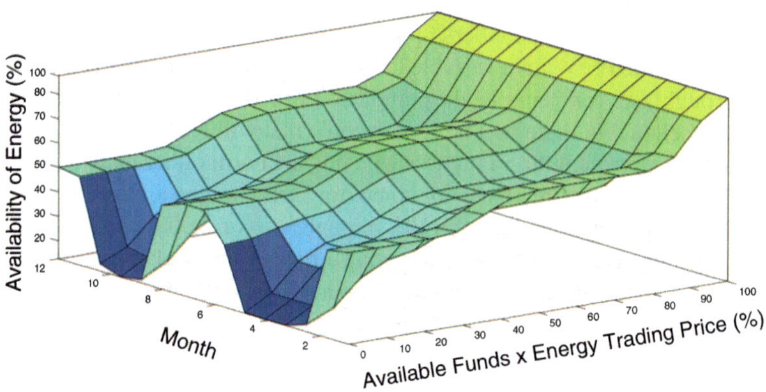


**Fig. 12.7** Graphical representation of the input/output mapping regarding the *availability FIS*

set which, after the defuzzification process, provides the final desired output. The benefit of such approach is its ability to encode apparently complex problems using just a few simple fuzzy rules, in contrast to exact approaches which usually require much greater effort.

To help visualize the aforementioned process, Fig. 12.7 depicts the mapping from the input to the output space regarding the *availability FIS*. The horizontal axes represent the inputs of the FIS while the vertical axes represent the output (availability of energy $A^E(t)$ (%)). The depicted surface incorporates the entire rule base of the *availability FIS*. The true elegance of an FIS is its capacity to encode intuitive linguistic terms into applicable information. For instance, as you can see from the plot, during the colder (Dec.–Feb.) and hotter (Jun.–Aug.) months of the year the availability of energy $A^E(t)$ is high in order to cope with the extreme weather conditions, while when we are low on funds (AF) or when the trading rate for purchasing energy ($P(t)$) is high then the availability of energy $A^E(t)$ is low in order to reflect our inability of purchasing energy from the grid.

An analogous process is performed by the *demand FIS*. The concept in this case is that the more dissatisfied the occupants are, the more "funds" they should have in their disposal in order to be able to afford to purchase energy. As you can see, in contrast to alternative approaches, ours is a human-centric one, since it configures its aggressiveness according to the occupants' requirements.

To summarize, the *availability FIS* assesses how affordable is to purchase energy according to the current rate, weather, and the available funds, while the *demand FIS* determines the amount of energy needed according to the occupants' thermal comfort. When these two FIS come to an agreement ($A^E(t) \simeq D^E(t)$), then we modify the thermostat's set-point to the agreed upon temperature. Hence, using intuitive and linguistic terms, we manage to implementing a flexible yet efficient decision-making mechanism for smart thermostats.

## 12.5 Communication Links

The communication infrastructure plays a key role at the IoT platform, since it provides the necessary information transfer gathered by the sender nodes and processed by local embedded processing nodes to the destination (e.g., actuators or next level of processing cores). Since different application domains impose variations in terms of data transferring problem, various protocols have been proposed at the context of IoT. Table 12.1 summarizes some representative approaches and provides a qualitative comparison in terms of various supported features.

Typically, such communication links can be classified either as constrained, or unconstrained networks. Specifically, an unconstrained network is characterized by high-speed communication links (e.g., wired network). On the other hand, constrained networks support relatively low transfer rates (typically smaller than 1 Mbps) and large latencies, as offered by, e.g., IEEE 802.15.4 (ZigBee) protocol. The differences between these two concern the hardware capabilities: in the former category the objective is to provide the better services by exploiting all the hardware capabilities offered; in the latter approach the devices are limited by low energy availability, since they are usually battery powered, bandwidth as low as 300 Kbps and computational power as low as few MHz. Due to the power saving constraints posed by the employed usecase, the communication link studied throughout this chapter (similar to the majority relevant implementations) relies on unconstrained network infrastructure.

Another crucial parameter for selecting the optimum communication infrastructure relies on the topology of the links (e.g., mesh, star, and point-to-point), as well as the maximum distance between nodes. Various constraints that are posed by the target application domain usually introduce guidelines about the selection of optimum communication infrastructure. Regarding the network, it is realized

**Table 12.1** Overview of available communication protocols

| | NFC | RFID | Bluetooth | Bluetooth LE | WiFi | Zigbee |
|---|---|---|---|---|---|---|
| Standard | ECMA-340 and ISO/IEC 18092 | ISO 15693 and ISO 14443 | IEEE 802.15.1 | IEEE 802.15.6 | IEEE 802.11a/b/g/n | IEEE 802.15.4 |
| Network | PAN | PAN | PAN | PAN | LAN | LAN |
| Topology | P2P | P2P | Star | Star | Star | Mesh, star, tree |
| RF frequency | 13.56 MHz | 13.56 MHz | 2.4 GHz | 2.4 GHz | 2.4/5.8 GHz | 868/915 MHz 2.4 GHz |
| Power | Very low | Very low | Low | Very low | High | Very low |
| Throughput | 106–424 Kbps | 400 Kbps | 723 Kbps | 1 Mbps | 11–105 MBps | 250 Kbps |
| Range | <10 cm | <3 m | 10 m | 5–10 m | 10–100 m | 10–300 m |
| Max. nodes | 2 | 2 | 8 | Application specific | 32 | 65,000 |
| Cost | Very low | Low | Low | Low | Medium | Medium |

with a two-level hierarchical approach, where the sensors located inside each room are supported through a local router, while the second level of abstraction provides packet routing through a central gateway (e.g., a gateway might collect information from the entire building). The smart thermostat industry historically has generated small amounts of data, as they were captured from the corresponding weather and activity sensors, respectively.

Apart from the topology, the desired functionality for data transfer is also affected by the employed communication protocol. For the scopes of our usecase, a new application-specific (fully customizable) communication protocol has been implemented. The functionality of this protocol is completely reactive, as it waits for the arrival of any packet to be processed. For every packet which is received, its type has to be analyzed. Our approach supports five packet types, as they are summarized in Table 12.2. Specifically, in case the packet type is *HELLO*, then the corresponding acknowledge (*ACK*) packet is created, which contains information about the employed DATA packet size (depending on the link's implementation), as well as its origin and destination nodes. The *START* and *STOP* packets denote the starting and finishing of data transfer between source and destination nodes. This is especially crucial since many of the available communication protocols (as they were discussed at Table 12.1) support only P2P links; thus an established link cannot be shared with other nodes. The weather/occupant's information is transmitted in *DATA* packets, while control messages (e.g., link configuration, desired link speed, selected encryption scheme, etc.) are sent in *CONTROL* packets. As a response to these packets, the node confirms their proper receive with an *ACK* packet.

Throughout this chapter, we applied the previously mentioned communication scheme as part of the underline IEEE 802.15.1 protocol in order to support the data transfers between distributed sensors and low-performance processing nodes.

**Table 12.2** Description of packets at the employed communication protocol

| Type of packet | Description |
|---|---|
| P_HELLO | This packet discovers the network |
| P_START | The P_START packet is used for establishing the connection |
| P_STOP | It is used for terminating the connection |
| P_CONTROL | Includes control messages for operating device and link |
| P_DATA | It contains data about the patient's heart rate, as well as all the necessary redundant data for error correction. Note that P_DATA packets can be adapted to carry even more information |
| P_ACK | This packet acknowledges the successful receipt of packet |

## 12.6 Evaluation

This section provides a number of experimental results that quantify the efficiency of the introduced framework. By appropriately computing the temperature set-points per thermal zone it is feasible to considerably improve the energy consumption and the occupants' thermal comfort. The experimental setup for our analysis consists of five buildings, as it was presented in Sect. 12.2. Without loss of generality both the proposed implementation and the reference control solutions adopt a 10-min time-step, i.e., each thermostat is configured once per 10 min. Additionally, we consider that people occupy the buildings only during the operating hours, while the distribution of people per room varies during the day as it is summarized in Table 12.3.

Each building is equipped with a number of weather sensors that monitor indoor/outdoor air temperature, humidity, and radiant temperature values. Furthermore photovoltaic (PV) panels are also employed in favor of minimizing the energy cost. Depending on the energy requirements the buildings can interact with the main grid in order to purchase or sell energy. The problem cannot be considered trivial due to the intermittent behavior of the solar energy, the uncertain dynamics of the buildings, and the need to meet the thermal comfort constraints for the micro-grid occupants. More thoroughly, the optimal operation for HVAC systems pre-assumes a mechanism that is able to handle in a closely coupled manner the increased number of thermal zones and cooling/heating strategies. This is a typical multi-objective problem, where the reduction of energy cost and the maximization of thermal comfort are in conflict with each other. Therefore, no single optimal solution can be found in these problems. Instead, a set of trade-offs that represent the best possible compromises among the objectives have to be computed. Regarding our experimentation, we consider that each of these cost metrics is of equal importance.

The target buildings are located in Crete (Greece), whereas our experimentation relies on real weather data [6]. Figure 12.8 plots the variation of the external air temperature and humidity values, as they have been acquired by the building's sensors. According to this figure, the variation of the external air temperature ranges between 14.6 and 37.2 °C, while the corresponding range for the external relative humidity and radiant temperature are 7.9–38.8 % and 0–47 KW/m$^2$, respectively. Therefore, proper selections of temperature set-points per thermal zone are absolutely necessary in order to minimize overall cost.

**Table 12.3** Summary of building properties

| Building | Surface area | Thermal zones | Operating hours | Warming-up pre-cooling | Random occupancy |
|---|---|---|---|---|---|
| #1 | 350 m$^2$ | 8 | 6:00am–9:00pm | No | Yes |
| #2 | 525 m$^2$ | 10 | 8:00am–9:00pm | Yes | Yes |
| #3 | 420 m$^2$ | 10 | 8:00am–5:00pm | Yes | Yes |
| #4 | 280 m$^2$ | 6 | 7:00am–8:00pm | Yes | Yes |
| #5 | 228 m$^2$ | 4 | 6:00am–6:00pm | No | Yes |

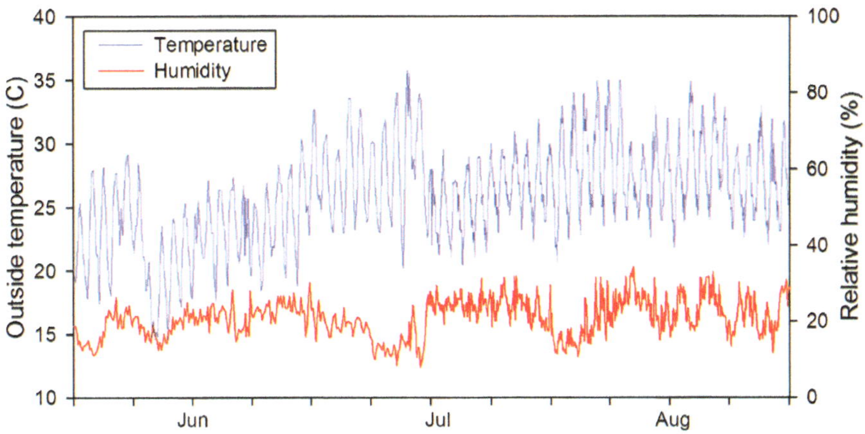


**Fig. 12.8** Variation of temperature and humidity values for the summer period

Next, we quantify the efficiency of proposed decision-making mechanism in term of computing accurately the desired temperature set-points. For demonstration purposes, two well-established control mechanisms are employed as reference solutions for this analysis. More specifically, we employ the *Pattern Search* [16] and the *Fmincon* [17] solves. Both of these techniques are implemented in Matlab's Optimization Toolbox as an open-loop optimization procedure. Note that throughout this analysis only the cooling operation of HVACs is considered. Due to the enormous computational complexity imposed by the aforementioned existing solvers, each month is studied as a separate sub-problem and is addressed individually.

A well-established metric for quantifying the efficiency of HVAC control mechanisms is the improvement of thermal comfort. For this purpose, we study this metric in terms of the predicted percentage of dissatisfied (PPD) people, which portrays the percentage of the buildings' occupants that are dissatisfied with the current thermal conditions. The results of this analysis regarding the two existing solvers, as well as the introduced decision making, are summarized in Fig. 12.9. For demonstration purposes, these results correspond to the hottest summer day (middle of August). According to this figure, we can conclude that the introduced decision-making mechanism results to the minimum PPD among the alternative solutions. We have to mention that early at the morning, all the solvers exhibit increased PPD values, since people enter into a "closed" building. However, even in this case, the proposed framework leads to the minimum overhead.

Along with the occupants' thermal comfort, we are equally interested in reducing the total energy consumption and thus the energy expenses as well. Figure 12.10 depicts the variation of energy cost regarding the alternative decision-making mechanisms for the same August's day. To be consistent, this analysis does not take into account the power saving from PV panels. This analysis indicates that our framework achieves to reduce energy demand compared to Fmincon solver (this solver exhibits similar performance in term of PPD). Although the pattern

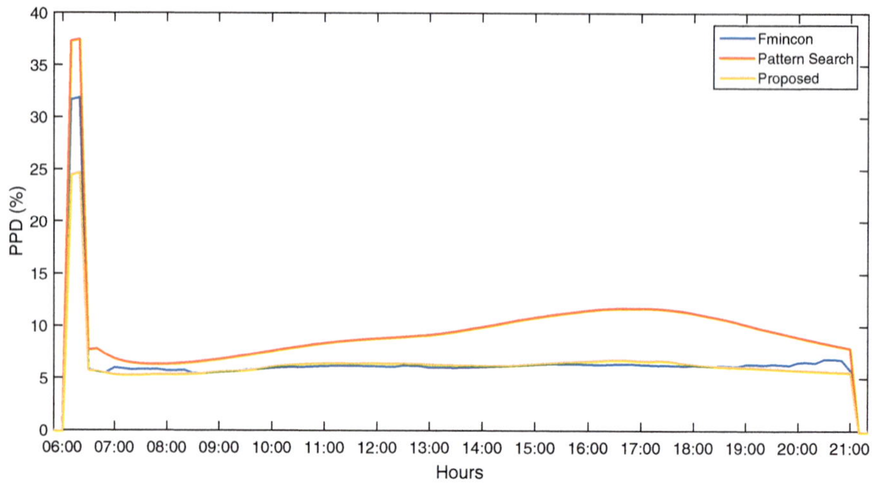


**Fig. 12.9** Variation of thermal comfort during an August's day

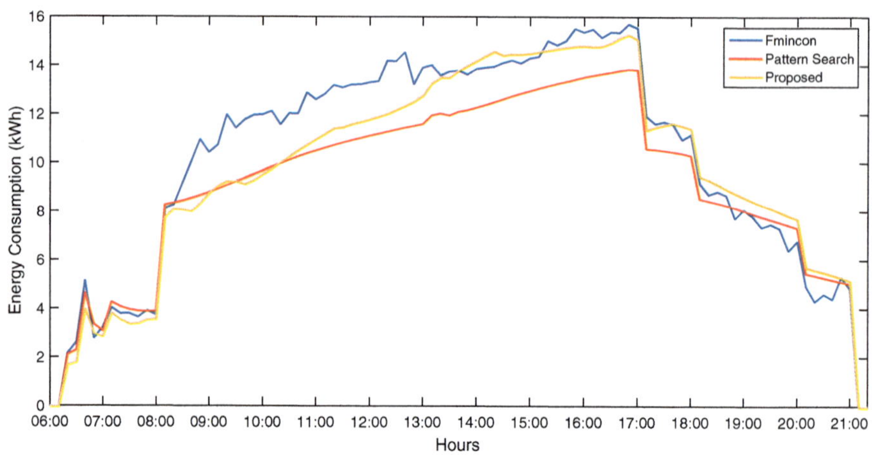


**Fig. 12.10** Variation of energy consumption during an August's day

search solution seems to improve further the energy consumption, it also leads to higher PPD values; thus, it could not be though as an overall good approach (by taking into consideration both of the cost metrics). Note that similar results for energy consumption and thermal comfort are also retrieved for the rest days of our experiment.

Finally, in order to discuss the overall efficiency of our decision-making framework, as compared to existing approaches, we quantify the total cost for the entire 3 month experiment (summer period). As we highlighted previously, the selected experiment duration (summer period) exhibits increased temperature and humidity

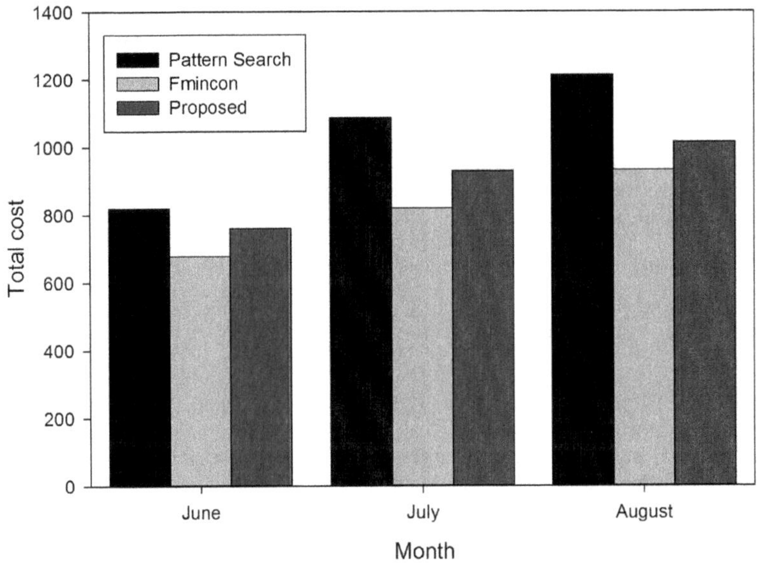


**Fig. 12.11** Overall cost of alternative decision-making mechanism for the summer period

values, which should be considered during the building cooling procedure. For this analysis, both occupant's thermal comfort and the total energy cost are equally considered. The results from this experiment are summarized in Fig. 12.11. For demonstration purposes, these results are plotted in normalized manner over the results derived from pattern search solver. Based on this analysis, we might conclude that our solution is by far more efficient as compared to pattern search solver (average improvement by 8–16 %), while the Fmincon solver achieves an additional average enhancement by 8 %. Apart from the efficiency of alternative solvers to compute the optimum (or optimal) temperature set-points, the computational complexity of these approaches is also of high importance. We have to notice that the proposed solution can compute the output much faster than the existing approaches.

## 12.7 Conclusions

A novel framework for implementing a flexible yet efficient decision-making mechanism for large-scale IoT platforms was introduced. To support this functionality, a number of connected smart thermostats that have the capability to monitor their own performance, to classify, to learn, and to take proper actions are employed in a building environment. The proposed solution evolves computational intelligence in order to maximize the occupant's thermal comfort without affecting the overall energy cost. Experimental results highlight the superior of the introduced framework compared to existing approaches. Also, it is important to highlight the considerable lower computational complexity imposed by the proposed solution, proving it ideal for employment as part of a smart thermostat.

**Acknowledgements** The research leading to these results has been partially funded by the European Commission FP7-ICT-2013.3.4, Advanced computing, Embedded Control Systems, under the contract #611538 (Local4Global, http://www.local4global-fp7.eu/). Also, this work presented is partially supported by the FP7-2013612069-2013-HARPA EU project.

## References

1. HiPEAC, HiPEAC Vision 2015 (2015) (available online: https://www.hipeac.net/publications/vision/)
2. J. Gantz, D. Reinsel, The digital universe in 2020: big data, bigger digital shadows, and biggest growth in the far east - United States, in *IDC iView: IDC Analyze the Future*, vol. 2007, pp. 1–16 (2012)
3. D. Estrin, D. Culler, K. Pister, G. Sukhatme, Connecting the physical world with pervasive networks. IEEE Pervasive Comput. **1**(1), 59–69 (2002)
4. ITRS, International Technology Roadmap for Semiconductors (2013) (available online: http:// http://www.itrs.net/)
5. C. Bertin, C. Guillon, K. De Bosschere, Compilation and virtualization in the HiPEAC vision, in *2010 47th ACM/IEEE Design Automation Conference (DAC)*, pp. 96–101 (2010)
6. National Renewable Energy Laboratory, Weather Data Sources (2016) (available online: https://energyplus.net/weather)
7. National Renewable Energy Laboratory, EnergyPlus™ (2016) (available online: https://energyplus.net/)
8. Modelica Association, Modelica (2016) (available online: https://www.modelica.org/)
9. L. Wei-Min, J. Doyle, $\mathbb{H}\infty$ control of nonlinear systems: a convex characterization. IEEE Trans. Autom. Control **40**(9), 1968–1675 (1995)
10. T. Baúar, P. Bernard, *$\mathbb{H}\infty$-Optimal Control and Related Minimax Design Problems: A Dynamic Game Approach* (Birkhäuser, Boston, 1995)
11. D. Mayne, J. Rawlings, C. Rao, P. Scokaert, Constrained model predictive control: stability and optimality. Automatica **36**(6), 789–814 (2000)
12. L. Magni, G. Nicola, L. Magnani, R. Scattolini, A stabilizing model-based predictive control algorithm for nonlinear systems. Automatica **37**(9), 1351–1362 (2001)
13. D. Bertsekas, *Nonlinear Programming*, 2nd edn. (Athena Scientific, Belmont, MA, 2004)
14. M. Trčka, J. Hensen, M. Wetter, Co-simulation of innovative integrated HVAC systems in buildings. J. Build. Perform. Simul. **2**(3), 209–230 (2009)
15. T. Nghiem, G. Pappas, Receding-horizon supervisory control of green buildings, in *American Control Conference (ACC)*, pp. 4416–4421 (2011)
16. C. Audet, J.E. Dennis Jr., Analysis of generalized pattern searches. SIAM J. Optim. **13**(3), 889–903 (2002)
17. R.H. Byrd, J.C. Gilbert, J. Nocedal, A trust region method based on interior point techniques for nonlinear programming. Math. Program. **89**(1), 149–185 (2000)

# Chapter 13
# Fuzzy Inference Systems Design Approaches for WSNs

**Sofia-Maria Dima, Christos Antonopoulos, and Stavros Koubias**

## 13.1 Data Classification in WSNs

The primary objective of data classification technique utilization in WSNs is application of sophisticated methods on excessive volumes of data aiming to achieve accurate classification results [1]. The increasing use of sensing units, the need for data acquisition and knowledge application upon acquired data, as well as the need for definite confirmation regarding the presence of an ambiguous event arise the necessity for adequate data mining techniques development, based on multiple modalities. The multi-modality events are based on the relational sequences of single-modality events and require adequate fusion process so as to derive a safe conclusion, able to activate respective actuation units. Furthermore it is argued that increased accuracy can minimize unnecessary communications between the actuation units, thus enabling the reduction of the communication overhead for sensor-actor communication [2], thus resulting into significant performance and behavior benefits for the whole system.

To implement such fusion processes, regions of interest are constructed and the local sensors' values are processed by a classification algorithm, which can be executed by a node or a subset of nodes (centralized/distributed approach) reducing the number of direct transmissions to the actuation units. Considering the centralized architecture, a central server maintains all sensors' readings. The sensors send their values to this server node (also denoted as cluster head in WSN networks), which is responsible to perform all aspects of the classification algorithm. In this way, the whole processing effort is placed on the cluster head node, which is responsible to extract from the data flows the high-level information needed for efficiently

S.-M. Dima • C. Antonopoulos (✉) • S. Koubias
Department of Electrical and Computer Engineering, University of Patras, Patras, Greece
e-mail: cantonop@ece.upatras.gr



G. Keramidas et al. (eds.), *Components and Services for IoT Platforms*,
DOI 10.1007/978-3-319-42304-3_13

monitoring of the system. On the distributed approach the CPU load is segmented, aiming to achieve balanced resource expenditure amongst nodes, primarily related to energy consumption [3].

Several data classification algorithms have being proposed and utilized towards event detection problem resolution with fuzzy logic systems and Bayes Networks [4, 5] comprising two prominent categories. Fuzzy logic is a well-adapted algorithmic framework for multi-sensor scenarios. It effectively tackles uncertainty and inaccuracy and it can be built based on experts' experience. It is also of significantly lower computational cost compared to other approaches such as support vector machines advocating respective utilization in WSNs and CyberPhysical systems. The use of fuzzy systems in classification problems has also been highlighted by several research efforts [5, 6]. The imprecise sensors' values, the complexity of accurately determining thresholds, when more than one parameters are involved, in an event detection scheme as well as the potentially conflicting rules, are some of the reasons highlighting the inadequacy of crisp values approach to describe WSN events. Fuzzy logic, on the other hand, can address these challenges more efficiently than crisp logic.

As a proof of concept, in this chapter we consider as a use case scenario a healthcare assessment application, able to evaluate the health status of a monitored person, while a fuzzy inference system (FIS) is utilized to accurately and reliably classify the health status. The healthcare application serves as a use case study. The healthcare domain is a critical, for many reasons, domain posing significant challenges. Respective difficulties pertain to the fact that heterogeneous sensor values are semantically related, and thus they have to be appropriately merged so as to derive a safe conclusion. There are two main factors of paramount importance in any WSN related application scenario. Firstly the detection of abnormal situations and the emergent response, and secondly the minimization of false alarms rate. In that respect, the use of fuzzy logic seems to offer a well-fitted solution [6]. In addition, fuzzy logic can efficiently handle sensors measurements' fluctuations (due to sensors' hardware imprecision), thus leading to a reduction of false alarms rate compared to conventional approaches like crisp logic. However the output of the fuzzy logic system requires that the pattern of the input values is always in time while only a marginal deviation between their timestamp and that of the FIS is acceptable. These two assumptions are at stake in case of wireless sensor networks, as varying and unreliable delay as well as the packet loss ratio (and in general networking related phenomena) can compromise reliable execution of the fuzzy system. Hence, the vital importance of health related applications poses the necessity for comprehensive study regarding efficient FIS under several networking conditions. In addition techniques that can reduce the effect of network quality of service to the algorithm's performance (i.e., centralized distributed approach) also represent critical objectives. The error prone nature of the wireless communication is a notorious constraint in all cases and even more in demanding scenarios such as healthcare applications considered here. Consequently, both the nature of the application (requiring advanced methods of data management, i.e., fuzzy logic system) and the nature of the environment (i.e., wireless sensor networks) pose numerous of challenging issues making it an ideal application scenario to be tackled through fuzzy logic based techniques.

### 13.1.1 Fuzzy Logic Critical Definitions

The main objective of this section is to highlight the most important fuzzy logic definitions and concepts required, so as to better introduce the design and implementation approaches presented in the following sections. Therefore it is not intended to provide a thorough or comprehensive introduction in the theory of fuzzy logic. In 1965, Zadeh [7] introduced Fuzzy Theory as a method for handling inaccuracy and uncertainty in many practical problems. The classic Set Theory is based on the fundamental principle that an item either does or does not belong to a specific set. In contrast, fuzzy logic expands the answer to the above question, as it accepts that an element can both belong to a set and not belong to the same set, with respect to its degree of involvement in it. Therefore, the fuzzy logic is a valuable extension of the classical binary (true or false) logic.

Consequently, the imprecision or ambiguity is the core of fuzzy sets and fuzzy logic. The introduced fuzzy sets are essentially a generalization of the classical sets. Furthermore, people usually do not think in terms of just symbols and numbers, but with vague/fuzzy terms. These fuzzy rules define categories which are not completely separated and well-defined sets. The alteration from one category to another is gradual, and the transition occurs when condition(s) related to a specified class occur.

In essence, fuzzy sets are functions that reflect a variable which is a member of the set, in the range [0, 1]. This value indicates the degree of correlation between the variable and the fuzzy set. When this degree is zero (0), it indicates that the value does not belong to the set, whereas when it is one (1), then it means that this value fully represents the fuzzy set. This degree is determined by the membership function of the fuzzy set, as a fuzzy set is fully described by its membership function. In practice the function participation can be extracted from:

1. Subjective assessments
2. Ad-hoc and simplified forms
3. Incidence and probabilistic
4. Natural measurements
5. Learning and adaptation processes (e.g., neural networks).

The simplest of these membership functions are formed by straight lines. Typical respective examples include the triangular and trapezoidal membership functions. Gaussian membership function is also widely used, and it can be structured in the form of a single or synthesis if different Gauss distributions. The Gaussian can be often used in fuzzy sets because of its normality. It also has the advantage of maintaining non-zero values at all points. There are also a lot of polynomial curves, which can be used as membership functions like Z or sigmoidal (S).

Another critical characteristic concerns the specialized variables encountered in fuzzy logic theory. Specifically, in such cases variables' values are not numbers, but words or phrases in a natural or artificial language. These values are represented by fuzzy sets and therefore the linguistic variables are also known as fuzzy variables. In order to understand a fuzzy logic based mechanisms it is also important to define fuzzy reasoning. Also denoted as fuzzy inference, it is a process through which vague conclusions can be extracted. In other words it represents a computing system, which evaluates fuzzy rules in the form: "if x a then y = b," via fuzzy procedures. A FIS is one of the most widespread applications of fuzzy logic. Such a system consists of the following parts:

1. A number of unclear rules of the form IF-THEN (fuzzy knowledge base).
2. A fuzzy reasoning engine.
3. A fuzzification unit, which converts the input data to fuzzy sets.
4. A defuzzification unit, which converts fuzzy conclusions/decisions in a specified format.

The fuzzy knowledge base, apart from fuzzy rules, usually contains a numerical database section which is required so as to extract processed results. The rules are usually taken by experts or/and by simulation procedures. The fuzzy reasoning engine is the core of the fuzzy system and contains the logical decision making.

The fuzzification unit performs the following tasks:

- Measures the (non-fuzzy) values of the system's
- Displays areas of change of the inputs' values at appropriate supersets
- Fuzzifies the incoming inputs' values. It converts them in a fuzzy linguistic form.

The defuzzification unit performs the following two operations:

- Displays areas of change of the outputs' variables in respective supersets
- Defuzzify the results given by the fuzzy reasoning engine. It converts them into deterministic (non-ambiguous) format, for further use by subsequent systems or decision processes.

In the processing level the variables expressed by classical numbers are introduced to the system. These variables are going through the fuzzification process, which translates classical data (crisp values) into fuzzy quantities. These fuzzy quantities are expressed in the form of linguistic variables. Then these fuzzy variables are inserted into the fuzzy rule engine, where through fuzzy rules, set by the stage of the knowledge base, fuzzy conclusions are produced. These conclusions are in the form of linguistic variables. At the last stage of the process, the last variables are converted into crisp values, via the defuzzification unit.

There are three prevalent methodologies that can be applied in the fuzzy logic: Mamdani systems, the diagrammatic method, and the Sugeno type systems [8].

Design and implementation techniques presented in this chapter follow the Mamdani system. To compute the output of this FIS given the inputs, one must go through six steps:

1. Determining a set of fuzzy rules
2. Fuzzifying the inputs using the input membership functions
3. Combine the fuzzified inputs according to the fuzzy rules to establish a rule strength
4. Finding the consequence of the rule by combining the rule strength and the output membership function (implication)
5. Combine the consequences to get an output distribution (aggregation)
6. Defuzzify the output distribution

Finally two prominent defuzzification methods are as follows:

1. *Maximum Defuzzification Method.* According to the method of clarifying maximum, the discrete value is corresponding to the maximum affinity value of the final result. If there is more than one of these values is taken as the case may either the average (middle-of-maximum) or the maximum value the (largest-of-maximum) or the minimum value (smallest-of-maximum).
2. *Centroid Defuzzification Method.* According to the method of centroid, the discrete value is resulting from the center of gravity of the final function of the fuzzy output parameter.

## 13.2 Design of a Fuzzy System

A critical parameter in knowledge acquisition is whether an expert is involved in the process. In case an expert is indeed participating, knowledge acquisition can be further characterized as either indirect or direct. In the former case the designer of the fuzzy system extracts required information by interacting with the expert or experienced operator. In contrast direct knowledge acquisition is achieve when the designer itself is an expert operator and thus can reflect the system's behavior correctly. On the other hand, when an expert is absent, the process is denoted as "automatic" and it involves the use of statistical modeling approaches and aims to extract knowledge from already acquired data [5, 9].

Although a complete FIS can be based solely on acquired data, and indeed various research papers try to address this problem through data modeling, it is noted that such approach is often prone to important shortcomings. On one hand, this approach cannot be applied when amount of data is not sufficient. On the other hand, utilized membership functions, thresholds, and extracted rules may not be able to realistically represent real-life scenarios and real phenomenon. Furthermore, the characteristics of the training set drastically influence quality of the extracted rules, while the quantity of considered cases of the phenomenon also determine the degree of generality regarding the induced rules.

Such issues can be tackled by experts' knowledge, complementary exploited to the extracted FIS derived from statistical data. This is because experts are familiar with the characteristics of fuzzy system variables and, based on the extended experience, can express them as linguistic rules. A domain expert is able to provide us with a global view of the system behavior, describing the most influential variables and using them in a few basic rules.

However, the two main challenges an expert is expected to address concern the definition of a variable and the variable interaction formalization. This is particularly important since interaction between knowledge and data can be achieved not only at various levels such as modeling but also at prediction or/and decision levels. A critical advantage of involving an expert concerns the fact that a linguistic rule definition can be made while respective data are absent.

Either expert rule and model validation with data (FETD—First Experts Then Data), or rule automatic generation and expert assessment of induced knowledge (FDTE—First Data Then Experts), is possible [9].

- FETD: Build an expert knowledge base (KB), and then complete it using the knowledge extracted from experimental data. Data are used to define the precise meaning, in the numerical space, of an expert linguistic concept.
- FDTE: Experts give a linguistic meaning to rules induced from data, thanks to system interpretability. Build an induced KB, and then look for an expert able to evaluate and refine it.

Specifically fuzzy partition and rule design is utilized where knowledge can complement automatic design, whereas member functions and rules are added in areas where no data are available, FIS parameter optimization, and system validation. In this way it is possible to both evaluate respective accuracy and analyze induced knowledge. Furthermore, cross-validation procedure can be employed regarding system's generalization, pinpoint areas of low performance, and analyze the aforementioned interaction between data and inference rules. As a downside of fuzzy formalization respective complexity increase must be highlighted, which is justified in most of cases considering the performance enhancements as well as resource conservation that can be offered.

## 13.3 Literature Review of Applying Fuzzy Logic in WSN Scenarios

In current literature there are several applications which utilize the fuzzy logic and targeting in WSN environments. Specifically, in [10], a fuzzy rule based system for fire detection is proposed, based on environmental values, while in [11] a fuzzy logic system for controlling home appliances is proposed, so as to create the smart home environment. The same use case scenario is considered in [6], where the authors offer a valuable analysis, targeting reduction of false alarms in fire-detection scenarios, under the utilization of the fuzzy logic framework. In addition, they proposed the use of spatial and temporal characteristics into the rule-base so as to increase the system's accuracy. Additionally, novel techniques are proposed that reduce the exponential growth of the rule-base without compromising the accuracy of event detection. As the authors claimed in their manuscript: the main goal of this work is threefold (a) recommend fuzzy logic as an accurate technique for detecting various events (b) propose new characteristics that will serve as inputs to the fuzzy engine so as to reduce the number of false alarms, and (c) propose new techniques for reducing the rule set length, so as to allow its storage to the limited sensor's memory. The evaluation analysis was performed utilizing the FuzzyJ Toolkit software, while the authors used the fire events public database from NIST so as to evaluate their system. At this point it is important to note that this software is not a network simulator so no realistic effect has been evaluated with respect to networking constrains such as sensors disconnection, packet loss, and latency. Moreover, an energy estimation on how this algorithm utilizes sensor's resources is also omitted. Furthermore, a congestion fuzzy logic estimation model for QoS in wireless sensor network is presented in [12], while the evaluation results confirm that the proposed FIS is able to efficiently sort out traffic and minimize packet loss, even in critical scenarios. However, although the fuzzy system is executed online, there isn't any specific application scenario under investigation, while the scheduling of the above system to the sensor nodes is not taken into consideration as well as the event triggering of this system. In addition, the authors of [13] present an activity recognition architecture utilizing multi-node collaboration and fuzzy logic, in order to produce a reliable recognition result from unreliable sensor data.

Moreover, in [14] it is argued that fuzzy logic can be effectively used to enable location estimation in mobile robots, while they emphasize on the computational requirements reduction and heuristic determination of combination rules. Nevertheless, none of the above papers present evaluation results with respect to required energy for executing the classification system or comparative metrics of the performance and the sensitivity of the fuzzy algorithm, under critical networking conditions. Also, in [15] the authors propose an effective fuzzy based faulty readings detection model in wireless sensor networks. The proposed model is able to detect the faulty readings and overcome the faulty nodes. More specifically, the difference between the sensed values of neighboring nodes was used as a parameter to determine faulty nodes and was represented by a linguistic variable in fuzzy logic. In the proposed framework, after error detection localization, confidence value of each node was obtained, which used to detect faulty readings. Also, the FIS was used to detect faulty readings in WSN. The application of the fuzzy logic in the healthcare domain has also been addressed by several studies in current literature. However, networking conditions are not taken into consideration in order to evaluate their impact to the final outcome of the system. More specifically, in [16] the authors have proposed a remote monitoring solution which allows the elderly people to live at home in safety. In the same way, Medjahed et al. have also presented in [17] a fuzzy logic system targeting recognition of daily activities, considering ubiquitous healthcare scenarios. However, both papers do not consider networking challenges, i.e., potential packet loss, or untimely arrivals of packets.

## 13.4 Designing a Health Status FIS for Wireless Sensor Networks Application Scenario

Driven by previous sections, the main objective of this section is to present an example of fuzzy logic application in the context of a realistic as well as demanding WSN application scenario. Both the design and implementation phases are discussed.

### *13.4.1 Challenges of Applying Data Mining Techniques in WSNs*

Healthcare assessment using specific indicators is a valuable tool for monitoring the elderly and detecting distress situations. In order to maximize classification accuracy, an intelligent system able to acquire and apply knowledge is of utmost importance. Currently, typical detection of critical situations is achieved using only one modality (i.e., health related sensors or sensors from individual's surroundings). However, exploitation of single-modality systems does not provide a comprehensive view of the individual's environment and situation. On the contrary, using multiple types of sensors allows the incorporation of various types of information so as to obtain higher level information. Moreover, in the case of an ambiguous situation it is essential to confirm the event detection, using several monitoring modalities. So the multi-sensor data fusion refers to the use of data by multiple sensors in a sophisticated process dealing with the correlation, aiming to achieve refined decision and classification and to improve results. Fuzzy logic is a well-adapted approach for multi-sensor scenarios.

However, a fundamental challenge in event detection is that ubiquitous sensors transmit data over an unreliable, error prone wireless medium often encountered in WSN/CPS environments. At the same time, these data should be processed rapidly in order to detect distress situations, while coupling with strict time constraints. Moreover, additional challenges are particularly related to the heterogeneity of the collected data, to factors of influence, and to the mutual dependencies between sensors values. Another critical issue is that a sensor's value includes a factor of uncertainty. These multidimensional and heterogeneous characteristics should be correlated, in order to derive high-level information. Thus, the use of data fusion is justified in CyberPhysical systems.

In WSNs, the wireless communication error prone nature, in combination with high performance demands, further augments uncertainty. This means that the requirements of the decision-making plane have to adjust with the requirements of the communication plane so as to take the appropriate action, while adhering demanding time constrains. For example, the time stamp that sensors' values arrive at the fusion center is not always the same, i.e., sensor's value may arrive after the fusion process or may be lost due to poor network's conditions. Traditional data mining approaches do not consider these problems. In WSNs applications the fusion is a run time process, that requires the availability of all inputs at this specific time stamp. Techniques for optimizing network performance are surely essential, but data fusion should suggest a safe action scenario also in case of network deterioration.

## 13.4.2 Use Case Scenario

All these challenges are going to be analyzed, taking into consideration a use case scenario for health monitoring [18]. A network driven event detection scheme, which is based on fuzzy logic techniques, is described in detail. More specifically, the main objective of the use case scenario is that sensors from a body area network and from monitoring individual's surroundings are processed in order to detect the individual's distress situations [19]. The accuracy of the implemented fuzzy system is evaluated assuming a reliable and robust communication method [18]. Furthermore, this fuzzy system is assumed to be deployed in a critical WSN environment. Therefore, the level of reliability of the communication model should be examined in order to derive a safe decision [20]. For these purposes, a second fuzzy system is implemented to provide the rate of network confidence, expressing the robustness and the reliability of the network. Hence, the proposed mechanism is composed of two main modules: assessment of health status and evaluation of network confidence and reliability.

### 13.4.2.1 Healthcare Assessment Fuzzy Inference System

The working set of input fuzzy variables in the proposed distress situation detection algorithm is comprised of: Sound Class (SC) recognition, Heart Rate (HR), Activity (A), Fall recognition (F), and Time (T) while health status (HS) is the resulting output variable. The membership functions are described in Table 13.1, while an indicative representation of the sound class and the health status is presented in Figs. 13.1 and 13.2. The sound signals are gathered and processed independently so as to recognize normal sounds (speech) or alarm sounds (screaming) and are labeled on a log scale according to their alarm level. The membership functions are defined on each input variable as trapezoidal, triangular, and singleton functions. According to [19], the limits of each membership function are determined. The thresholds for the individual's parameters are defined by physicians and can be easily adjusted. The

**Table 13.1** Membership functions of the fuzzy variables

| | |
|---|---|
| Heart rate | Alarm: trapezoidal (0, 0, 30, 50) + trapezoidal (140, 160, 199, 199) |
| | Possible alarm: trapezoidal (30, 50, 60, 80) + trapezoidal (110, 130, 140, 160) |
| | Normal: trapezoidal (60, 80, 110, 130) |
| Sound class | Possible-alarm: triangular (7.5, 15, 22.5) |
| | Alarm: trapezoidal (12.5, 22.5, 30, 30) |
| | Normal: trapezoidal (5, 5, 10, 10) |
| | No-signal: trapezoidal (1, 1, 1, 1) |
| Activity | Immobile: trapezoidal (0, 0, 7, 10) |
| | Rest: trapezoidal (7, 10, 20, 30) |
| | Normal: trapezoidal (20, 30, 40, 60) |
| | Tense: trapezoidal (40, 60, 100, 100) |
| Fall detection | No-fall: singleton (0) |
| | Fall: singleton (1) |
| Time | Day: Trapezoidal (6, 8, 16, 18) |
| | Night: trapezoidal (0, 0, 6, 8) + trapezoidal (16, 18, 24, 24) |
| Health status | Alarm: Gaussian (18, 11.98) |
| | No-alarm: Gaussian (18, 87.35) |

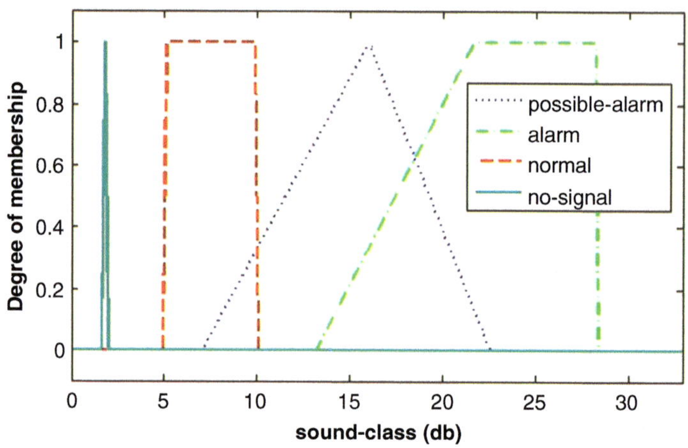


**Fig. 13.1** Membership function of the sound class 1

core of the fuzzy system is the rule based system, which leads the inference system to produce the fuzzy outputs. Using the number of linguistic variables for each input, the total number of rules is calculated to be 192 ($4 \times 3 \times 4 \times 2 \times 2$). A subset of them is selected experimentally so as to prevent degradation of the system's accuracy and predictability, while cutting down the total number of rules, which facilitate complexity and network resource expenditure reduction. The fundamental rules (20 in total) have been extracted using medical studies [21] as references and have been evaluated by physicians.

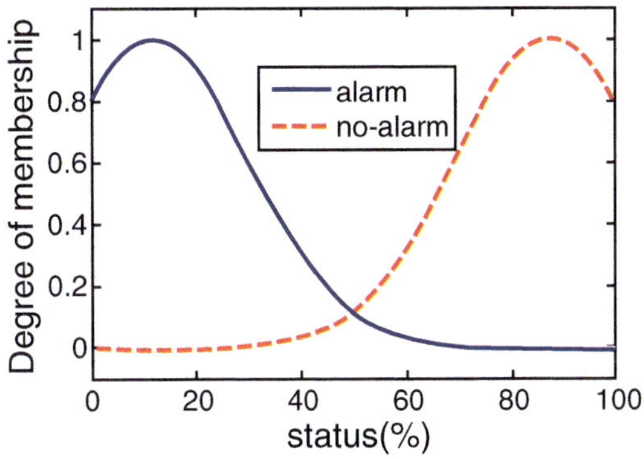


**Fig. 13.2** Membership function of the health status output

By evaluating this set of rules using random attributes within the predefined limits, merging rules, and using physician's expertise we extracted the following 19 rules (Table 13.2).

The utility of fuzzy logic can be shown with an indicative example. Assuming the following values for each fuzzy input: Sound Class: 6.24 (limits 0–30), Heart Rate = 100 beats per minute, Activity = 56 (limits 0–100), Time = 9 a.m, and Fall = 0 (0: false, 1: true). In such a scenario, since the value of heart rate is quite high the crisp logic will conclude that there is an abnormal situation. However, according to fuzzy logic the well-being is 80.2 % (Fig. 13.3), clearly contradicting the result of the crisp logic.

#### 13.4.2.2 WSN QoS Assessment Fuzzy Inference System

Taking into account the demanding scenarios, a critical aspect not typically considered is that as the health status identification is an event-driven application, it requires accurate and time constrained data delivery. Otherwise the fuzzy system may use the previous value of a sensor or a mean value of the previous readings. However, as it is extracted through respective measurements, in such a case the output of the fuzzy system is strongly affected by the network performance capabilities and robustness parameters. This is a significant input related to network conditions that we have to take into consideration, so as to trigger the most suitable action scenario. As an example, the detection of a distress situation during monitoring elderly people should be recorded in a medical database, so as to update his medical history. If all inputs are not updated due to poor network conditions, this alarm situation will have low confidence rate. Therefore the output should not be taken into consideration. Data accuracy is a metric of the service quality that the network offers to applications. In our use case scenario, data accuracy is critical as a metric of confidence level. Data accuracy (DA) is related to the delay (D), the packet loss ratio (PL), and the packet delay variance (jitter) (J).

**Table 13.2** Rule engine in healthcare assessment fuzzy inference system

| IN | R1 | R2 | R3 | R4 | R5 | R6 | R7 | R8 | R9 | R10 | R11 | R12 | R13 | R14 | R15 | R16 | R17 | R18 | R19 |
|---|---|---|---|---|---|---|---|---|---|---|---|---|---|---|---|---|---|---|---|
| Heart alarm | x | ✓ | x | x | x | x | x | x | x | x | x | x | x | x | x | x | x | x | x |
| Heart normal | x | x | x | x | x | x | x | ✓ | ✓ | ✓ | x | ✓ | ✓ | x | x | x | ✓ | ✓ | x |
| Heart possible alarm | ✓ | x | ✓ | x | ✓ | x | ✓ | x | x | x | ✓ | x | x | x | ✓ | ✓ | x | x | ✓ |
| Sound no Signal | x | x | x | x | x | x | x | x | x | x | ✓ | x | x | x | x | x | x | x | x |
| Sound alarm | ✓ | x | x | ✓ | x | x | x | x | x | x | x | x | x | x | x | x | x | x | x |
| Sound normal | x | x | x | x | x | x | x | p | x | x | x | x | x | x | x | x | x | x | x |
| Sound possible alarm | x | x | x | x | ✓ | ✓ | x | x | x | ✓ | x | x | x | x | x | x | x | x | x |
| Activity immobile | ✓ | x | x | x | ✓ | x | x | x | x | x | ✓ | x | x | x | x | x | x | x | ✓ |
| Activity normal | x | x | x | x | x | x | x | ✓ | x | x | x | x | x | x | x | ✓ | x | ✓ | x |
| Activity rest | x | x | ✓ | x | x | x | x | x | x | ✓ | x | ✓ | x | x | x | x | x | x | x |
| Activity tense | x | x | x | ✓ | x | x | ✓ | x | x | x | x | x | x | x | ✓ | x | ✓ | x | x |
| Fall | x | x | x | x | ✓ | x | x | x | x | ✓ | x | x | ✓ | x | x | x | x | x | x |
| Not fall | ✓ | x | x | x | ✓ | x | x | x | ✓ | ✓ | x | x | x | x | x | x | x | x | x |
| Day | x | x | x | ✓ | x | ✓ | x | x | x | x | x | x | x | x | x | x | x | x | x |
| Night | x | x | ✓ | x | x | x | x | x | x | x | x | ✓ | x | ✓ | x | x | x | x | x |
| Health status | A | A | A | A | A | A | NA | NA | NA | NA | A | NA | NA | A | A | A | NA | NA | A |

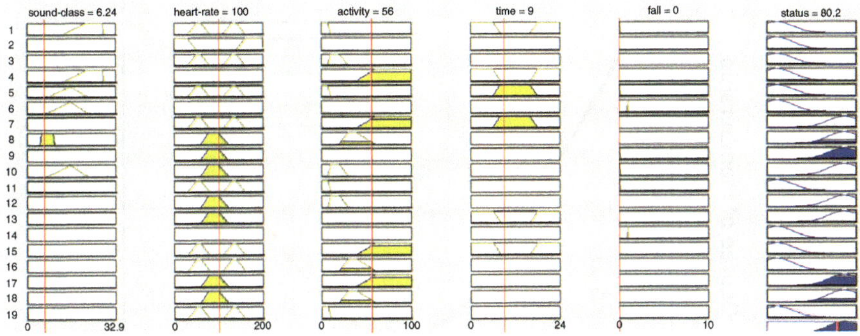


**Fig. 13.3** Evaluation of FIS in a specified input pattern

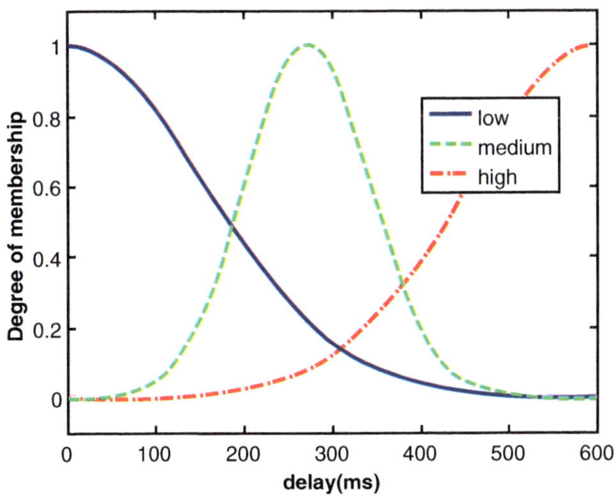


**Fig. 13.4** Fuzzy input delay of network reliability system

The limits of each network reliability related input as denoted in Figs. 13.4, 13.5, and 13.6 are configured in order to satisfy the minimum requirements of healthcare applications as described in [22] and [23]. For this reason, the packet loss ratio cannot exceed 3 % at health applications. Otherwise the application is out of its main goal. The data accuracy output is depicted in Fig. 13.7. Gaussian membership functions are used because they provide a smooth transition between the different levels. In addition, Gaussian membership functions facilitate continuously differentiable hyper surfaces of a fuzzy model. These functions facilitate theoretical analysis of fuzzy systems as they have derivatives of any grade. At the same time, it is noted that network conditions considered above are rather demanding as expected to apply in industrial or medical environments. These parameters are strongly affected

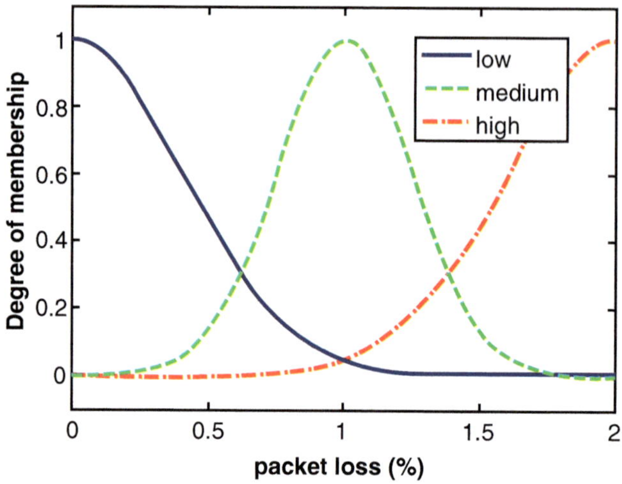


**Fig. 13.5** Fuzzy input packet loss of network reliability system

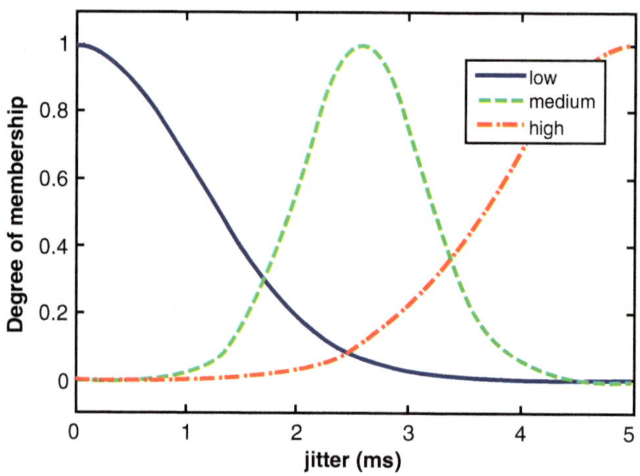


**Fig. 13.6** Fuzzy input jitter of network reliability s

by the communication protocols and traffic flow. Therefore, in cases of very low network reliability, suggestions for changing network parameters are essential. For the implementation of this system a subset of rules has been used (Table 13.3).

#### 13.4.2.3 Evaluation of Both Systems

The aforementioned fuzzy systems have been implemented based on Mamdani's inference system and the centroid defuzzification method has been used [18].

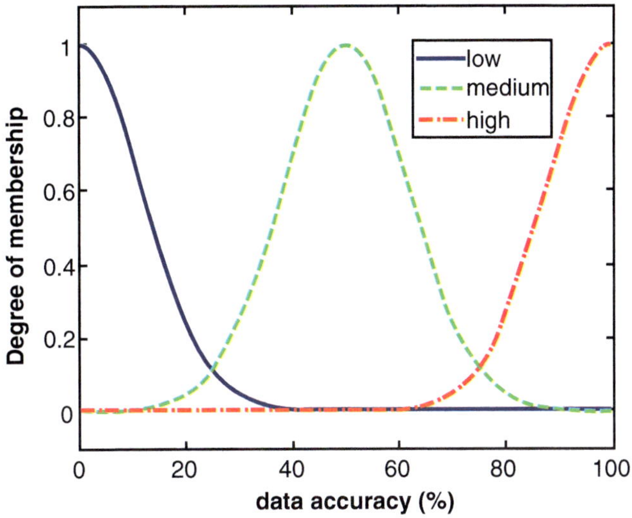


**Fig. 13.7** Fuzzy output data accuracy of network reliability system

**Table 13.3** Rule engine in WSN QoS assessment fuzzy inference system

| | | | | | | | | |
|---|---|---|---|---|---|---|---|---|
| Delay | L | L | L | M | L | M | M | H |
| Packet loss | L | M | L | L | M | L | M | x |
| Jitter | L | L | M | L | M | M | M | H |
| Data accuracy | H | M | H | M | M | M | M | L |

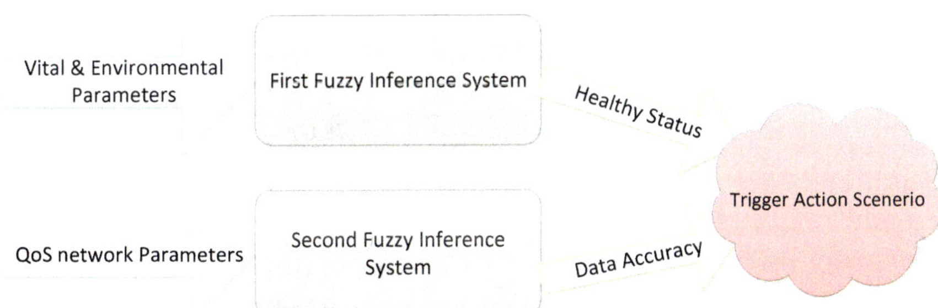


**Fig. 13.8** The combination of two systems

The defuzzified values of the aforementioned systems should be co-examined, as depicted in Fig. 13.8 in order to verify that the application's goal is achieved. The evaluation results obtained from the two FIS for the health assessment and the evaluation of the Network Reliability FIS are represented as follows in Tables 13.4 and 13.5.

**Table 13.4** Evaluation analysis of health assessment system

| Sound class (db) | Heart-rate (pulses/min) | Activity motion (%) | Time (h) | Fall/No fall (1/0) | Status (%) |
|---|---|---|---|---|---|
| 7.5 | 170 | 35 | 11 | 0 | 19.45 |
| 7.5 | 65 | 35 | 11 | 0 | 74.518 |
| 27 | 130 | 2 | 22 | 0 | 19.45 |
| 18 | 90 | 2 | 22 | 0 | 88.19 |
| 0 | 135 | 15 | 22 | 0 | 19.45 |
| 0 | 145 | 15 | 22 | 0 | 20.547 |
| 0 | 170 | 15 | 22 | 0 | 19.45 |
| 0 | 80 | 10 | 15 | 0 | 88.19 |
| 13 | 130 | 4 | 9 | 0 | 21.16 |
| 13 | 80 | 4 | 9 | 0 | 88.19 |
| 13 | 80 | 4 | 9 | 1 | 21.16 |
| 25 | 60 | 4 | 22 | 1 | 50 |
| 7.5 | 70 | 70 | 9 | 0 | 77.615 |
| 9 | 128 | 70 | 9 | 0 | 79.952 |
| 7.5 | 85 | 35 | 9 | 0 | 88.19 |
| 12 | 63 | 35 | 9 | 0 | 72.509 |
| 28 | 85 | 35 | 9 | 0 | 80.185 |
| 0 | 85 | 70 | 4 | 1 | 50.056 |
| 15 | 130 | 0 | 9 | 1 | 19.736 |

**Table 13.5** Evaluation analysis of network reliability FIS

| Delay (ms) | Packet loss (%) | Jitter (ms) | Data accuracy (%) |
|---|---|---|---|
| 20 | 0.98 | 0.65 | 53.9 |
| 60 | 0.4 | 0.65 | 81.735 |
| 4 | 0.3 | 0.3 | 96.583 |
| 60 | 0.8 | 0.65 | 57.2438 |
| 120 | 0 | 0.65 | 77.7857 |
| 120 | 0.3 | 3 | 78.0099 |
| 300 | 0.3 | 3 | 54.3374 |
| 300 | 1.2 | 3 | 50 |
| 450 | 1.8 | 3 | 12.3777 |
| 50 | 1.8 | 3 | 50.4189 |
| 50 | 0.3 | 3 | 86.5609 |
| 500 | 0.35 | 2 | 71.198 |
| 550 | 1.8 | 2 | 10.5892 |
| 200 | 1.8 | 1.6 | 33.4226 |
| 75 | 0.5 | 1.6 | 69.11 |

Data accuracy can be classified in regions so as to represent the network confidence level. The thresholds that are used for this categorization are based on application's requirements and can adaptively change online. As an example, we assume that the categorization of data accuracy is based on two thresholds 25 % and 80 %. In that respect, if the output of the network reliability fuzzy system has a value below 25 %, then the application will not satisfy its fundamental characteristics. Such an indication is critical in taking the decision for the optimum action and should be considered as the triggering event for the reconfiguration of network conditions and/or the reexamination of all nodes' existence (i.e., a sensor station may be energy-depleted). Another reaction would be the modification of the sampling periods and/or the initiation of a new synchronization sequence. On the contrary, if the network confidence is high (>80 %), the health status will have an acceptable value and thus appropriate actions should be recommended based on his vital conditions. In this case, the sampling periods should also change at the highest rates so as to monitor continuously, the critical situation. If the data accuracy level is within the two thresholds, this means that the network confidence is average. If this is due to network deterioration, this value will exceed the lower threshold and downgrade further the network's status. Otherwise, transient network failures may be responsible for the low level of data accuracy. However, this does not affect the application's goal and will be corrected immediately. In both cases the actor should control the transmission process on demand so as to recheck data accuracy. If data accuracy insists on fluctuating in average levels, then this implies deterioration of the network, so the operations of the first case scenario should be performed.

### *13.4.3 Centralized Implementation of the HealthCare Assessment FIS in a WSN Platform*

In the proposed use case scenario the set of input fuzzy variables (Sound Class, Heart Rate, Activity, Fall Detection, and Time) and the resulting output (Health Status) have been presented. The described FIS is implemented in a WSN embedded platform, and is evaluated under a networking simulation scenario [24, 25].

More specifically, the input variables are collected by the related sensor nodes and are used as input to the FIS. Mamdani's model is considered for this fuzzy system. These values are transmitted to the Fuzzy Cluster Head Node (FSHN), which is responsible for executing the whole FIS. After all vital and environmental inputs have been fuzzified, the minimum implication method is applied to each rule, taking the fuzzified inputs and applying them to the antecedent of the fuzzy rule in order to derive its weight. The outputs of all rules are aggregated using the maximum method [26]. The aggregated output fuzzy set is defuzzified using the centroid of gravity method [26] and the output is a crisp number. The aforementioned procedure is presented in Algorithm 1. The aforementioned functions of fuzzy logic (aggregation, implication, and centroid of graph) can be utilized to any FIS (e.g., environmental or industrial applications).

**Algorithm 1: Pseudocode of the fuzzy process**

```
Input: Hrt, Snd, Act, Fall, Time,
HrtMembFunc, SndMembFunc, ActMembFunc,
TimeMembFunc, FallMembFunc, HealthMembFunc, RuleEngine
Output: Health Status
FHrt = Fuzzify (Hrt)//Heart Fuzzification
FSnd = Fuzzify (Snd)//Sound Fuzzification
FAct = Fuzzify (Act)//Activity Fuzzification
FTime = Fuzzify (Fall)//Fall Fuzzification
FFall = Fuzzify (Time)//Time Fuzzification
for (i in Len (RuleEngine) do
weightrule[i] = findweight (RuleEngine
(i),FHrt,FSnd, FAct,FTime,FFall)
end
{ for (i in Len (RuleEngine)do                                  }
{    if (antencedent(RuleEngine(i) == notalarm)                 }
{   then                                                        }
{       if(weightrule(i) ≥ maxnotalarm)then                     }
{            maxnotalarm = weightrule(i)                        }
{     end                                                       } implication
{   else                                                        }
{      if(weightrule(i) ≥ maxalarm)then                         }
{         maxalarm = weightrule(i)                              }
{     end                                                       }
{   end                                                         }
{ end                                                           }
area = aggregation (HealthMembFunc, maxalarm, maxnotalarm)
Health Status = centroidofgraph (area)//defuzzification
```

The algorithm presented above can be executed in a resource rich node which is called FCHN. The heart, sound, activity, time, and fall sensor node (HSN, SSN, ASN, TSN, and FSN) are responsible for acquiring the respective data values. According to this centralized approach, these values are transmitted to the FCHN, which also serves as the fusion cell, which performs the FIS algorithm. The role of each node in the centralized scenario is described in (Algorithm 2). However, if we assume that a realistic size sensor network consists of tens or hundreds of nodes, restrictive limitations must be taken into consideration. The number of messages sent to the cluster head and the processing load of the data mining tasks, require increased processing capabilities and increase the probability of network congestion as well. Hence a decentralized application of a data mining algorithm is needed, in order to achieve more efficient operation and minimize network performance degradation. Driven by this observation the next chapter aims to address this necessity offering both a general analysis of respective approaches and presenting a specific design and implementation with respect to the Health Status FIS utilized as proof of concept throughout this chapter.

**Algorithm 2: Pseudocode of node functionality in the centralized approach**

```
i ← NodeID
if (i ! = FCHN) then
    Sense its value;
    Send to the FCHN;
else
    Receive the incoming packets;
    Execute the FIS periodically;
end
```

### *13.4.4 Decentralized Implementation of the HealthCare Assessment FIS in a WSN Platform*

Besides a purely centralized approach a distributed version is also considered. In the distributed approach, tasks are assigned to the nodes based on criteria such as, data dependencies, nodes' sensing capabilities, and locality and energy resources. Hence for the Health Status application scenario, the sensing tasks and the respective fuzzification processes can be considered as independent tasks that can be executed on the WSN node. In this way, we accomplish the on-site processing of the data. The next step of the considered application demands the evaluation of the coefficient rules. This task depends on the fuzzified inputs.

Hence it is decided to relocate it closer to the sensing nodes, and specifically on one of them. In this case this node doesn't have to transmit its own data, and thus the number of transmissions is reduced. The node that is assigned with the additional task of evaluating the coefficient of each rule was selected based on the average distance among the others and its energy resources. The other two remaining tasks: the implication and the aggregation method are the most computational intensive. However the aggregation method needs as input the weight of each rule. In Heath Status application, this means that 19 floating numbers, one for each rule need to be transmitted. If we merge these values in one packet, the packet size is increased and this affects the duration that a node occupies the medium as well as the probability of packet collision.

Moreover if these two tasks are not assigned in different nodes, the main processing workload will remain in the cluster head. For this reason we assume the node that calculates the rule of each weight should also apply the implication method. In this way, only the output of this method need to be transmitted to the cluster head, which is going to apply the aggregation method and to defuzzify the fuzzy output by finding the center of the gravity. A centralized and a distributed topology of our network is depicted in Fig. 13.9. In the centralized approach the five nodes are responsible only for the sensing and transmitting the values to the base station. Therefore, the base station will take over the reception of the sensing data

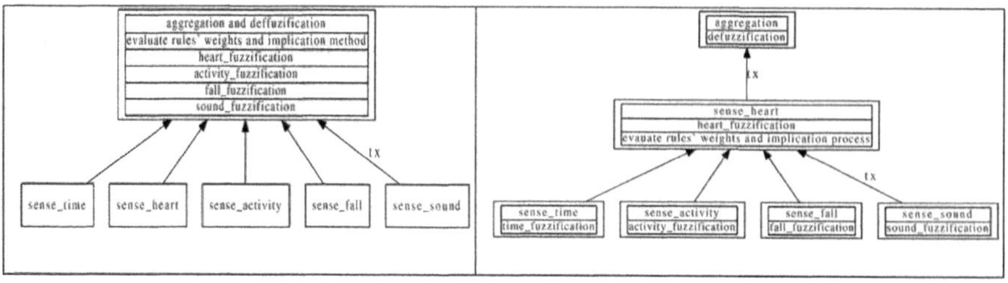


**Fig. 13.9** Centralized and distributed approach

and their processing. In the distributed approach, as we have already mentioned, one node from the FSNs is responsible for evaluating the rules' weights and for applying the implication method. This node is called coef_FSN. We assume that for a given topology the heart sensor is closer to the other fuzzy nodes. Hence for this topology the heart sensor is the coef_FSN. The remaining fuzzy related nodes sense their values and fuzzify them according to the membership functions, while the FCHN takes as input the two weights (maxalarm – maxnotalarm, see Algorithm 1). The functionality of each fuzzy related node in the distributed approach can be derived from the Algorithm 3.

**Algorithm 3: Pseudocode of node functionality in the distributed approach**

i ← NodeID
**if** (i == FSN and i != coef_FSN) **then**
    Fuzzification ();
    Send to the coef_FSN;
**else if** (i == coef_FSN) **then**
    Evaluate the weights of each rule ();
    Apply the implication ();
**else if** (i == FCHN) **then**
    Apply the aggregation;
    Defuzzify the fuzzy output;
**else**
    Send to the FCHN;

## 13.5 Evaluation Analysis of the Centralized and the Distributed Approach

As described in the previous sections the proposed FIS for event detection in the healthcare domain has been implemented in a centralized and distributed way, utilizing the well-known WSN embedded platform of TelosB (http://www.sentilla.com/moteiv-transition.html) nodes. The two approaches are evaluated on top of the Rime communication stack provided by the Contiki OS (http://www.sics.se/contiki/) and the network is simulated in the Cooja simulator using emulated TelosB nodes in order to offer a comparative analysis between them, and highlight the pros and cons under various networking conditions.

### *13.5.1 Evaluation Setup*

The number of nodes is considered twenty-five. In each case one node is considered the cluster head node (FCHN) and five nodes are related to the fuzzy process (Fuzzy Sensor Nodes FSN). FSNs values represent heart rate, sound class, activity rate, fall, and time, respectively. The remaining nodes (19 in total) are the Non-FSNs, whose purpose is to simulate a realistic scenario, where numerous nodes coexist, and expectedly some of them are irrelevant to the FIS objective and thus effectively disturb the direct transmissions of the FSNs to the FCHN. One-hop communication model is considered.

The packet generation rate of the FSNs nodes (tf1) is tuned from 0.25 to 3 s, while the Traffic Factor of the remaining nodes (tf2) is tuned from 0.25 to 4 s. The fuzzy timer (ft) parameter represents the execution frequency of the fuzzy algorithm, indicating the time between two consecutive executions. It must be noted that the input data considered are the latest received FSN packets. Therefore, depending on algorithms frequency and channels errors, the data utilized may not be up to date, due to network delay or packet loss, revealing another critical trade-off. The fuzzy timer is predefined at 4 s.

The most indicative evaluation metrics are presented as follows:

- Packet loss ratio
- Energy consumption profile
- Percentage of not updated fuzzy processes: The algorithm's performance depends on the freshness of input data. However this requirement is challenged by the considered wireless medium. Therefore, an important evaluation metric is the number of executions of the fuzzy algorithm with not updated input data. If the last generated fuzzy inputs have not been received by the FCHN or the coef_FSN when the fuzzy timer expires, then the fuzzification will be executed with old inputs (i.e., not updated).

Particularly in Fig. 13.10 we observe that the packet loss in the distributed approach is about 20 % when tf2 has a value of 0.25 s, while it varies from 3 to 8.5 % when the network is quite relaxed (tf2 = 4 s). Concurrently, high packet loss ratios occur when the Non-FSNs generate and transmit packets in a higher rate introducing congestion in the network. The traffic that the FSN nodes introduce to the rest of the network is negligible and do not contribute the overall packet loss ratio. In the more relaxed scenarios, where the tf2 is more than 1 s the packet loss decreases to lower than 4 %. Respectively in the centralized scenario we observe packet loss around 75 % in the more stressed scenarios, while it varies from 55 to

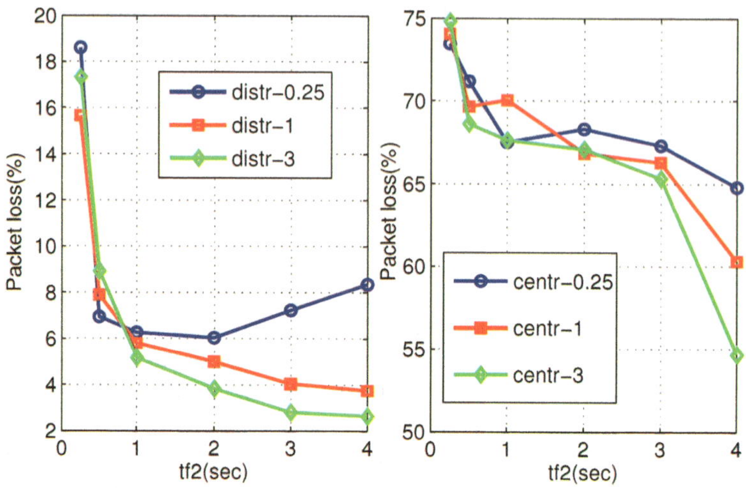


**Fig. 13.10** Packet loss ratio in the centralized and distributed approach

65 % when the generation rate decreases (tf2 = 4 s). In the centralized approach the packet loss is higher because all the nodes send directly their data to the FCHN. As anticipated, the distributed implementation balanced the traffic flows among the network as the FSNs nodes forward their fuzzified results to the coef_FSN, while the Non-FSNs send directly their data to the FCHN. Moreover an important computational task of the FIS has been executed to the FSNs and to the coef_FSN, and thus the FCHN is not overloaded with the reception of all the incoming packets and the whole processing. As a result, the congestion effects were minimized for all the Traffic Factors. Additionally in the distributed approach, the number of transmissions decreases since the generation packets from the coef_FSN are not transmitted to the FCHN. They are fuzzified locally and are used as input to the implication method, with the received fuzzified values from the 4 remaining FSNs.

Figure 13.11 shows the ratio of the energy consumed by the CPU and the radio over the total energy spent for each of the 6 FSNs for both centralized and decentralized approach. The energy consumption gain of the distributed scenario can be easily observed. The results of the consumed energy in these nodes are similar while varying tf1 value, however, these figures consider a tf1 value of 1 s. Node 6 is the FCHN in both cases, while node 1 has been selected as the coef_FSN in the distributed implementation. On the one hand, in the centralized approach the FCHNs processing load is overloaded since it is responsible for executing the FIS while the FSNs nodes only sense and transmit their value. On the other hand, in the distributed approach all the FSN nodes are responsible for the sensing and the fuzzification of the sensed values, while the coef_FSN is responsible for getting this data, for applying the implication method and for sending the two weights (alarm, not-alarm) to the FCHN. Finally, the FCHN is responsible for the reception of the Non-FSNs and the coef_FSN packets, and the defuzzification

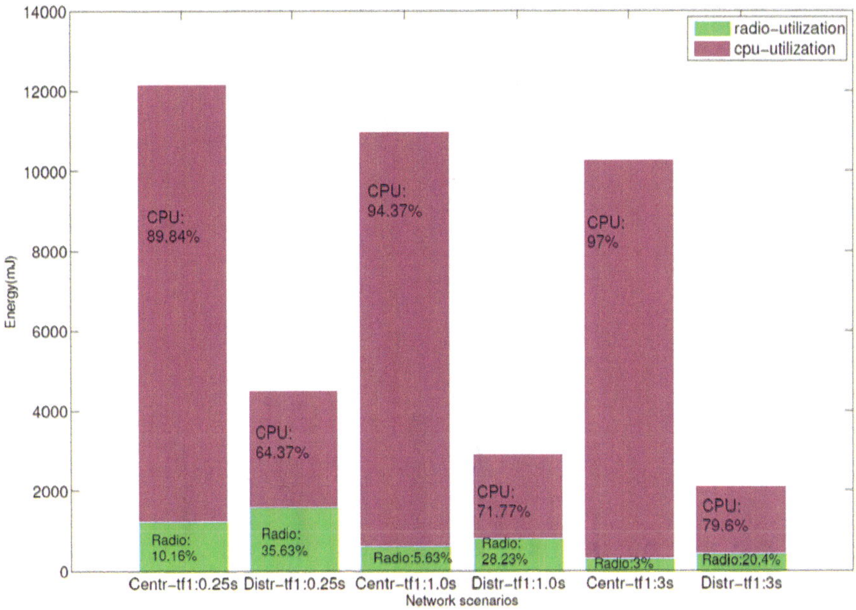


**Fig. 13.11** Percentile CPU utilization and radio energy

method. As we can see the CPU utilization is almost evenly balanced among all the nodes in the distributed implementation. Especially nodes 1 and 6 represent the two major processing elements and this is obvious to their CPU utilization. However the CPU utilization has been balanced among the nodes, we should assure that the total energy is not increased. From Fig. 13.12, it is observed that the overall energy is significantly reduced in the decentralized scenarios even if there is a slight increase in the radio consumption. For instance, the total energy that is consumed in the 6 nodes in the distributed approach has been decreased by a factor of 62.5 % compared to the centralized approach. The reason for the radio energy consumption increase is the output of each task of the FIS has a bigger payload than the payload of the raw sensor values and therefore the FSNs transmit longer packets. It should be noted that the CPU utilization includes also the CPU energy for executing the code for transmission and receive and for handling the interrupts between the monothread. This means that the CPU utilization is not only the energy for executing the FIS, which leads to a significant difference in the CPU utilization, as shown in Fig. 13.12. The processes handling required by the Contiki OS and passing required information between the communications and processing threads lead to the observed CPU utilization increase.

The next metric is application oriented, and depicts how the network health affects the FIS' performance. High ratios of packet loss lead to FIS's execution with outdated values (not updated fuzzy processes). In Fig. 13.13 the number of the fuzzy processes that were executed with invalid inputs (old values) is in all

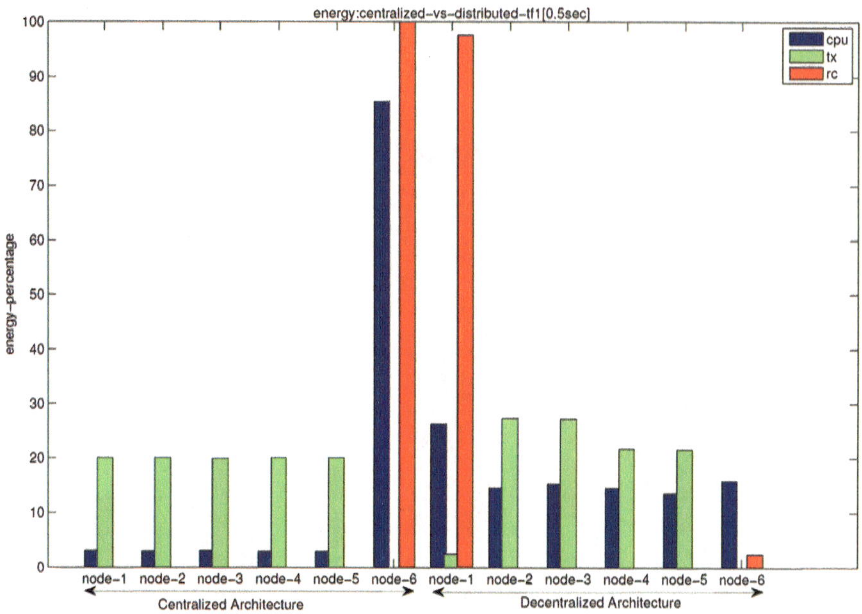


**Fig. 13.12** Percentile energy per node

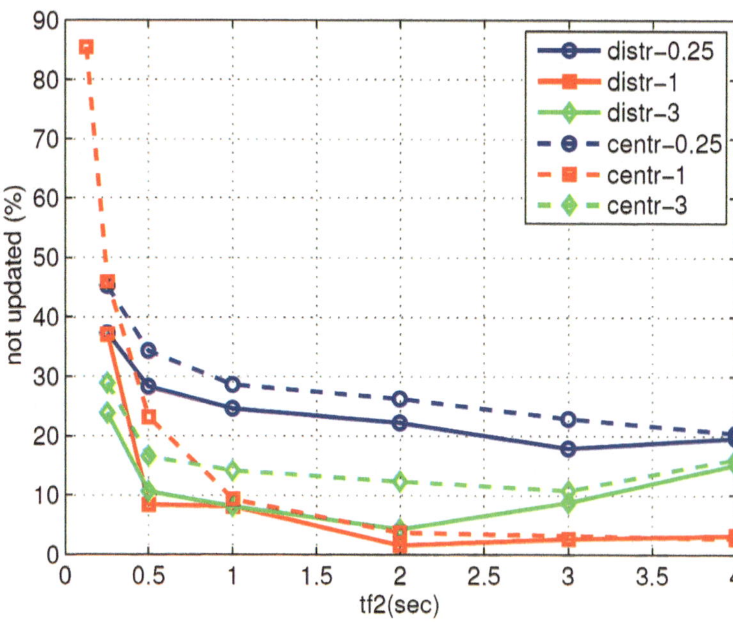


**Fig. 13.13** Percentage of not updated fuzzy processes

scenarios considerably lower in the distributed scenario. The increase of tf2 has a significant inversely analogous result on NU fuzzy processes. A decrease of the NU fuzzy process with respect to tf2 can be noted up to 0.5 s. Further increase of tf2 (i.e., reducing the Non-FSN packet workload) has rather diminishing effect on NU fuzzy process, as indicated by the slope of the respective graphs. This is because the networking conditions are more relaxed in this case. In the most stresses scenarios ($tf2 = 0.25$ s) the difference in the ratios of the not updated fuzzy process varies from 5 to 7 % when tf1 value varies from 3 to 0.25 s, respectively. This behavior is almost similar for the remaining tf2 range, except of the most relaxed scenarios ($tf2 > 3$ s) where the difference between the two implementations is negligible.

## 13.6 Conclusions

Having analyzed the most prominent approaches in designing FIS in WSNs applications, and highlighted all respective main characteristics of each approach, this section aims to summarize our survey and extract useful conclusions.

The increasing use of sensing units, the need for data acquisition and knowledge application upon acquired data, as well as the need for confirming the presence of an ambiguous event arise the necessity for adequate machine learning techniques development. Traditional machine learning algorithms are characterized by high performance demands. However the requirement of accurate classification results in WSNs applications has to compromise with the limited resources as well as the wireless communication error prone nature. Hence new techniques for applying event detection algorithms in WSNs are needed. Fuzzy logic can tackle these challenges.

This chapter offers a theoretical background on fuzzy logic framework and describes the steps for designing a FIS. A case study of healthcare assessment is described in detail. Vital parameters from individual's body area as well as environmental parameters are considered as inputs to the FIS, which evaluates the individual's well-being. The FIS performance is evaluated under Matlab framework assuming ideal networking conditions, and then varying the network confidence rate. The network confidence rate is extracted using a second FIS, which evaluates the data accuracy considering as inputs QoS metrics. As a general conclusion the FIS performance is strongly affected by the network confidence rate, and the outcome from both FISs should be merged so as to extract a safe conclusion about the criticality of a situation.

In addition this chapter presents implementation of the aforementioned health assessment FIS in a WSN embedded platform (TelosB) and the evaluation analysis under a simulated WSN infrastructure. The nodes related to the FIS input parameters (FSNs) send their data to the FCHN, which is responsible for the FIS execution. However, this burdens the CPU utilization of the FCHN. Thus, a distributed implementation of the FIS is presented, aiming to split FIS in subtasks, and allocate them to the FSNs. Comparative evaluation analysis is offered, indicating the advantages of the distributed approach.

We hope that this effort can serve as a roadmap for future research efforts that use fuzzy logic in real WSN deployments.

## References

1. L. Gu, D. Jia, P. Vicaire, T. Yan, L. Luo, A. Tirumala, Q. Cao, T. He, J.A.Stankovic, T. Abdelzaher, B.H. Krogh, in *Lightweight Detection and Classification for Wireless Sensor Networks in Realistic Environments*. Proceedings of the 3rd International Conference on Embedded Networked Sensor Systems, SenSys'05, ACM (2008) pp. 205–217. doi:10.1145/1098918.1098941
2. I.F. Akyildiz, I.H. Kasimoglu, Wireless sensor and actor networks: research challenges. Ad Hoc Netw. **2**(4), 351–367 (2004). doi:10.1016/j.adhoc.2004.04.003
3. Y. Tian, Y. Gu, E. Ekici, F. Ozguner, in *Dynamic Critical-Path Task Mapping and Scheduling for Collaborative In-Network Processing in Multi-Hop Wireless Sensor Networks*. 2006 International Conference on Parallel Processing Workshops, 2006, ICPP 2006 Workshops (2006), pp. 8–222.
4. B. Krishnamachari, S. Iyengar, Distributed Bayesian algorithms for fault-tolerant event region detection in wireless sensor networks. IEEE Trans. Comput. **53**(3), 241–250 (2004)
5. K.-X. Thuc, K. Insoo, A collaborative event detection scheme using fuzzy logic in clustered wireless sensor networks. AEU Int. J. Electron. Commun. **65**(5), 485–488 (2011)
6. K. Kapitanova, S.H. Son, K.-D. Kang, Using fuzzy logic for robust event detection in wireless sensor networks. Ad Hoc Netw. **10**(4), 709–722 (2012)
7. L. Zadeh, The concept of a linguistic variable and its application to approximate reasoning. Inform. Sci. **8**, 199–249 (1975). doi:10.1016/0020-0255(75)90036-5
8. A. Hamam, N.D. Georganas, in *A Comparison of Mamdani and Sugeno fuzzy Inference Systems for Evaluating the Quality of Experience of Hapto-Audio-Visual Applications*. IEEE International Workshop on Haptic Audio visual Environments and Games, 2008. HAVE 18–19 Oct 2008 (2008), pp. 87–92, doi:10.1109/HAVE.2008.4685304
9. J.M. Alonso, L. Magdalena, S. Guillaume, in *Designing Highly Interpretable Fuzzy Rule-Based Systems with Integration of Expert and Induced Knowledge*. 12th International Conference on Information Processing and Management of Uncertainty in Knowledge-Based Systems (IPMU) (2008), pp. 682–689. ISBN 978-84612-361-7
10. P. Manjunatha, A.K. Verma, A. Srividya, in *Multi-Sensor Data Fusion in Cluster Based Wireless Sensor Networks Using Fuzzy Logic Method*. The Third international Conference on Industrial and Information Systems (2008)
11. L. Zhang, H. Leung, K. Chan, Information fusion based smart home control system and its application. IEEE Trans. Consum. Electron. **54**(3), 1157–1165 (2008)
12. S.A. Munir, Y. Wen Bin, R. Biao, M. Man, in *Fuzzy Logic Based Congestion Estimation for QoS in Wireless Sensor Network*. IEEE Wireless Communications and Networking Conference (2007)
13. M. Marin-Perianu, C. Lombriser, O. Amft, P.J.M. Havinga, G. Troster. in *Distributed Activity Recognition with Fuzzy-Enabled Wireless Sensor Networks*. Distributed Computing in Sensor Systems, 2008. Lecture Notes in Computer Science (2008)
14. A. G. Dharne, Jaeyong Lee, S. Jayasuriya, in *Using Fuzzy Logic for Localization in Mobile Sensor Networks: Simulations and Experiments*. IEEE American Control Conference (2006)
15. A. Barati, S.J. Dastgheib, A. Movaghar, I. Attarzadeh, in *An Effective Fuzzy Based Algorithm To Detect Faulty Readings In Long Thin Wireless Sensor Networks*. International Conference on Future Communication Networks (ICFCN) (2012)

16. D. Istrate, J. Boudy, H. Medjahed, J.L. Baldinger, Medical remote monitoring using sound environment analysis and wearable sensors. Biomed. Eng. Carlos Alexandre Barros de Mello (Ed.), InTech, DOI: 10.5772/7852. http://www.intechopen.com/books/biomedical-engineering/medical-remote-monitoring-using-sound-environment-analysis-and-wearable-sensors
17. H. Medjahed, D. Istrate, J. Boudy, B. Dorizzi, in *Human Activities of Daily Living Recognition Using Fuzzy Logic for Elderly Home Monitoring*. IEEE International Conference on Fuzzy Systems (2009)
18. S.-M. Dima, C. Antonopoulos, J. Gialelis, S. Koubias, in *A Network Reliability Oriented Event Detection Scheme for Wireless Sensors and Actors Networks*. The International Conference on Industrial Technology, IEEE (2012)
19. H. Medjahed, D. Istrate, J. Boudy, B. Dorizzi, in *A Fuzzy Logic System for Home Elderly People Monitoring (EMUTEM)*. Proceedings of the 10th WSEAS international conference on Fuzzy systems (2009). pp. 69–75
20. P. Rafiee, G. Latif Shabgahi, Evaluating the reliability of communication networks(WAN) using their fuzzy fault tree analysis – a case study. J. Math. Comput. Sci. **2**(2), 262–270 (2011)
21. M. Mahfouf, M.F. Abbod, D.A. Linkens, A survey of fuzzy logic monitoring and control utilisation in medicine. Artif. Intell. Med. **21**(1–3), 27–42 (2001)
22. O. Gama, P. Carvalho, J.A. Afonso, P.M. Mendes, Quality of service in wireless e-emergency: main issues and a case-study. Paper presented at the 3rd Symposium of Ubiquitous Computing and Ambient Intelligence 2008
23. S. Ullah, B. Shen, S. Riazul Islam, P. Khan, S. Saleem, K. Sup Kwak, A study of MAC protocols for WBANs. Sensors **10**(1), 128–145 (2010)
24. S. Dima, D. Tsitsipis, C. Antonopoulos, J. Gialelis, S. Koubias, in *FLOGERA: A Fuzzy Logic Event Recognition Algorithm in a WSN Environment*. 2012 8th International Wireless Communications and Mobile Computing Conference (IWCMC), (2012), pp. 850–855. doi: 10.1109/IWCMC.2012.6314315
25. S.-M. Dima, C. Panagiotou, D. Tsitsipis, C. Antonopoulos, J. Gialelis, S. Koubias, Performance evaluation of a WSN system for distributed event detection using fuzzy logic. Ad Hoc Netw. **23**, 87–108 (2014)
26. R.R. Yager, D.P. Filev, *Essentials of Fuzzy Modeling and Control* (Wiley, New York, 1994)

# Part V
# Use Cases for IoT

# Chapter 14
# IoT in Ambient Assistant Living Environments: A View from Europe

**Christos Panagiotou, Christos Antonopoulos, Georgios Keramidas, Nikolaos Voros, and Michael Hübner**

## 14.1 Introduction to AAL IoT Approach

The Internet of Things (IoT) paradigm, as already manifested, is leading to smart objects being capable of identifying, locating, sensing and connecting, and thus opening possibilities for new forms of communication between people and things, as well as amongst things. This evolution benefits new application scenarios that promote the ambient assisted living (AAL) environments. AAL systems encompass ICT in order to support elderly people in their daily routine thus prolonging an independent and safe lifestyle.

It is commonly accepted that intelligent, customizable, and multifaceted home monitoring and control comprise one of the most rapidly evolving ICT domains. To address respective challenges, novel technical approaches are required to cover all aspects of home environment monitoring. On one hand, the significance of such endeavors is highlighted by the high interest that home monitoring receives both from academia and from industry [1–4] mainly under the term AAL environment systems. On the other hand, relative views are further emphasized by several studies concerning the population's everyday life characteristics.

Focusing on the European view on the issue, according to the 2012 EU Ageing Report, the part of European population aged over 65 will almost double by 2060—rising from 87.5 million in 2010 to 152.6 million—while the number of older people (aged 80 years and above) is projected to increase even more, almost tripling from

C. Panagiotou (✉) • C. Antonopoulos • G. Keramidas • N. Voros
Technological Educational Institute of Western Greece, Embedded System Design and Application Laboratory (ESDA), Antirio, Greece
e-mail: chpanag@gmail.com

M. Hübner
Chair for Embedded Systems for Information Technology (ESIT), Ruhr Universität Bochum, Bochum, Germany

 
G. Keramidas et al. (eds.), *Components and Services for IoT Platforms*,
DOI 10.1007/978-3-319-42304-3_14

23.7 million in 2010 to 62.4 million in 2060 [1]. At the same time, critical advances in the medical domain offer valuable tools to the senior population allowing them to stay active for as long as possible. However, extending the life time expectation of the general population is unavoidably accompanied by increased frequency of chronic disorders manifestation. Such occurrences greatly degrade life quality and at the same time emphasize the need for advanced AAL technologies and deployments. It is important to note that respective needs apply both to the user and their equally old spouses or relatives that undertake the responsibility of care provision, effectively extending the range of people requiring AAL services.

Additionally, moving beyond specific chronic disorders, AAL includes other equally important aspects under the umbrella of new solutions towards improving the quality of life, supporting health needs, facilitating concepts such as "aging well" or "healthy aging," and maintaining the social participation of the elders. All the above clearly indicate a new rapidly growing market for health devices and services that support seniors in their home environments. Provision of such services must cover a wide range of requirements including medication schedule reminders, monitoring of vital signs, and dispatching of medical alerts in the case of emergency situations such as falls and accidents, affective monitoring, urging people to stay active and creative and many more.

Based on the aforementioned EU report, the number of elderly people in need of constant care and attention will increase significantly in the near future. Another observation advocating the extensions and enhancement of AAL technologies relates to the fact that contemporary approaches and services are and will be rendered totally inadequate. Specifically, hospitalization, nursing homes, or employing personal care givers at home for a continuously increasing population percentage are not a viable solution due to practical [2], economical [5], and psychological [4] reasons. Another perspective of paramount importance, yet most of times ignored by existing approaches is that AAL services (and not only for elderly people but also in general) are a highly personalized, multiparametric and multifaceted challenge not able to be addressed solely by existing approaches.

AAL IoT solutions, on the other hand, acknowledge and even capitalize on the fact that what a person may need or expect from home monitoring can be quite diverse. On one hand, a typical category is the users suffering from specific medical conditions. In that respect, the actual condition (e.g., kinetic difficulties, heart conditions, brain conditions, etc.) and the needs for each specific user are diverse (or diversified) and thus a "one size fits all" solution is not adequate. On the other hand, AAL need to also cover cases where home monitoring is needed because he/she spends a lot of time on its own and a general monitoring on the user's well-being is desired. Furthermore, home monitoring requirements gaining active interest focus on safety issues (e.g., the user forgot to turn off the oven or close a window/door before going to sleep). Such cases although not directly related to medical conditions also comprise services of high added value for the general population [6].

## 14.2 AAL Involved Technologies

For all the above reasons, AAL represents a fundamental research domain that has drawn massive attention. Moving to the more technical aspects, AAL rather than being a technology itself, AAL encompasses IoT approach, embedded networked devices and sensors, algorithms, and diverse ICT technologies to support people in their preferred environment. Such interdependencies and collaborations amongst different ICT domains further enhance the added value of respective research efforts and deployments. Another aspect differentiating AAL development from other ICT advancements concerns the requirement to pose minimum intrusion thus yielding significant added value of allowing the person to be self-reliant and independent. Furthermore, as AAL approaches are applied into more and more scenarios they also move beyond the human factor and focus on a wide range of characteristics of the environment itself characterizing typical IoT platforms.

Specific areas towards which AAL technologies extend include security, surveillance, energy consumption, and remote access attracting increasing interest. Additionally, in many cases people (especially elderly) tend to spend significant part of each day away from home. Consequently, a home monitoring system able to detect events related to safety and security (i.e., a window/door left open, the oven is left on, etc.) could be an invaluable tool [7, 8]. Power consumption monitoring pertaining to security, well-being, economy as well as carbon footprint control services comprise another critical aspect of nowadays AAL enabled by IoT capabilities with energy demand-response programs support [9, 10] representing also research objectives of high added value. In order for the aforementioned services to be implemented significant advancements in ICT fields such as wireless sensor networks (WSN) regarding ultra-low power consumption embedded systems and miniaturization [11, 12] are required. Based on the respective observations over the last years both the number of available WSN platforms offering different capabilities and the range of different sensors' support continuously increase moving towards a holistic home environment platform. However, high degree of heterogeneity in all prominent technological areas still remains a critical challenge prohibiting widespread utilization [13, 14]. Finally, one of the latest trends in AAL technologies receiving active interest from the European research community (e.g., Horizon 2020 research projects calls) is help provision with daily activities, based on monitoring activities of daily living (ADL) and issuing reminders [15], as well as helping with mobility and automation [16]. Exploiting novel sensor modality such as video processing, audio processing as well as processing approaches such as fuzzy logic and affective modeling offer enhanced monitoring and control capabilities. Finally, such technologies promote sociability amongst users sharing common interest, activities, and hobbies or with their family, friends, doctors, and caregivers [17, 18] highlighting further the interconnection with IoT approaches.

## 14.3 Research Efforts: Paving the Way of AAL IoT

This section describes mainly active projects funded by the EU regarding the AAL domain under the IoT paradigm. The projects are presented under the view of the objectives and the expected impact addressed by common or similar calls which will be used as a common ground. The majority of the described projects (five out of six) have been accepted in "PHC-19-2014—Advancing active and healthy ageing with ICT: service robotics within assisted living environments" call [19] while the last one has been funded under the "FP7-ICT-Challenge 5: ICT for Health, Ageing Well, Inclusion and Governance" call.

In the PHC-19 call, the EU identifies and highlights the needs that the increasing ageing European population has to deal with. Namely, issues regarding risk of cognitive impairment, frailty, and social exclusion with considerable negative consequences for their independence, quality of life, including that of those who care for them, and for the sustainability of health and care systems. The call focuses on the development of robotic services applied in the context of AAL and particularly for supporting ageing population. The key concepts that need to be addressed regard modularity, cost effectiveness, reliability, flexibility, applicability to realistic settings, and acceptability by the end users.

The outcomes of the projects funded under the PHC-19 call are expected to contribute with respect to validation of the benefits offered by the developed robotic services, the reduction of the time spent in institutional medical care, and the improvement of quality of life.

**RAMCIP Project** The RAMCIP (Robotic Assistant for MCI Patients at home, Grant Agreement no: 643433) [20] scope is to research and develop real robotic solutions for assistive robotics for the elderly and those suffering from mild cognitive impairments (MCI) and dementia. The desired functionalities are given through the development of a service robot with high-level cognitive functions. These functions are driven through modeling and monitoring of the home environment, the user and other co-located persons, allowing the robot to take optimal decision regarding when and how to provide assistance, in a proactive and discreet way [21].

The proposed system's core is an autonomously moving service robot with an on-board touch screen and a dexterous robotic hand and arm. The objectives of the project can be summarized in developing a service robot that will be capable of understanding actions, complex activities, and behavior of multiple persons in the user's home and provide proactive, discreet, and optimal assistance to the user. Furthermore, the project deals with establishing advanced communication channels between the user and the robot. Another key objective of the project is the development of advanced physical interaction between the robot and the home environment in order to support assistance activities that involve physical interaction between the robot and the user. These key functionalities that the RAMCIP robot will deliver are briefly presented in Fig. 14.1.

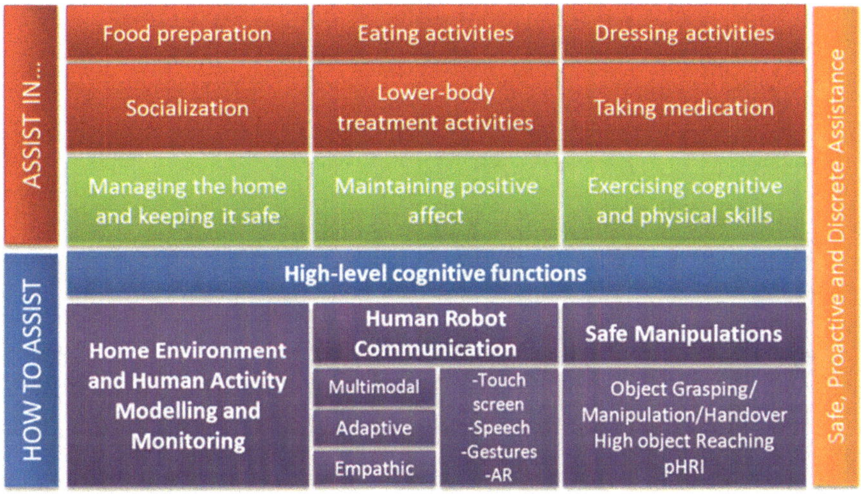


**Fig. 14.1** The RAMCIP vision

The innovation points of the RAMCIP project can be summarized as:

- Advanced cognitive functions of the service robot driving proactive and discreet assistance provision through user and environment monitoring and modeling.
- Advanced human–robot communication channels based on multimodal interfaces (speech, gestures, touch screen, and augmenting reality (AR) displays).
- Dexterous robotic manipulation introduced in the context of service robots for AAL environments focused on human–robot interaction (HRI).

The promising functionalities that the project envisages to impose to current robotic technologies will deliver new case studies for the IoT field. The solutions that the project presents regarding the HRI aim to improve the system's acceptability and provide a more user-centric approach. The evaluation of the project's advances will be performed in two pilot sites in Poland and Spain.

**Growmeup Project** The Growmeup project (Specification of User Needs Analysis and Design of Behaviour Patterns Model, Grant Agreement no: 643647) [22], as the rest of the PHC-19 call projects, presents its approach on providing affordable robotic services in favor of the elders and their well-being. The project focuses on five innovation key points and particularly [23]:

1. To enhance HRI dialogue.
2. Provide a robot with learning capabilities, i.e., learn user habits, preferences, routines, etc.
3. Interaction of robots through cloud infrastructure.
4. Detection of external devices that extend robot's functionalities through a scalable and adaptable architecture.
5. Deliver robot capable to manipulate uncertainty of user's behavior and adapt to its environment.

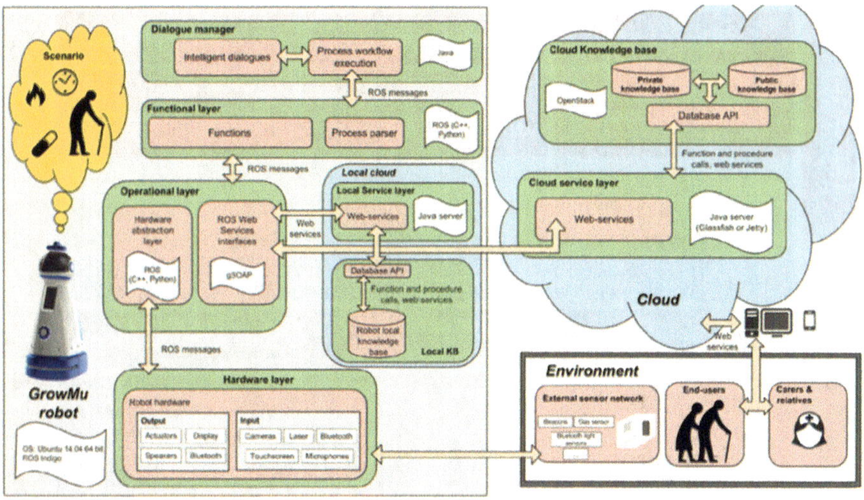


**Fig. 14.2** Growmeup architecture

The Growmeup system consists of four main components. These components along with their sub-systems are presented in Fig. 14.2. The core component of the system is the cloud infrastructure that supports information exchange between service robots. This structure equips the robot with capabilities to adapt its services and abilities and therefore enhance its functionalities autonomously. The cloud infrastructure is composed of the knowledge base, the computational expensive algorithm, and the application services. The interconnection with the rest components is accomplished through web services as it is also shown in Fig. 14.2.

The smart environment is the second main component of the system. It refers to external network of sensors and actuators that are deployed in the user's environment. These external devices provide functionalities such as temperature measurements, door status, fire detection, object localization based to iBeacons, etc.

Finally, the last two components of the system are the human and robot actors. The human actors apart from the elder it includes care givers and relatives. Humans are interacting with the robot directly or indirectly through the cloud infrastructure. Figure 14.2 gives some insights in the robot's structure, technologies, and interfaces.

The Growmeup project realizes the IoT paradigm through the cloud infrastructure that constitutes the link among the objects, the user, and the services. The service is an application that runs on the cloud and is accessible through web services. The main contribution of the services to the overall system functionality is to:

- provide support for daily home activities
- provide interfaces for exchanging and storing data
- execute the algorithms and deliver their outcomes.

The service approach of the system through the cloud structure is expected to deliver the required flexibility as addressed by the call objectives. Moreover, the interfaces that deliver the interaction with external devices will provide modularity that is required to make the system flexible and adaptable to user needs and characteristics. Safe results about the cost effectiveness of the system cannot be made since the system is yet under development. Finally, the last objectives of the call that refer to the reliability, the applicability to realistic setting, and the acceptance by the end users will be investigated during the trials of the project.

**MARIO Project** The MARIO project (Managing Active and healthy aging with use of caRing servIce robots, Grant Agreement no: 643808) [24] addresses the difficult challenges of loneliness, isolation, and dementia in older persons through service robots. The main objectives of MARIO pertain to an improved interaction with end users and assisted living environments in order to enable iterative development and preparation for post-project uptake. The project aims to support and receive "robot applications" similar to the paradigm offered by community for smartphones so as to empower development and creativity, enable the robot to perform new functionalities over time, and support discovery and improve usefulness for end users through cost effective solutions.

From the ICT point of view, the project incorporates advances in machine learning techniques and semantic analysis methods to make the robot more personable, useful, and accepted by end users [25].

The challenges that the project tries to address as highlighted in [26] concern:

1. Effectively assessing persons with dementia in the community provided they are suitable users for the robot (especially in countries with less developed community care structures).
2. Finding optimum trade-off between the needs and preferences of caregivers and end users while most efforts so far are focused solely on the caregivers.
3. Identifying meaningful activities and potential forms of achieving social connectedness that will add genuine value to the daily life of a person with dementia.
4. Designing a dementia-friendly interface that allows for a sufficiently comprehensive range of functionalities without being overwhelming or disorienting for a person with dementia will be essential for the successfully implementation of a genuinely empowering use of the technology.
5. Realistic assessment of the benefits of robot assisted care over traditional approaches to care, without implicit endorsement of a technological imperative.

MARIO claims to offer opportunities to progress beyond the current state of the art in the five following innovation areas:

1. Integration of robot semantics with existing structured and unstructured data, leveraging on current data integration practices such as the linked data principles, W3C semantic web standards Resource Description Format (RDF), SPARQL, and rule interchange format (RIF), semantic web-oriented machine reading, and ontologies.

2. "Entity-centric" knowledge management: each entity and its relations have a public identity that provides a first access to the knowledge consumed by robots. Such identity is given by resolvable URLs that use simple Web and Internet protocols to provide useful knowledge as a representative of real world entities.
3. Introduction to reading/listening semantic-web-oriented machine in robots.
4. Development of an ontology network using the Ontologies for Robotics and Automation. The network will evolve and expand by integrating ontologies emerging from interaction with assisted human, sensors, or with other robots.
5. Ability to advance robot knowledge by learning new ontology patterns from its experience with users and the robot network eventually in place.

MARIO tries to address the needs for modularity and flexibility through smartphone's application community. Developed applications can be tailored to user's needs and characteristics. In that sense, MARIO is expected to deliver flexibility, modularity, and increased system's acceptance by the users. From the IoT perspective, MARIO invests on the semantic interoperability through the usage of ontologies that deliver data and their semantics at the same time leaving the communication technologies aside. As expected, concrete conclusions about the system's overall performance will come up after the respective evaluation process.

**ENRICHme** The ENRICHme project (ENabling Robot and assisted living environment for Independent Care and Health Monitoring of the Elderly, Grant Agreement no: 643691) [27] aims to improve the quality of life of elderly people suffering of MCI in the context of AAL for the ageing European population at greater risk of cognitive impairment, frailty, and social exclusion, using a service robot within an assisted living environment.

The project utilizes robotic approaches in order to develop engaging solutions for HRI that automatically learn from long-term experience and adapt to the elderly person in terms of perceptual and cognitive capacity. In that context, it introduces three levels of intelligence: robot, ambient, and social intelligence. These three key concepts are realized in the project through a robot that encompasses safe navigation, monitoring, and interaction with the person along with home automation and ambient sensing for long-term monitoring of human motion and daily/weekly/monthly life activities (in relation to the use of RFID tagged objects around the environment). Figure 14.3 demonstrates an abstract view of the system's architecture. The figure highlights the components of the intelligent environment of the elder (ambient, robot, and social intelligence) along with their interconnections.

Two in-depth validation trials will be carried for the overall scientific assessment of the ENRICHME system aiming at identifying the potential impact of the system in supporting the active and healthy ageing of European citizens. Data on system sensibility and sensitivity will be compared with those already available in the literature and current elderly housing facilities. Physiological indexes will be analyzed versus other parameters (e.g., motion, behavioral, clinical, video analysis,

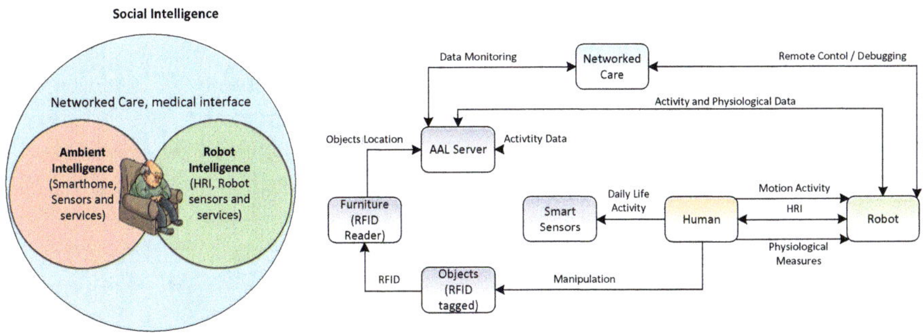


**Fig. 14.3** The ENRICHme system

and ethnographic observation) to check if the collected information coincides with and brings new insights into clinical and HRI situations.

**RADIO** The RADIO project (Robots in Assisted Living Environments: Unobtrusive, efficient, reliable and modular solutions for independent ageing, Grant Agreement no: 643892) [28] develops an integrated smart home/assistant robot system, attempting to increase the levels of acceptance and unobtrusiveness. RADIO objectives are focused on using the integrated smart home/assistant robot system as the sensing equipment for health monitoring with sensors that do not need to be discrete and distant or masked and cumbersome to install. Instead, sensors will be realized as a natural component of the smart home/assistant robot functionalities. The key ICT advancements of the RADIO project can be summarized in:

1. Integration of unobtrusive medical sensors
2. Heterogeneous networking and architectural solutions in the robotic and smart home domain
3. Design for usability by users not familiar to technologies
4. Scalable architecture for wide area application and heterogeneous components

The conceptual architecture schema as depicted in Fig. 14.4 highlights the three major interconnected planes of the system. Namely, the RADIO system consists of the end user's environment, the care giver's environment, and the care institution environment. Each plane contains several sub-modules that deliver their functionalities to the respective plane [29].

In the context of RADIO, the smart home service and the IoT platform are viewed as a single solution, however, the IoT platform has ICT resources that can be exploited aside from the smart home service, and hence by other ICT resources of the RADIO solution. Those resources are exposed as a RESTful API. The core services currently provided by the IoT platform are Sensor Service that is mainly responsible for managing the data and the status of the sensors and actuators of the respective deployment; Event Service that provides real time communication between the IoT platform and the smart home. RADIO Ecosystem services of

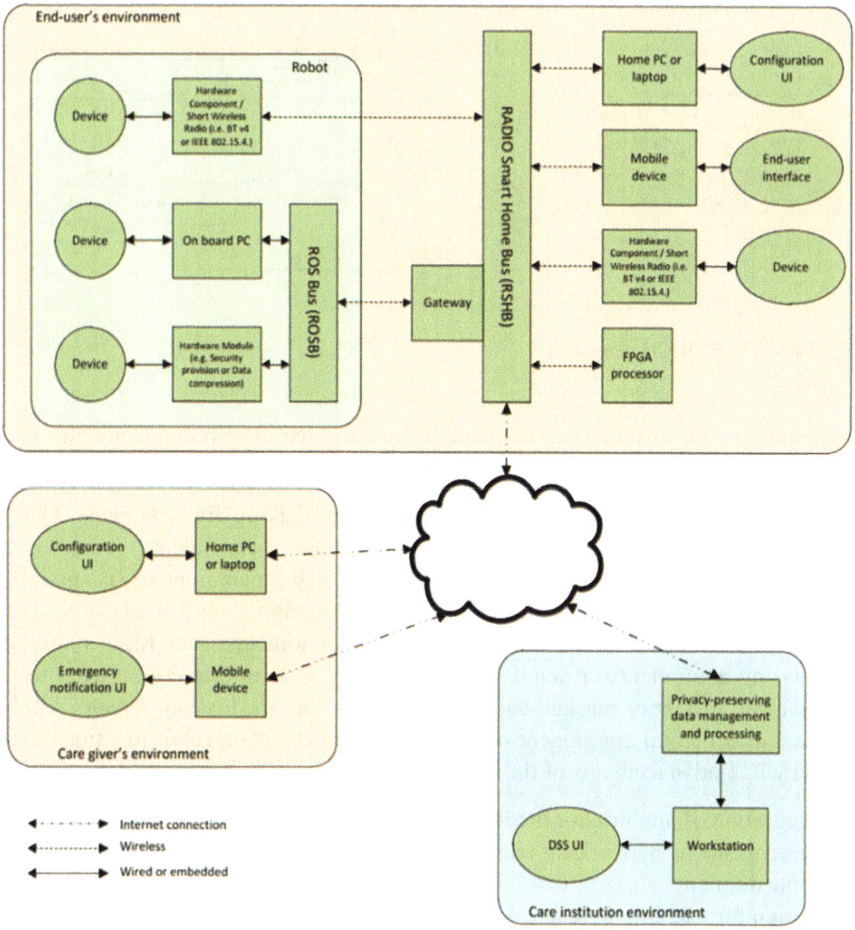


**Fig. 14.4** Conceptual architecture of the RADIO system

automation and notifications will build upon those services, thus, in the following, we provide some context of how those services currently operate.

**ROBOT-ERA** The objective of the Robot-Era project (Implementation and integration of advance Robotic systems and intelligent Environments in real scenarios for the ageing population, Grant Agreement no: 288899) [30] is to develop, implement, and demonstrate the general feasibility, scientific/technical effectiveness, and social/legal plausibility and acceptability by end users of a plurality of complete advanced robotic services. These services will be integrated in intelligent environments, which will actively work in real conditions and cooperate with real people and between them to favor independent living, improve the quality of life, and the efficiency of care for elderly people.

ROBOT-ERA aims to enhance the quality and the acceptability of current robotic services and deliver new functionalities and services with increased performance. Its scope is pursued by exploiting previous knowledge on robotics and ambient intelligence technologies, cognitive-inspired robot learning architectures, elderly user-needs, and methodology of design for acceptability. These keystones are fundamental for the real deployment in order to find the most appropriate and successful solutions and overcome actual barriers to the exploitation of robotic services [31, 32].

The system intends to provide a personalized medical support to the users in a continuous manner, by leveraging the use of companion robots, distributed sensors in the environment, and a cloud platform. Exploiting the cloud robotics paradigm, computation-intensive tasks are assigned to the cloud. Therefore, access is achieved to a vast amount of data and the storage capabilities are increased. The indoor user localization and environmental monitoring algorithms are developed and offloaded onto the cloud considering the Software as a Service (SaaS) paradigm. This architecture improves the robot's functionalities without increasing the on-board computational load. Furthermore, different users and robotic agents are able to use the proposed services not being preoccupied about software maintenance. The list of the implemented SaaS that the project offers is reported below:

1. User indoor localization based services
2. Care reminding services
3. Environmental monitoring services

Modern cloud approaches are also applied in the ROBOT-ERA project through the SaaS model. SaaS, as leveraged by ROBOT-ERA, has become a key requirement for the evolving IoT solutions. The applicability of the system along with acceptability rate from the users will be tested during the pilots that will be performed in Peccioli (Italy) and Örebro (Sweden).

## 14.4 Maturity Analysis of AAL IoT Developments and Platforms

Having analyzed the technical and the societal objectives for each EU funded project, in this section we illustrate the in-field trials and planned use-cases of each project. For a fair and meaningful comparison, we narrow down the set of projects to those that are based on mobile robotic platforms for AAL environments. This leaves at our disposal five projects: RADIO [28], RAMCIP [20], growmeup [22], Mario [24], and ENRICHME [27]. Note that in this selection we include only active projects.

The purpose of this section is to reveal how the use-cases are selected, what is the common ground and the differences between the selected use-cases among the projects and eventually if there are use-case scenarios that are overlooked by the projects' workplan. The reason that we concentrate in the selection of the use-cases and the in-field trials is that these aspects of the projects represent a first-class metric for validating the technology readiness level (TRL) produced in each case. Below is our characterization for each project.

Finally, we have to mention at this point that our assessment is restricted to the information that is available in the corresponding projects' websites (including related information such as press releases, public deliverables, and project presentations). Any aspects of the projects that are not covered in their websites and probably described in the detailed technical annexes of the projects are out of scope for this chapter.

**RADIO** The RADIO project sets forward a novel approach to acceptance and unobtrusiveness in AAL environments. The underlying idea is that the sensing mechanism is not discrete but it acts as an obvious and accepted part of the user's daily life. More specifically, the sensing mechanism consists of an integrated smart home and the assistant robot system. According to RADIO, the robotic system must be considered as an unobtrusive "pet" and in this way, sensors do not need to be discrete and distant or masked and cumbersome to install. In other words, the sensing mechanism must be perceived as a natural component of the smart home/assistant robot functionalities.

The target population of RADIO technology is elderly people without any significant physical or mental impairment. By collecting and analyzing behavioral data, doctors can diagnose cognitive impairment symptoms early and take timely remedial actions. As such, the end users are able to maintain their independent life, prolong the time spent at home, and improve their quality of life. In RADIO, the robot also acts as the contact point for home automation conveniences, such as lights that can be switched on and off. Additional activities of daily life (ADL) and mood recognition methods ranges from the time spent out of the house or carrying out a given activity in it, to sleeping patterns, to recognizing whether the end user has changed clothes, washed, or other crucial indications required by the doctors. Finally, the RADIO robot is designed to recognize the mood of the users by means of both speech and facial characteristics analysis.

The RADIO approach will be evaluated at actual elderly care facilities and under realistic conditions at clinical partners' premises. Moreover, technical testing is also planned in two full-scale smart home apartments with multiple rooms. In particular, the RADIO consortium includes three clinical piloting partners that will run four sets of experiments in three phases. The first phase, called formative phase, will be carried out in an early stage of the project with elderly end users in a hospital dedicated to neuro-motor rehabilitation of patients. The second phase, called intermediate phase, will run in the same hospital, while the third phase, called summative phase, includes two sets of experiments: one at a healthcare provider and geriatric center and one at the private homes (outside of any medical care institutions) of end users who have volunteered to participate in the RADIO piloting plan.

Overall, it can be considered that the RADIO project includes a rich piloting plan that is sufficient enough to ensure a mature end solution. This is because the clinical objectives of the projects will be tested in three different environments with different parameters and requirements, while technical objectives will be further validated in two smart home environments. What is also interesting is that the piloting plans are spread across the whole duration of the project allowing a complete and iterative evaluation of the RADIO system. Apart from the above objectives, installing the RADIO in different environments and premises will also validate the flexibility and ease of installation of the prototype.

**RAMCIP** The RAMCIP project follows a different mentality. It is based on robotic assistant at home; in contrast to the RADIO approach in which the robot acts more as an "observer." As noted, the target population in RAMCIP is patients with MCI and dementia. The approach relies on a robotic platform consisting of an autonomous moving robot with an on-board touch screen and dexterous robotic hand and arm. The robot is considered as a service-centric machine with cognitive functionality. This functionally is driven through modeling and monitoring of the home environment, the user and other co-located persons, allowing the robot to take optimal decision regarding when and how to provide assistance.

In essence, the RAMCIP project aims to develop a service robot (i.e., a home assistance to the user) capable to understanding the behavior and activities of the monitored persons. The project also deals with providing entertainment functions to the users, e.g., access to social media, teleconferences with relatives or caregivers as well as other, more practical assistive functionalities such as bringing objects to the user or picking up objects from the floor.

The RAMCIP project includes a detailed demonstration/validation plan with multiple pilot trials. In the foreseen trials, elderly people will be assessed by medical doctors and psychologists in order to be categorized into two groups: the control group and the clinical group. The patients suffering from MCI and Alzheimer's disease (AD) will be assigned to the clinical group. The control group will consist of normal elderly persons and this group will be used to assess the normal person–robot interaction. With this approach, potential problems related with normal aging process will be defined and separated from MCI/AD dedicated problems. This is of paramount importance and within the RAMCIP project, a detailed analysis is performed to study the inclusion/exclusion criteria of each group.

The evaluation of the project's advances will be performed in two pilot sites. More specifically, in two neurological instructions specializing in several neurodegenerative diseases such as Alzheimer's disease, dementia, multiple sclerosis, and epilepsy. It can be considered that the field areas covered by the piloting partners are capable to specify/validate/demonstrate the user-centric technology of RAMCIP project. Moreover, the project set forward a unique approach to improve the social life of the patients. This has a real social impact since this is the main and most important problem in the target patients group of the project.

**GrowMeUp** Similar to the RAMCIP project, the GrowMeUp target is to build a service-centric robotic platform aiming to increase the years of independent and active/autonomous living, thus the quality of life of older persons. In contrast to the RAMCIP project, the target population group is people with light physical or mental health problems (at the age of 65+) who live alone at home and can find pleasure and relief in getting support or stimulation to carry out their daily activities over the ageing process.

In more details, in the GrowMeUp approach the robotic system will be able to learn the older persons needs and habits over time and enhance ('grow up'/scale up) its functionality to compensate for the elder's degradation of abilities, to support, encourage, and engage the older persons to stay longer active, independent, and socially involved while carrying out their daily life at home.

The envisioned technology heavily relies on machine learning mechanisms that will enable the GrowMeUp service robot to extend and increase its knowledge continuously over time. In addition, cloud based technologies are utilized to share and distribute the gained knowledge with multiple other robots, so that other robots making use of the cloud, can learn from the experience of each other and thus increase their own functionality/competencies and at the same time reduce learning effort. The daily activities support will be provided in a human like way characterized by behavior and emotional understanding, intelligent dialoguing, and personalized services provision.

The GrowMeUp prototypes will be tested in one big healthcare site that includes fully operational apartments and in the private homes of elderly people managed by a home care service provider. The two trials have been carefully selected so as elderly persons can live with the greatest possible independence and activity. The trial workplan is organized in two main phases: the pre-trials and the final-trials phases. Both phases will last for a remarkably long duration (3 and 9 months, respectively) revealing that the project follows an iterative and thorough evaluation plan. In other words, the project workplan includes ample room to identify new user needs and requirements, to suggest corrective measures to any encountered problem, and finally to propose refinements or use-case scenarios' adaptations.

It is important to note that the GrowMeUp trials involve a multidisciplinary team including technical directors, social workers, social animators, psychologists, nurses, and direct action helpers. Also during the final trials, for each one of the elderly, a virtual care team will be assigned consisting of minimum three carers and one younger person (family relative or volunteer at the end user institution in the age of 15–18). As such, it can be considered that the project has a sustainable exploitation strategy since the generated technology will be validated by a large and multidisciplinary team in real environments and for a long duration.

**MARIO** The MARIO project will also build a service oriented robotic platform. The aim of the project is to provide a solution to loneliness, isolation, and dementia in older persons. The underlying idea of the project is to increase the self-perception and brain stimulation of monitoring elders through the robot. Similar to the previous projects, MARIO includes an extensive and long running piloting plans that are overlapped with the validation and R&D activities of the project.

There are three piloting partners in the project namely, a nursing unit, a social care and community support unit, and a hospital unit specializing in comprehensive geriatric assessment.

Overall, the project has a clear commercialization strategy and specific effort is devoted to bring the MARIO solution to the end users through market deployment, e.g., by addressing licensing and integration aspects. As also noted, this direction is also supported by the fact that MARIO will set forward a framework for receiving "robot applications" similar to the app community for smartphones. As such, this will enable the robot to perform new functionalities over time, improve usefulness for the end users while lowering costs.

**ENRICHME** As in the previous three projects, the ENRICHME project relies on an interactive, service-oriented, and mobile robot. The target population group is elderly people suffering from MCI. As noted, the project relies on three main pillars: robot, ambient, and social intelligence. These three concepts are realized through a robot with safe navigation, monitoring, and interaction with the person along with home automation and ambient sensing for long-term monitoring of human motion and daily/weekly/monthly life activities. In essence, a prime target of ENRICHME is to act as an enabling tool to enrich the social activities of the patients and as a result their quality of life.

The project includes two piloting phases: testing in AAL laboratories and validation at the elderly facilities. The first one will be performed in two different sites and its aim is to provide user feedback to the development activities of the project in order to improve the final prototype. The second one will be performed in three different sites and its aim is to offer the final assessment of the ENRICHME prototype. It is important to note that all trial sites will be provided by the project partners. Another important aspect is that the final validation phase will run for 12 months in three different sites simultaneously.

**A Critical View** Although the analyzed projects have different technical and societal objectives and in the most of the cases different target population groups, the common ground in all of them is that they include a rich and long running piloting plan. More specifically, in all cases the piloting/trials approach includes multiple phases that are spread across the whole duration of the project. This common to all projects approach ensures, to the extent that it is possible, that the projects will deliver a mature and close to market technology.

Another common parameter that must be highlighted is that the consortia include the correct balance between ICT and medical/clinical experts. The largest number of the medical/clinical partners is in the ENRICHME projects (five partners in total). Among all projects, almost 35 % (on average) of the partners belong to the medical/clinical field. This is of added value, since the end solutions delivered by each project will be assessed by real-life end users, namely not only by elderly persons, doctors, and nurses but also by social workers, psychologists, and any kind of direct action helpers and caregivers.

While at the time of writing the mentioned projects are running for almost 1 year, it would be interesting to repeat the current analysis at the end of the lifetime of the projects. This will allow us to put the results and achievements of each project side-by-side and draw more concrete results and conclusions about this relatively new field of research, i.e., robotic assistance in AAL environments. Such evaluation will be interesting to be performed in all aspects of the projects (apart from the achievement of the technical objectives) including commercial achievements, achievements in medical assessment technology, etc.

## 14.5 Conclusions

This chapter presented how the IoT paradigm is realized on the AAL domain through a thorough study of EU funded projects. Initially, it has been highlighted the societal impact of the AAL advancements and how they are benefitted by the IoT evolution. Current trends are captured by restricting the survey on recent projects that five of them are still active and one of them has been recently completed. The projects have been described based on the objectives they address as defined by the respective calls. The main scope of the chapter was to identify and highlight the technological pillars that IoT paradigm shares with the AAL domain along with the respective ICT and societal advancements. A big part of the chapter was dedicated on the presentation of the use-cases and the planned in-fields trials of each project. Thereby, a primary evaluation of each project's TRL level has been attempted.

## References

1. European Commission, *The 2012 Ageing Report. Economic and Budgetary Projections for the 27 EU Member States (2010–2060)* (European Economy, Feb 2012)
2. T. Panagiotakopoulos, A. Theodosiou, A. Kameas, Exploring ambient assisted living job profiles. Proceedings of the 6th International Conference on Pervasive Technologies Related to Assistive Environments, 2013
3. Centers for Disease Control and Prevention, *Chronic Disease Overview: Costs of Chronic Disease*, http://www.cdc.gov/chronicdisease/
4. A.M. Holmes, P. Deb, The effect of chronic illness on the psychological health of family members. J. Ment. Health Policy Econ. **6**, 13–22 (2003)
5. Centers for Disease Control and Prevention, *Chronic Disease Overview: Costs of Chronic Disease*, http://www.cdc.gov/nccdphp/overview.htm
6. G. Van Den Broek, F. Cavallo, C. Wehrmann, *AALIANCE Ambient Assisted Living Roadmap. Ambient Intelligence and Smart Environments* (IOS Press, Amsterdam, 2010)

7. N. Xian, X. Chen, C. Liu, Development of safety monitoring system based on WSN. Proceedings of the International Conference on Electrical and Control Engineering, 2011
8. O. Garcia-Morchon, P. Res, H. Baldus, The ANGEL WSN security architecture. Proceedings of the Third International Conference on Sensor Technologies and Applications, 2009
9. L.D. Ha, S. Ploix et al., Tabu search for the optimization of household energy consumption. Proceedings of the International Conference on Information Reuse and Integration, 2006
10. H. Albadi, E.F. El-Saadany, Demand response in electricity markets: An overview. Proceedings of the Power Engineering Society General Meeting, 2007
11. Y. Zhang, L. Sun et al., Ubiquitous WSN for healthcare: Recent advances and future prospects. J. Internet Things (2014)
12. D. Ruffieux, M. Contaldo et al., Ultra low power and miniaturized MEMS-based radio for BAN and WSN applications. Proceedings of the ESSCIRC, 2010
13. V. Potdar, A. Sharif, E. Chang, Wireless sensor networks: A survey. Proceedings of the International Conference on Advanced Information Networking and Applications Workshops, 2009
14. D. Culler, D. Estrin, M. Srivastava, Overview of sensor networks. IEEE Comput. **37**(8), 41–49 (2004)
15. M.E. Pollack, L. Brown et al., Autominder: An intelligent cognitive orthotic system for people with memory impairment. Proceedings of the Robotic Autonomous Systems, 2003
16. M. Spenko, H. Yu et al., Pamm—A robotic aid to the elderly for mobility assistance and monitoring for the elderly. IEEE Trans. Neural Syst. Rehabil. Eng. **15**(3), 327–335 (2006)
17. E.D. Mynatt, J. Rowan et al., Digital family portraits: Supporting peace of mind for extended family members. Proceedings of the SIGCHI Conference on Human Factors in Computing Systems, 2001
18. F. Vetere, H. Davis et al., The magic box and collage: Responding to the challenge of distributed intergenerational play. Int. J. Hum. Comput. Stud. **67**, 165–178 (2009)
19. PHC-19-2014, Advancing active and healthy ageing with ICT: Service robotics within assisted living environments, http://ec.europa.eu/research/participants/portal/desktop/en/opportunities/h2020/topics/668-phc-19-2014.html
20. RAMCIP Project, Robotic assistant for MCI patients at home, http://www.ramcip-project.eu/ramcip/
21. RAMCIP Project, Report on end user requirements and use cases break down (Public Deliverable, 2016)
22. Growmeup Project, Specification of user needs analysis and design of behaviour patterns model, http://www.growmeup.eu/
23. Growmeup Project, Specification of overall system architecture and security and privacy infrastructure (Public Deliverable, 2016)
24. MARIO Project, Managing active and healthy aging with use of caring service robots, http://www.mario-project.eu/
25. M. Mongiovì, D.R. Recupero et al., Semantic reconciliation of knowledge extracted from text through a novel machine reader. Proceedings of the International Conference on Knowledge Capture, 2015
26. H. Felzmann, K. Murphy et al., Robot-assisted care for elderly with dementia: Is there a potential for genuine end-user empowerment? Proceedings of the International Conference on Human-Robot Interaction Workshops, 2015
27. ENRICHme Project, Enabling robot and assisted living environment for independent care and health monitoring of the elderly, http://www.enrichme.eu/
28. RADIO Project, Robots in assisted living environments: Unobtrusive, efficient, reliable and modular solutions for independent ageing, http://www.radio-project.eu/
29. C. Antonopoulos, G. Keramidas et al., Robots in assisted living environments as an unobtrusive, efficient, reliable and modular solution for independent ageing: The RADIO perspective. Proceedings of the International Symposium in Applied Reconfigurable Computing, 2015

30. ROBOT-ERA Project, Implementation and integration of advanced robotic systems and intelligent environments in real scenarios for the ageing population, http://www.robot-era.eu/
31. M. Bonaccorsi, F. Lauraa et al., Design of cloud robotic services for senior citizens to improve independent living in multiple environments. J. Artif. Intell. Soc. Econ. (2015)
32. P. Dario, A synoptic glance and main objectives. Robot-Era presentation, 2013

# Chapter 15
# Software Design and Optimization of ECG Signal Analysis and Diagnosis for Embedded IoT Devices

**Vasileios Tsoutsouras, Dimitra Azariadi, Konstantina Koliogewrgi, Sotirios Xydis, and Dimitrios Soudris**

## 15.1 Introduction and Theoretical Background

Internet of Things is perceived as a key application to introduce smart, sensor-based, inter-connected devices to many aspects of all parts of life. A most promising domain for the deployment of such kind of inter-woven systems is considered to be the healthcare one. The added values of combining these domains are numerous since patients will be offered better treatment and quality of life while the already stressed healthcare system will be alleviated since remote monitoring will reduce the need of patient hospitalization.

From the embedded IoT designer point of view, the design and implementation of a medical analysis flow creates new challenges since there are strict requirements considering real-time processing, system predictability, security, and low power consumption. Consequently, bio-signal processing flows have to be meticulously designed and optimized both on algorithm and implementation level. The aim of this study is to provide the basis for such an exploration based on a case study of arrhythmia detection using the electrocardiogram signal as input. This first section introduces the necessary background information both on medical concepts and algorithms which will be further utilized when exploring different design choices.

V. Tsoutsouras (✉) • D. Azariadi • K. Koliogewrgi • S. Xydis • D. Soudris
Microprocessors and Digital Systems Laboratory, Electrical and Computer Engineering Department, National Technical University of Athens, Zografou Campus, Athens, Greece
e-mail: billtsou@microlab.ntua.gr; dimitra@microlab.ntua.gr

 
G. Keramidas et al. (eds.), *Components and Services for IoT Platforms*,
DOI 10.1007/978-3-319-42304-3_15

### *15.1.1 The Electrocardiogram Signal*

The electrocardiogram signal, most commonly referred to as ECG signal, is widely accepted as one of the most fundamental bio-signals for monitoring and assessing the health status of a patient. It is produced by recording the electrical activity of the heart over a period of time using electrodes placed on a patient's body. The cardiac muscle contracts in response to electrical depolarization of the muscle cells, and the sum of this electrical activity is what the electrodes record. The intensity, duration, and shape of the waveform reflecting depolarization and repolarization of the atria and ventricles.

Figure 15.1 illustrates a typical example of an ECG waveform. Different parts of this waveform can be distinguished and these parts are utilized by medical experts to understand the development of the ECG signal. More specifically, the P wave indicates that the atria are contracting, pumping blood into the ventricles, and is usually 0.08–0.1 s in duration. The QRS complex represents ventricular depolarization and contraction, with duration normally 0.06–0.1 s. The T wave represents ventricular repolarization and is longer in duration than depolarization. Sometimes a small positive U wave may be seen following the T wave, which represents the last remnants of ventricular repolarization.

By convention, electrodes are placed on each arm and leg, and six electrodes are placed at defined locations on the chest. These electrode leads are connected to a device that measures potential differences between selected electrodes to produce the characteristic ECG tracings. The limb leads are called leads I, II, III, AVR, AVL, and AVF. The chest leads are called V1, V2, V3, V4, V5, and V6. In this study, the ECG signals we examine are modified limb lead II, a bipolar lead parallel to the standard limb lead II, but acquired using electrodes placed on the torso, a requirement for long-term ECG monitoring.

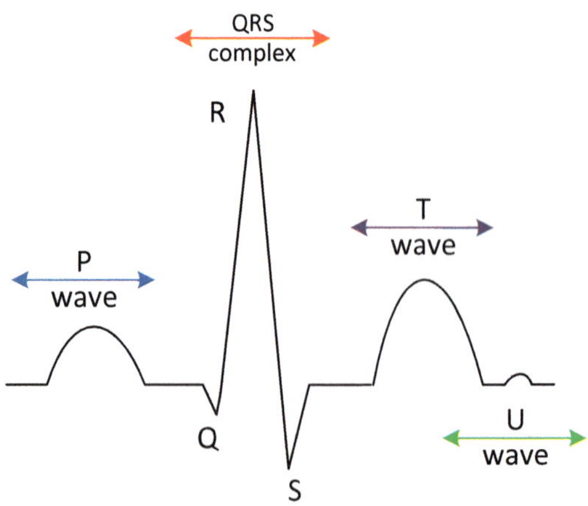


**Fig. 15.1** ECG waveform [1]

### 15.1.2 Arrhythmia Detection and MIT-BIH Database

One of the most common heart malfunctions is arrhythmia [2]. In this case the heartbeats differently compared to its normal rhythm and while the phenomenon is not always critical, in other cases it can lead to strokes or heart failure. Given the severity of the latter cases, arrhythmia and its detection using the ECG signal is a well-studied domain. Recently, new techniques from the machine learning domain have been adapted [3, 21, 22] and as a part of this work we utilize such a machine learning-based ECG analysis flow to be incorporated in the IoT node under design.

To efficiently design and employ machine learning techniques an enriched and descriptive input data set of phenomenon under analysis is needed. For the purposes of this study, data from the MIT-BIH arrhythmia database [4, 20] were used. The database is composed of 48 half-hour excerpts of two-channel (two leads) ambulatory ECG recordings, obtained from 47 subjects. Of these, 23 recordings were chosen at random from a collection of over 4000 24-h ambulatory ECG recordings, serving as a representative sample of routine clinical recordings. The remaining 25 recordings were selected from the same set to include a variety of rare but clinically important phenomena [2] (complex ventricular, junctional, and supraventricular arrhythmias), which would not be well represented in a small random sample. The subjects included 25 men aged 32–89 years and 22 women aged 23–89 years. Approximately 60 % of the subjects were inpatients and 40 % outpatients.

A crucial advantage of the MIT-BIH database is that it also provides annotations for each record, where cardiologists have designative a label for each individual heartbeat included in the record. There are approximately 110,000 annotations.Two arrhythmia groups are examined in this analysis, "Normal" (N) and "Abnormal," which includes all the different types of arrhythmia as they have been indicated by doctors.

### 15.1.3 Discrete Wavelet Transform

To efficiently process the ECG signal, a means of extracting features of it is necessary. A widely used such algorithm is the discrete wavelet transform (DWT) [3, 5]. The wavelet transform (WT) is similar to the Fourier transform, with the extension that it is capable of providing the time and frequency information simultaneously. This is essential when analyzing signals whose frequency response varies in time, such as the ECG signal, where the time localization of the frequency spectral components is required.

The DWT analyzes the signal at different frequency bands with different resolutions by decomposing the signal into a coarse approximation and detail information. During DWT decomposition, the time domain signal is successively low-pass and high-pass filtered by the use of two sets of functions: scaling functions and wavelet functions.

The original signal $x[n]$ is first passed through a halfband high-pass filter $g[n]$ and a low-pass filter $h[n]$. After filtering, half the frequencies of the signal have been removed, thus half the samples can be discarded according to Nyquist's rule. The signal is downsampled by 2, simply by discarding every other sample. This constitutes one level of decomposition and can mathematically be expressed as follows:

$$y_{\text{high}}[k] = \sum_n x[n] * g[2k - n]$$

$$y_{\text{low}}[k] = \sum_n *h[2k - n]$$

where $y_{\text{high}}[k]$ and $y_{\text{low}}[k]$ are the outputs of the highpass and lowpass filters, respectively, after downsampling by 2.

This operation halves the time resolution since only half of each filter output now characterizes the entire signal. At the same time, the frequency resolution is being doubled, as each filter output now spans only half the frequency of the input. The above procedure can be repeated for further decomposition.

### *15.1.4 Support Vector Machines*

Support vector machines (SVMs) are machine learning algorithms, which analyze data and recognize patterns based on the statistical learning theory. They are used for classification and regression analysis. Given a set of training examples, each labelled as belonging to one of two categories (supervised learning), an SVM training algorithm builds a model that assigns new examples into one category or the other, making it a non-probabilistic binary linear classifier. Extending this notion, in this work they will be employed as means of distinguishing whether a heartbeat is exhibiting arrhythmia or not.

In a binary classification problem where input data belong to two classes, a hyperplane can be defined which is the geometrical division or separation between the two classes. The SVM algorithm will try to deduce the optimal hyperplane, called maximum margin hyperplane that best divides the classes. This is achieved by discovering the hyperplane that has the largest distance to the nearest training data point of any class, since in general the larger the margin, the lower the generalization error of the classifier. The decision of the optimal hyperplane is fully specified by a (usually small) subset of the data which defines the position of the separator. These points are referred to as the support vectors and hence the name of the classifier.

While the original problem may be stated in a finite-dimensional space, it often happens that the categories to discriminate are not linearly separable in that space. For this reason, it is proposed that the original finite-dimensional space be mapped into a much higher-dimensional space, making the separation possible in that space. To achieve this mapping, a kernel function is used. Commonly chosen kernel functions are non-linear ones such as choices for a kernel function such as a polynomial one with order greater than one, Gaussian radial basic function (RBF) which is used in this work and hyperbolic tangent (tanh).

## 15.2 Design and Exploration of ECG Analysis Flow

The first essential part of creating an ECG analysis IoT node is defining the exact components of the flow and exploring different design alternatives which are better suited for the current case study. This exploration is performed on high, algorithmic level and is executed offline on devices with high computational capabilities in order to explore the biggest possible subset of the design space within a reasonable time frame. Consequently, in this section, the overview of the ECG analysis flow is presented and then the focus is changed to each individual sub-part and the strategies to best define its structure.

### *15.2.1 Overview of the Proposed ECG Analysis Flow*

Figure 15.2 illustrates the structure of the proposed analysis steps for heartbeat classification. A digitized ECG signal is applied as the input to the system. A filtering unit is used as a preprocessing stage, to remove baseline wander and noise from the ECG signal. The filtered signal is then passed to the heartbeat detection unit, which attempts to locate all the heartbeats contained in the input ECG signal.

Next in the flow, is the segmentation unit, where the input signal is segmented into single heartbeats, accordingly with the information extracted from the previous stage. In order to achieve greater classification performance, a feature extraction unit is included. There, for each produced heartbeat, a feature vector is extracted, containing a smaller number of elements than the ECG samples forming the heartbeat. This vector serves as input for the classification stage, where the heartbeat is classified as either "N" (normal) or "ABN" (abnormal) by a single classifier.

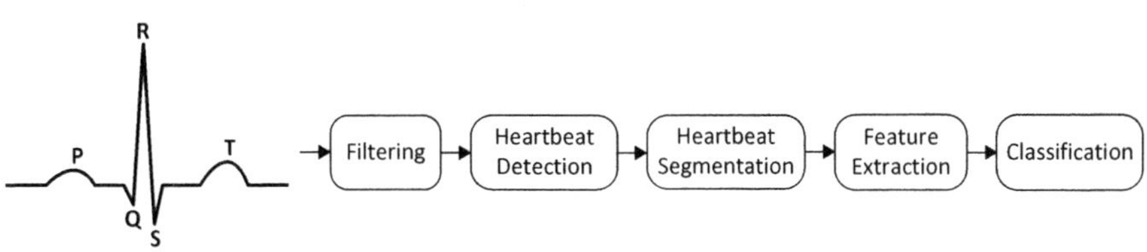


**Fig. 15.2** Proposed ECG analysis flow

Along with the MIT-BIH annotated ECG records, a library of software for physiologic signal processing and analysis is provided by PhysioNet [6]. The initial stage of this study was implemented in Matlab environment, and so the WFDB Toolbox for Matlab[7] was used for the reading and processing of the database.

## *15.2.2 Building Blocks of the ECG Analysis Flow*

### 15.2.2.1 Filtering

The power line interference and the baseline wandering are significant noise sources that can strongly affect the ECG signal analysis. The Matlab function *filter*() was used to filter the signals with a band-pass filter of bandwidth 1–50 Hz. The signals were band-pass filtered at 1–50 Hz [8].

### 15.2.2.2 Heartbeat Detection

Firstly, the wqrs function[9] is applied to the signal, which gives us the locations of all QRS complexes found in the signal. This information along with the ECG signal are the inputs to ecgpuwave function, which gives us the exact position of all the R peaks found in the signal. QRS detection, especially detection of R wave in heart signal, is easier than other portions of ECG signal due to its structural form and high amplitude.

Each R peak detection corresponds to the detection of a single heartbeat. The ecgpuwave program introduced in [10] has been validated on the Common Standards in Electrocardiography Multilead database [11] and the MIT-BIH QT database [12] and its performance in detecting significant points in ECG signals is comparable to those given by experts.

### 15.2.2.3 Heartbeat Segmentation

Having located the R peaks of each heartbeat waveform, we can proceed to segment the ECG signal into single heartbeats. To do that, we have to decide on a window width, which having as center the detected position of the R peak, will cover the whole of the heartbeat waveform. We choose a window width of 257 samples, as suggested in [3].

### 15.2.2.4 Feature Extraction

As a feature extraction mechanism we use DWT [13], since it has been proven to produce very accurate results. The wavelet base for the DWT is Daubechies of order

2 (db2) [14] and we perform four levels of decomposition as proposed in [3]. The Matlab function *wavedec*() was used to perform the wavelet analysis on the signal, and functions *appcoef*() and *detcoef*() were used to extract the approximation and detail coefficients, respectively.

The DWT is used to compute compressed parameters of the heartbeat data which are called features. These parameters characterize the behavior of the heartbeat. The method of using a smaller number of parameters to represent the heartbeat is particularly important for recognition and diagnostic purposes. The four levels of decomposition produce eight sets of coefficients each one for four levels of detailed and four levels of approximate coefficients. Since the heartbeat on which the DWT is applied consists of 257 samples, the number of wavelet coefficients for the first, second, third, and fourth level are, respectively, 130, 66, 34, and 18. Thus, 494 wavelet coefficients are obtained for each heartbeat. The final feature vector that serves as input to the classification stage resulted from a design space exploration performed on all combinations of these eight sets of coefficients, as will be discussed later on.

#### 15.2.2.5 Classification

The last stage of the structure consists of a binary classifier, which labels each heartbeat as either "Normal" or "Abnormal." In this study, we focus on using an SVM classifier [15], mainly due to its ability to support non-linear classification with efficient accuracy and computation cost. The Matlab interface of LIBSVM [16], a library for SVMs, was used for the implementation of the SVM classifier.

As kernel function, we use the radial basis function (RBF). This kernel non-linearly maps samples into a higher-dimensional space so it, unlike the linear kernel, can handle the case when the relation between class labels and attributes is non-linear, with smaller complexity in the model selection than the polynomial kernel does. There are two parameters, $C$ and $\gamma$, which need to be tuned, so that the model which best predicts the unknown data is selected. Parameter $\gamma$ is set to $1/k$, where $k$ means the number of attributes in the input data, and parameter $C$ is set to 1, which is the default option recommended by LIBSVM.

### *15.2.3 Design Space Exploration on SVM Classifier*

The training process of the SVM model can be done offline. Having defined a training data set and a testing data set for evaluation, we perform Design Space Exploration on the different choices of input feature vectors of the ECG signal in order to designate the best SVM model for arrhythmia detection (different feature sets generate different SVM models).

A high level view of the process to decide upon the feature vector which gives the model with the best performance results is depicted in Fig. 15.3. The training phase

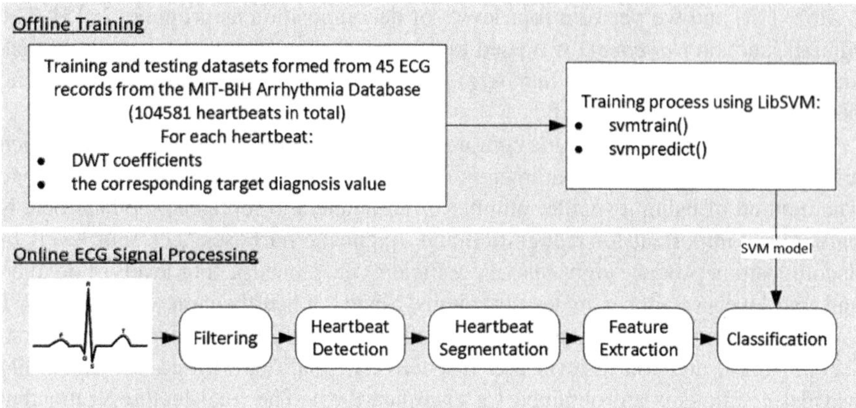


**Fig. 15.3** Offline training and online classification

is executed offline in a machine with high computational capabilities. Its result is a number of model exhibiting trade-offs regarding their classification accuracy and computational requirements. This section elaborates on the required steps to set up and perform the exploration of the design space.

#### 15.2.3.1 Creating and Input Data Set for SVM Design Space Exploration

In order for the different SVM models to be produced, an input data set has to be appropriately formulated using the 45 selected ECG records from the MIT-BIH database. This input data set will be afterwards divided to produce its training and testing sub-parts. As the previous flow indicates, all records were firstly filtered, the R-peaks were located, and then the records were segmented into single heartbeats, forming a set of 104,581 heartbeats. As stated before, for each heartbeat we need a vector of attributes and a target value. The vector of attributes used is the feature vector which contains the DWT coefficients of the heartbeat and is formed at the feature extraction stage. As target value, we use the label given for each heartbeat in the annotation files.

The problem detected at this point of the analysis is that there is a mismatch in the heartbeats detected in a recording by the functions provided in the toolkit and the heartbeats annotated by the doctors. In Fig. 15.4, the black circles indicate the R peaks defined by the doctors. The green circle shows a correctly detected R peak, a "True" beat, which is a detection close to an R peak annotation. A faulty detected R peak, a "False" beat, is one which is far from the corresponding R peak annotation (red circle). "Missed" beat is considered an R peak which the detector failed to detect (the second black circle).

We overcome this problem by forming a procedure that allows us to match the correctly detected heartbeats with their corresponding labels in the annotation files. We use the distance from the actual R peak to decide on the type of detection, comparing it to a defined threshold $T$. A statistical analysis over the whole database

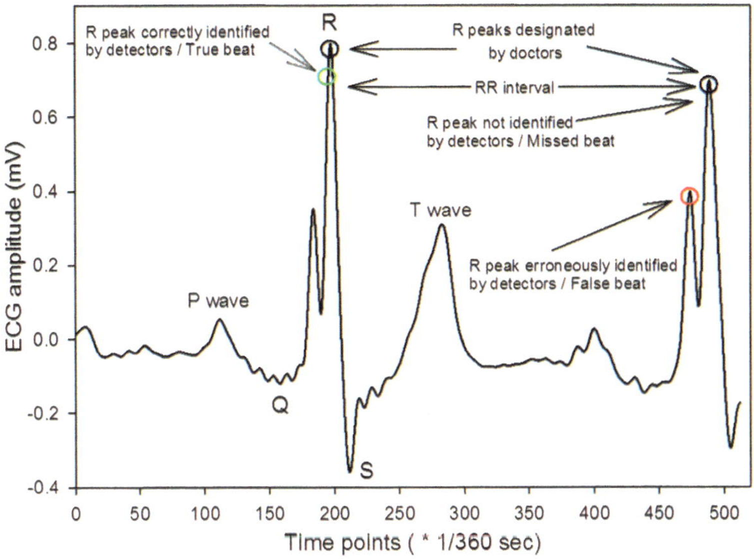


**Fig. 15.4** Correct vs. faulty QRS detection [17]

was conducted to estimate the duration of the QRS complex, and its median value was found approximately 0.11 s. With a sampling frequency $Fs = 360$ Hz, the QRS complex is approximately 40 data samples wide. We define the threshold $T$ as half the duration of the QRS complex, that is, 20 samples.

Our input consists of two time series; one consists of the detected heartbeats using the provided R peak detection functions (*D_times*) while the other consists of the annotated heartbeats provided by medical experts (*A_times*). Beginning from the first elements of each time series, the procedure goes as follows: we compare the detected beat *i_D* with the annotated beat *i_A*. If their absolute distance is less or equal than $T$, then we have matched a detected heartbeat with an annotated one. Consequently, the detected beat under examination is characterized as "True," and both time series indexes are increased in order to move to the next heartbeats. If the annotated heartbeat proceeds the detected for more than $T$ samples, then it is characterized as a "Missed" beat, and its index is increased to move on to the next annotated beat. If the detected heartbeat proceeds the annotated for more than $T$ samples, then it is characterized as a "False" beat, and its index is increased to examine the next detected beat.

This procedure allows us to match the true detected heartbeats with their corresponding labels in the annotation files. As for the falsely detected heartbeats, we set the target value as "Abnormal." This seems as a logical assumption, but to validate this choice, we train a classifier with only true detected heartbeats and evaluate it under false heartbeats testing scenarios. The result was that about 86 % of the heartbeats were indeed classified as Abnormal. Of course, the missed heartbeats are not taken into consideration, since they are not detected during the heartbeat detection stage.

**Algorithm 1** Derived R Matching

```
procedure MATCHING(Dtimeseries, Atimeseries)
    Truebeats, Falsebeats, Missedbeats ← []

    iD, iA ← 0
    lenD ← length(Dtimeseries)
    lenA ← length(Atimeseries)
    while iD <> lenD AND iA <> lenA do
        if abs(Dtimes(iD) − Atimes(iA)) <= T then
            Truebeats ← Truebeats + Dtimes(iD)
            iD ← iD + 1
            iA ← iA + 1
        else if Dtimes(iD) − Atimes(iA) > T then
            Missedbeats ← Missedbeats + Atimes(iA)
            iA ← iA + 1
        else if Atimes(iA) − Dtimes(iD) > T then
            Falsebeats ← Falsebeats + Dtimes(iD)
            iD ← iD + 1
        end if
    end while

    return Truebeats, Falsebeats, Missedbeats
end procedure
```

Finally, we have a data set of 104,581 heartbeats (100,231 true heartbeats and 4350 false heartbeats), along with a vector of their corresponding target values, and a matrix with the DWT coefficients for each heartbeat. Half of the heartbeats are used as the training data set for the SVM model, and the rest of the heartbeats are used as the testing data set.

#### 15.2.3.2 Design Space Exploration for SVM Tuning

A design space exploration is performed over all combinations of the eight sets of DWT coefficients produced in the feature extraction stage. To reduce exploration time, DWT is calculated only once, producing a vector containing all sets of coefficients for each heartbeat detected in the database. An iteration takes place, choosing a different combination of these sets to serve as feature vector, and producing a different classifier. The goal is to produce a classifier with

- maximized accuracy $\left(\frac{\text{Number of correctly classified points}}{\text{Total number of points}}\right)$
- minimized computational cost

In this initial stage, the computational cost was not measured as the absolute execution time of classification on the platform used, but it was rather computed theoretically as the product of the number of support vectors and the size of feature vector, since this is the number of multiplications required for a heartbeat to be classified [15].

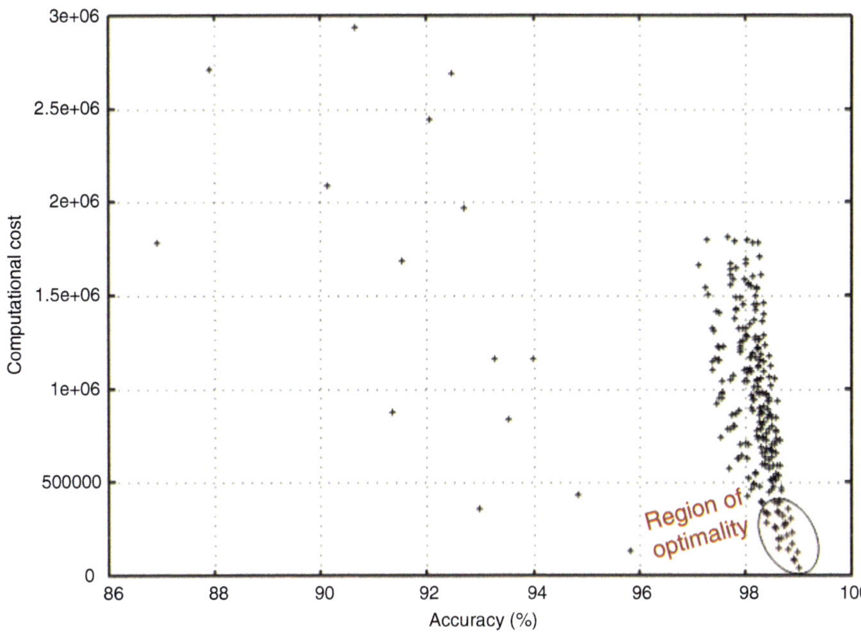


**Fig. 15.5** Design space exploration of SVM classifier

Figure 15.5 presents the results of the design space exploration, regarding accuracy and computational cost. As we can see, in almost all cases the accuracy is above 97 %, with only a few exceptions. The best result comes from the feature vector which only contains the approximate coefficients of the fourth level of decomposition. In this case, we have an accuracy of 98.9 %, a feature vector of size 18, and 2493 support vectors.

## 15.3 Analysis Flow on Embedded IoT Platform

Up to this point the focus was towards the design and optimization of the different parts of the ECG analysis and arrhythmia detection flow. This process has been performed using high level tools such as Matlab, which inherently provide implementations of the various functional blocks required to perform both building and testing of the various components of the ECG processing flow.

The proceeding goal is having designated these components to further define ways in order to port them to target embedded IoT device which is much more constrained in terms of available off the self source code implementations and computational resources. Consequently, this section summarizes all the necessary decisions to support the aforementioned requirement.

### 15.3.1 Employed IoT Platform

Figure 15.6 demonstrates an abstract view of a complete IoT-based ECG monitoring design. Electrodes are used to sense the ECG voltage on the skin, amplifiers to increase the magnitude of the signal, and ADC to get digital samples. The signal is then sent to a wearable IoT-based device to be processed. The data transfer can be conducted via a Bluetooth module. The output of the computing devise, the heartbeat diagnosis in the case of ECG signal classification, is finally sent via Internet connection to a remote device for further analysis and storage.

We implement the ECG signal analysis and classification algorithm developed on the Galileo board, Intel's proposed IoT device [18]. The Galileo is the first product to feature the Intel Quark SoC X1000, a chip designed for small-core products and low power consumption, and targeted at markets including the Internet of Things and wearable computing. The Quark SoC X1000 is a 32-bit, single core, single-thread, Pentium (P54C/i586) instruction set architecture (ISA)-compatible CPU, operating at speeds up to 400 MHz. The use of the Pentium architecture gives the Galileo the ability to run a fully fledged Linux kernel. What's more, an on-board Ethernet port provides network connectivity, while also the underside provides a mini-PCI Express slot, designed for use with Intel's wireless network cards to add Wi-Fi connectivity to designs (Fig. 15.7).

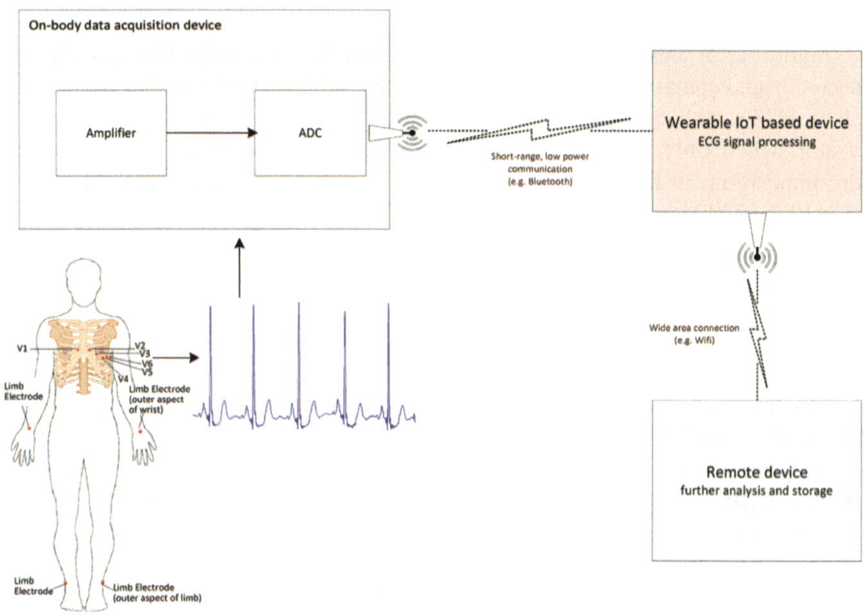


**Fig. 15.6** Iot-based ECG monitoring design

**Fig. 15.7** Intel Galileo board [18]

The Galileo board's technical specifications:

- **Operating system:** Yocto Project-based Linux
- **Processor:** Single-Core 400 MHz Intel Quark X1000
- **Memory:** 256 MB RAM
- **Dimensions:** 107 x 74 x 23 mm
- **Weight:** 50 g (excluding PSU)
- **GPIO:** 14x Digital Input/Output Pins, 6x Analogue Input Pins
- **Networking:** 1x Wired 10/100 Ethernet, Optional PCIe Wireless
- **Expansion:** USB 2.0 Host, Micro-SD Card

The Quark X1000 features:

- Up to 400 MHz clock speed
- 16 KB L1 Cache
- 512 KB SRAM
- Single core, single thread
- Integrated SDIO, UART, SPI, USB, I2C, Ethernet, RTC

### *15.3.2 Mapping High Level Functionalities to Low Level Source Code*

The online ECG signal processing and classification is the real-time part of the analysis that will be implemented on the embedded platform, while the offline training procedure (see Fig. 15.3) is considered to already have been implemented on some other platform. So far all the analysis was implemented in high level using Matlab. At this point, the different stages of the analysis flow have to be implemented in C, as described below, and combined into a single program. The program reads sample by sample a digitized (at 360 samples per second) ECG signal. The analysis flow is executed for every set of 3000 samples that is read which is a limitation imposed by the source code responsible for R-peak detection (Sect. 15.1.1) as it has been provided by Physionet [4].

#### 15.3.2.1 Filtering

The input signal, consisting of 3000 samples, is firstly filtered. For the filtering stage, the same FIR filters that were used in the Matlab environment are implemented in C.

#### 15.3.2.2 Heartbeat Detection

In the Matlab implementation of heartbeat detection two functions provided by PhysioNet in the WFDB Toolbox, *wqrs*() and *ecgpuwave*(), are used. PhysioNet also provides the source codes of these two functions, wqrs in C and ecgpuwave in Fortran. Since we only have wqrs in C, we alter the algorithm to only apply wqrs to the signal, with no substantial divergence in the output of this stage. Wqrs is a QRS complex detection function, that returns the approximate position of the QRS complex onset.

Instead of passing this information to ecgpuwave, in order to get the exact position of the R peak and then apply the 257-sample window on the next stage of the heartbeat segmentation, we keep the QRS onset information as the heartbeat reference point, and adapt the window to cover the heartbeat waveform. Wqrs function analyzes an ECG signal, detecting QRS onsets, using a non-linearly scaled ECG curve length feature. The algorithm averages the first 8 s of the length-transformed input to determine on some initial thresholds that it uses. This is the reason we choose a 3000-samples (8.33 s) waveform as the minimum input of our program.

#### 15.3.2.3 Heartbeat Segmentation

For the implementation of the heartbeat detection stage, we adapt the window to cover the PR and the QT intervals (Fig. 15.1). Based on statistics over the ECG waveform [19], we decide on a window of 86 samples before the QRS onset, and 170 samples after the QRS onset (window width of 257 samples). Comparing the output of the two implementations up to this stage, we see that the resulted heartbeat is satisfactory in both the Matlab and the C implementation, as demonstrated in Fig. 15.8 for a random ECG waveform. The heartbeats detected in the 3000-sample signal are passed one by one to the next stages.

#### 15.3.2.4 Feature Extraction

For the stage of feature extraction the Matlab function *wavedec*() is implemented in C. We use Matlab function *wfilters*() to acquire the wavelet decomposition low-pass and high-pass filters associated with wavelet Daubechies of order 2 ("db2"). We apply symmetric-padding (boundary value symmetric replication) to the signal, since it is the default DWT extension mode of Matlab. Then, the convolution of

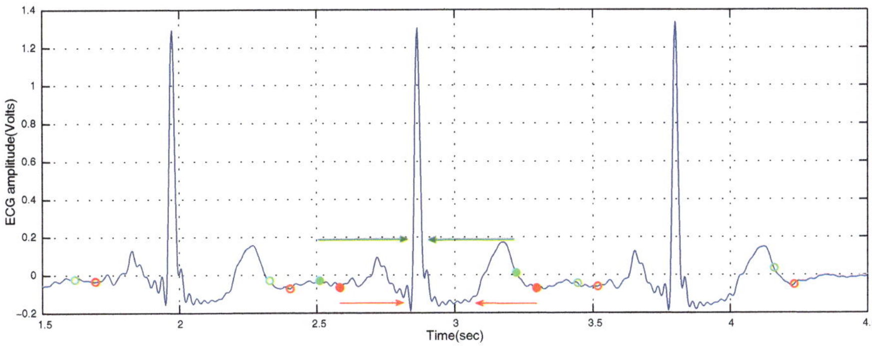


**Fig. 15.8** Heartbeat segmentation output for random ECG waveform: the *green circles* show the beginning and end of the heartbeat as determined by the Matlab implementation, and the *red circles* indicate the beginning and end of the heartbeat as determined by the implementation in C

the signal with each filter is implemented to produce the approximation and detail coefficients for each of four levels of decomposition, according to the following process:

Given a signal $s$ of length $n$, the DWT consists of $\log_2 n$ stages at most. The first step produces, starting from $s$, two sets of coefficients: approximation coefficients CA1 and detail coefficients CD1. These vectors are obtained by convolving $s$ with the low-pass filter Lo_D for approximation, and with the high-pass filter Hi_D for detail, followed by dyadic decimation (downsampling). The length $N$ of each filter is equal to the order of the Daubechies wavelet. The signals $F$ and $G$ are of length $n + 2N - 1$ and the coefficients cA1 and cD1 are of length $floor((n - 1)/2) + N$. The next step splits the approximation coefficients cA1 in two parts using the same scheme, replacing $s$ by cA1, and producing cA2 and cD2, and so on (Fig. 15.9).

This stage is implemented parametrically in terms of the coefficients that compose the feature vector of the heartbeat used in the classification stage. For each decomposition level, the coefficients produced are stored in a matrix if they are part of the final feature vector, or else they are only used for the computation of the next decomposition level. The process stops whenever all required coefficients have been produced.

#### 15.3.2.5 Classification

For the final stage of classification, we use the LIBSVM library which includes the source code of *svmtrain*() in C. We convert the SVM model under examination produced by the same function in the Matlab environment into the format that is used in the C implementation, using a generator. In the beginning of the program, the function *load_svm_model*(), which is included in the LIBSVM library, is called to load the SVM model from that file that we have created. The SVM model is

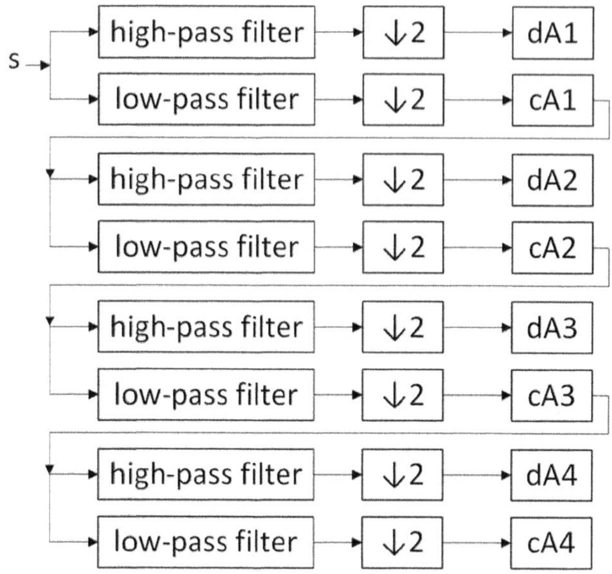


**Fig. 15.9** Four levels of DWT

only created once in that initialization section of the program, and used in the classification stage for each heartbeat.

In Fig. 15.10 the overall exploration framework is illustrated. This includes the exploration phase conducted in the Matlab environment, meaning the DSE which resulted in a set of Pareto solutions. In the customization phase, having selected a desirable solution, the feature vector is set accordingly and the corresponding SVM model is produced by a generator which converts the Matlab SVM model to the appropriate format for the C implementation. Finally, the runtime phase is the implementation of the algorithm in C, for the configuration selected, conducted on the Galileo board.

### *15.3.3 Supporting Back-End Server Communication*

The complete data flow of the IoT-based ECG monitoring design must support data communication to a back-end server which will perform data aggregation and monitoring of the entire IoT-based system. In the presented case study, the Galileo Board, which serves as the wearable IoT device, is connected via Ethernet to the network in order to transmit the output data of the application over Internet to the database of the back-end server. The information sent consists of the output of the classification stage, which indicates the detection of either a "Normal" or an

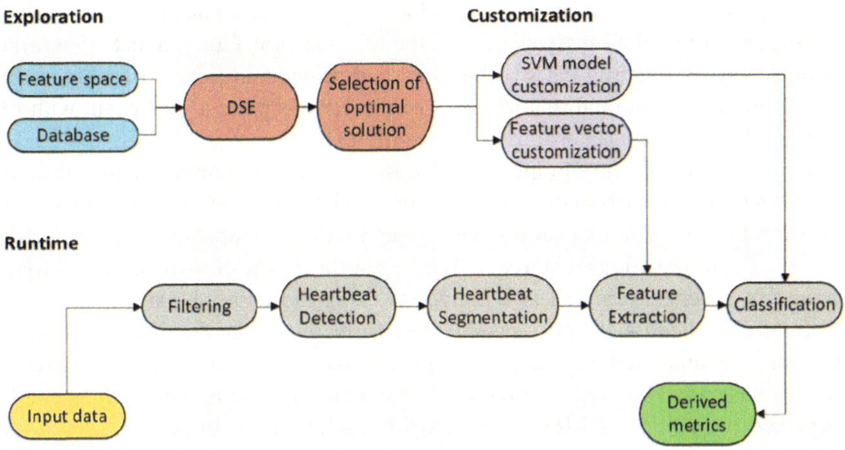


**Fig. 15.10** Structure of algorithm implemented in C

"Abnormal" heartbeat in the ECG signal. The transmission is performed in real-time, after the detection and classification of a new heartbeat.

In the software stack of the IoT node, libcurl library was used to perform HTTPS Post requests of the ECG monitoring application to the remote web server. Libcurl is a C-based client-side URL transfer library that supports many Internet protocols including HTTP, HTTPS, DICT, FILE, FTP, SFTP, SCP, Telnet, and TFTP. SSL certificates are also supported, ensuring that data exchanged remain private and integral. Secure data transmission is especially critical in medical IoT monitoring, as the information transmitted is of high clinical significance, while also medical confidentiality must be preserved. Furthermore, the library is IPv6 compatible which is important for IoT as it simplifies the assignment of a global IP address to an IoT device thus enabling efficient peer-to-peer communication.

An example back-end service was also developed, featuring an Apache/MySQL/PHP software stack in order to handle HTTPS requests and store the received data to a database.

### *15.3.4 Evaluation of ECG Analysis Flow on the IoT Node*

From the offline training procedure, we get the SVM model which is being used at the classification stage of the analysis flow (see Fig. 15.3). The configurations of this model derived from the design space exploration for the SVM tuning. The metrics taken into consideration for the selection were, as mentioned there, accuracy and computational cost.

In this section, we evaluate the behavior and performance of the various components of the ECG analysis flow on the IoT platform. Our goal is to determine the most time-consuming parts of the flow and provide an insight of whether the trend of the computational requirements of the various parts is consistent with the theoretical expectation according to their complexity.

During the DSE, the computational cost was not measured as the absolute execution time of heartbeat classification on the platform used, but it was estimated as the product of number of support vectors and size of feature vector, since it is the number of multiplications required for a new heartbeat to be classified, as mentioned already. In order to verify these estimations using the absolute execution time of heartbeat classification on the embedded platform, and make the final SVM model selection, we sampled the available configurations as a result of the DSE. In overall, using as basic discriminant the predicted computational intensiveness of the model we picked the 10 lighter, 10 heavier, and 11 configurations from in-between. The selected configurations are summarized in Table 15.1 where the columns include the subset of DWT coefficient used as input by the model (Sect. 15.2.2.4), the number of support vectors of the model, the size of its input feature vector and its accuracy, and estimated needs in computations.

The results of the time averages and the percentage of total execution time, for each stage of the analysis flow, for the selected configurations, are shown in Table 15.2. The time averages for the filtering and the heartbeat detection stage correspond to the 3000 sample input, while the time averages of the feature extraction and the classification stage are per heartbeat detected. As input we used 101*3000 samples (303,000 out of 650,000 samples contained in each ECG signal record) from 43 records (out of 48 total records of MIT-BIH database).

In the total of 4343 sets of 3000 samples provided as input to each configuration, 45,529 heartbeats were detected each time, and it was seen that approximately 10 heartbeats were contained in each set of 3000 samples. Therefore, for the estimation of the percentage of the total execution time of each stage, it was assumed that for every 3000 sample input, 10 heartbeats are detected. In the best cases examined, the total execution time is much less than the approximate 9 s required for the 3000 samples of the signal to be read. This means that the ECG signal analysis and classification proposed can be performed in real-time.

As we can see in Fig. 15.11, the scaling of execution time of the classification stage for the different configurations is as expected from the theoretical approach, meaning the computations required in each configuration.

In Fig. 15.12, we present the average % of total execution time for each stage of the algorithm, of the 10 best configurations examined, the 11 configurations from the in-between and the 10 most computationally demanding configurations. In the last case, the classification stage takes up more than 90 % of the total execution time.

**Table 15.1** Selected configurations from DSE

| Configuration | DWT coefficients | Number of support vectors | Feature vector size | Accuracy (%) | Computational cost |
|---|---|---|---|---|---|
| 1 | cA4 | 2493 | 18 | 98.99 | 44,874 |
| 2 | cA3 | 2490 | 34 | 98.93 | 84,660 |
| 3 | cA4, cD4 | 2513 | 36 | 98.90 | 90,468 |
| 4 | cA3, cA4 | 2408 | 52 | 98.99 | 125,216 |
| 5 | cA3, cD4 | 2696 | 52 | 98.80 | 140,192 |
| 6 | cA4, cD3 | 2913 | 52 | 98.62 | 151,476 |
| 7 | cA3, cA4, cD4 | 2449 | 70 | 98.88 | 171,430 |
| 8 | cA2 | 2920 | 66 | 98.62 | 192,720 |
| 9 | cA3, cD3 | 3017 | 68 | 98.67 | 205,156 |
| 10 | cA4, cD3, cD4 | 2931 | 70 | 98.62 | 205,170 |
| 11 | cA2, cA3, cA4 | 2595 | 118 | 98.84 | 306,210 |
| 12 | cA4, cA3, cA4, cD4 | 2650 | 136 | 98.79 | 360,400 |
| 13 | cD3, cD4 | 8378 | 52 | 94.83 | 435,656 |
| 14 | cA3, cA4, cD1 | 3258 | 182 | 98.36 | 592,956 |
| 15 | cA4, cD2, cD3, cD4 | 3861 | 184 | 98.00 | 710,424 |
| 16 | cA3, cA4, cD1, cD3, cD4 | 3510 | 234 | 98.26 | 821,340 |
| 17 | cA1, cA2, cA3, cA4, cD3, cD4 | 3122 | 300 | 98.60 | 936,600 |
| 18 | cA2, cD1, cD3, cD4 | 4240 | 248 | 97.70 | 1,051,520 |

(continued)

**Table 15.1** (continued)

| Configuration | DWT coefficients | Number of support vectors | Feature vector size | Accuracy (%) | Computational cost |
|---|---|---|---|---|---|
| 19 | cD2, cD3 | 11,635 | 100 | 93.25 | 1,163,500 |
| 20 | cA2, cA3, cD1, cD2, cD3 | 3885 | 330 | 98.02 | 1,282,050 |
| 21 | cA1, cA3, cD1, cD2, cD4 | 3955 | 378 | 97.81 | 1,494,990 |
| 22 | cA1, cA2, cA3, cA4, cD1, cD2, cD3, cD4 | 3592 | 496 | 98.22 | 1,781,632 |
| 23 | cA1, cA2, cA3, cD1, cD2, cD3 | 3876 | 460 | 98.13 | 1,782,960 |
| 24 | cA1, cA2, cD1, cD2, cD3 | 4204 | 426 | 97.78 | 1,790,904 |
| 25 | cA1, cA2, cA3, cD1, cD2, cD3, cD4 | 3761 | 478 | 98.01 | 1,797,758 |
| 26 | cA1, cD1, cD2, cD3, cD4 | 4760 | 378 | 97.28 | 1,799,280 |
| 27 | cA1, cA2, cD1, cD2, cD3, cD4 | 4077 | 444 | 97.65 | 1,810,188 |
| 28 | cD1, cD3, cD4 | 10,815 | 182 | 92.70 | 1,968,330 |
| 29 | cD1, cD2, cD4 | 11,413 | 214 | 92.04 | 2,442,382 |
| 30 | cD1, cD2, cD3, cD4 | 10,854 | 248 | 92.45 | 2,691,792 |
| 31 | cD1, cD2, cD3 | 12,791 | 230 | 90.64 | 2,941,930 |

**Table 15.2** Results from the implementations on Intel Galileo

| | Filtering | | Beat detection | | Feature extraction | | Classification | |
|---|---|---|---|---|---|---|---|---|
| Configuration | Average (in sec) | % of total ex. time | Average (in sec) | % of total ex. time | Average (in sec) | % of total ex. time | Average (in sec) | % of total ex. time |
| 1 | 0.77127 | 70.01 | 0.01691 | 1.53 | 0.00210 | 1.91 | 0.02925 | 26.55 |
| 2 | 0.77807 | 61.88 | 0.01707 | 1.36 | 0.00192 | 1.53 | 0.04430 | 35.23 |
| 3 | 0.62330 | 60.49 | 0.01373 | 1.33 | 0.00171 | 1.66 | 0.03762 | 36.52 |
| 4 | 0.62322 | 54.80 | 0.01381 | 1.21 | 0.00166 | 1.46 | 0.04836 | 42.53 |
| 5 | 0.62386 | 52.48 | 0.01379 | 1.16 | 0.00171 | 1.44 | 0.05341 | 44.92 |
| 6 | 0.62326 | 50.45 | 0.01380 | 1.12 | 0.00173 | 1.40 | 0.05811 | 47.03 |
| 7 | 0.65056 | 48.14 | 0.01451 | 1.07 | 0.00178 | 1.32 | 0.06685 | 49.47 |
| 8 | 0.68187 | 45.72 | 0.01504 | 1.01 | 0.00147 | 0.99 | 0.07799 | 52.29 |
| 9 | 0.62327 | 44.76 | 0.01396 | 1.00 | 0.00158 | 1.13 | 0.07394 | 53.10 |
| 10 | 0.62284 | 44.03 | 0.01385 | 0.98 | 0.00173 | 1.22 | 0.07604 | 53.76 |
| 11 | 0.62648 | 36.43 | 0.01390 | 0.81 | 0.00171 | 0.99 | 0.10622 | 61.77 |
| 12 | 0.62396 | 33.13 | 0.01388 | 0.74 | 0.00171 | 0.91 | 0.12284 | 65.23 |
| 13 | 0.62414 | 25.97 | 0.01386 | 0.58 | 0.00170 | 0.71 | 0.17479 | 72.74 |
| 14 | 0.71297 | 23.17 | 0.01594 | 0.52 | 0.00199 | 0.65 | 0.23284 | 75.67 |
| 15 | 0.62899 | 20.30 | 0.01380 | 0.45 | 0.00177 | 0.57 | 0.24380 | 78.68 |
| 16 | 0.62782 | 18.38 | 0.01395 | 0.41 | 0.00176 | 0.51 | 0.27556 | 80.69 |
| 17 | 0.65934 | 16.99 | 0.01470 | 0.38 | 0.00184 | 0.47 | 0.31877 | 82.15 |

(continued)

**Table 15.2** (continued)

| Configuration | Filtering | | Beat detection | | Feature extraction | | Classification | |
|---|---|---|---|---|---|---|---|---|
| | Average (in sec) | % of total ex. time | Average (in sec) | % of total ex. time | Average (in sec) | % of total ex. time | Average (in sec) | % of total ex. time |
| 18 | 0.62769 | 15.04 | 0.01393 | 0.33 | 0.00175 | 0.42 | 0.35130 | 84.20 |
| 19 | 0.73301 | 12.75 | 0.01622 | 0.28 | 0.00185 | 0.32 | 0.49811 | 86.65 |
| 20 | 0.65047 | 12.75 | 0.01433 | 0.28 | 0.00173 | 0.34 | 0.44180 | 86.63 |
| 21 | 0.71573 | 11.15 | 0.01593 | 0.25 | 0.00206 | 0.32 | 0.56645 | 88.28 |
| 22 | 0.62523 | 09.02 | 0.01401 | 0.20 | 0.00181 | 0.26 | 0.62759 | 90.52 |
| 23 | 0.62374 | 09.28 | 0.01376 | 0.20 | 0.00169 | 0.25 | 0.60676 | 90.27 |
| 24 | 0.62314 | 09.29 | 0.01374 | 0.20 | 0.00171 | 0.25 | 0.60542 | 90.25 |
| 25 | 0.69495 | 08.90 | 0.01530 | 0.20 | 0.00204 | 0.26 | 0.70780 | 90.64 |
| 26 | 0.62363 | 09.45 | 0.01389 | 0.21 | 0.00179 | 0.27 | 0.59465 | 90.07 |
| 27 | 0.63113 | 08.91 | 0.01399 | 0.20 | 0.00183 | 0.26 | 0.64231 | 90.64 |
| 28 | 0.62258 | 08.50 | 0.01389 | 0.19 | 0.00177 | 0.24 | 0.66672 | 91.06 |
| 29 | 0.62295 | 06.95 | 0.01380 | 0.15 | 0.00177 | 0.20 | 0.83134 | 92.70 |
| 30 | 0.70772 | 06.32 | 0.01552 | 0.14 | 0.00199 | 0.18 | 1.04530 | 93.36 |
| 31 | 0.65894 | 05.89 | 0.01472 | 0.13 | 0.00176 | 0.16 | 1.05047 | 93.83 |

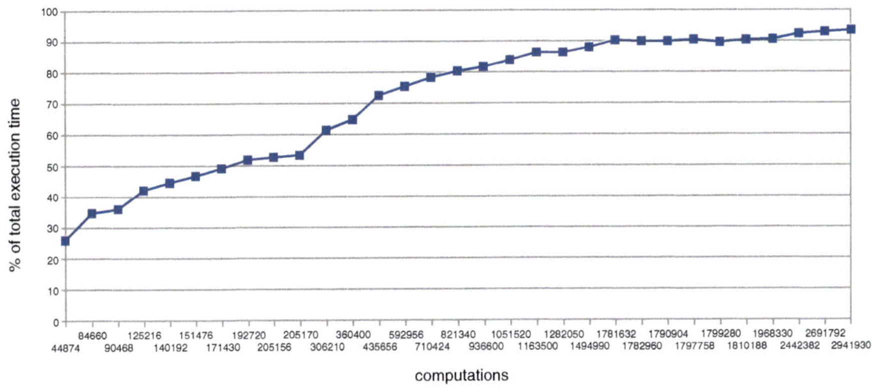


**Fig. 15.11** Scaling of execution time in accordance with the computations required in each configuration

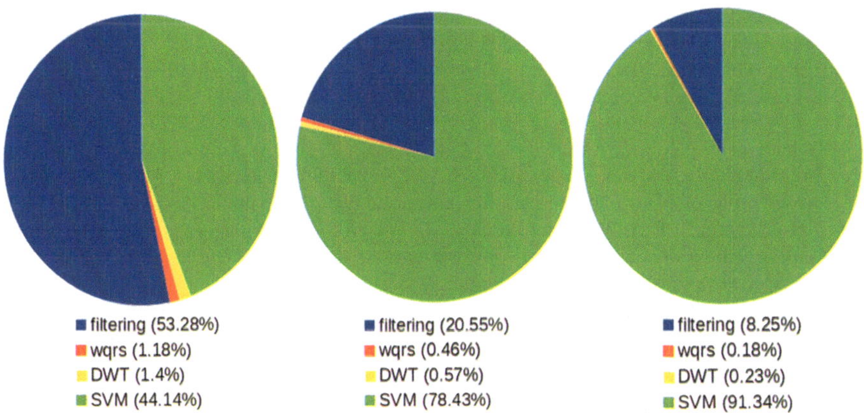


**Fig. 15.12** Average percentage of total execution time for each stage for the three categories of examined SVM models

# References


1. Electrocardiography, https://en.wikipedia.org/wiki/Electrocardiography
2. F.I. Marcus, J.N. Ruskin, B. Surawicz, Arrhythmias. J. Am. Coll. Cardiol. **10**(2), 66A–72A (1987)
3. E.D. Übeyli, ECG beats classification using multiclass support vector machines with error correcting output codes. Digital Signal Process. **17**(3), 675–684 (2007)
4. R. Mark, G. Moody, *MIT-BIH Arrhythmia Database Directory* (Massachusetts Institute of Technology, Cambridge, 1988). Available online from: http://www.physionet.org/physiobank/database/mitdb/
5. R.J. Martis, U.R. Acharya, L.C. Min, ECG beat classification using PCA, LDA, ICA and discrete wavelet transform. Biomed. Signal Process. Control **8**(5), 437–448 (2013)

6. A.L. Goldberger, L.A. Amaral, L. Glass, J.M. Hausdorff, P.C. Ivanov, R.G. Mark, J.E. Mietus, G.B. Moody, C.-K. Peng, H.E. Stanley, Physiobank, physiotoolkit, and physionet components of a new research resource for complex physiologic signals. Circulation **101**(23), e215–e220 (2000)
7. M.U. Guide, The Mathworks, Inc., vol. 5, p. 333, Natick, MA (1998)
8. Ojha, D. K., & Subashini, M. (2014). Analysis of Electrocardiograph (ECG) Signal for the Detection of Abnormalities Using MATLAB. World Acad. Sci. Eng. Technol. Int. J. Med. Health Pharm. Biomed. **8**(2), 114–117.
9. W. Zong, G. Moody, Wqrs-single-channel QRS detector based on length transform. Physionet (2003). http://wwwphysionetorg/physiotools/wag/wqrs-1htm
10. P. Laguna, R. Jan, E. Bogatell, D. Anglada, QRS detection and waveform boundary recognition using ecgpuwave. PhysioToolkit Open Source Software for Biomedical Science and Engineering, in *Proceedings of the 5th EAI International Conference on Wireless Mobile Communication and Healthcare*. ICST (Institute for Computer Sciences, Social-Informatics and Telecommunications Engineering), pp. 50–53 (2002)
11. P. Laguna, R. Jané, P. Caminal, Automatic detection of wave boundaries in multilead ECG signals: validation with the CSE database. Comput. Biomed. Res. **27**(1), 45–60 (1994)
12. R. Jané, A. Blasi, J. García, P. Laguna, Evaluation of an automatic threshold based detector of waveform limits in Holter ECG with the qt database. Comput. Cardiol. **24**, 295–298 (1997)
13. I. Daubechies, The wavelet transform, time-frequency localization and signal analysis. IEEE Trans. Inform. Theory **36**(5), 961–1005 (1990)
14. I. Daubechies et al., *Ten Lectures on Wavelets*, vol. 61 (SIAM, Philadelphia, 1992)
15. C. Cortes, V. Vapnik, Support-vector networks. Mach. Learn. **20**(3), 273–297 (1995)
16. C.-C. Chang, C.-J. Lin, Libsvm: a library for support vector machines. ACM Trans. Intell. Syst. Technol. (TIST) **2**(3), 27 (2011)
17. Tsoutsouras, V., Azariadi, D., Xydis, S., & Soudris, D. (2015, December). Effective Learning and Filtering of Faulty Heart-Beats for Advanced ECG Arrhythmia Detection using MIT-BIH database. In *Proceedings of the 5th EAI International Conference on Wireless Mobile Communication and Healthcare*. ICST (Institute for Computer Sciences, Social-Informatics and Telecommunications Engineering), pp. 50–53
18. M.C. Ramon, *Intel Galileo and Intel Galileo Gen 2* (Springer, New York, 2014)
19. G.D. Clifford, F. Azuaje, P. McSharry, *Advanced Methods and Tools for ECG Data Analysis* (Artech House Publishing, London, 2006)
20. G.B. Moody, R.G. Mark, The impact of the mit-bih arrhythmia database. IEEE Eng. Med. Biol. Mag. **20**(3), 45–50 (2001).
21. P. de Chazal, M. O'Dwyer, R.B. Reilly, Automatic classification of heartbeats using ECG morphology and heartbeat interval features. IEEE Trans. Biomed. Eng. **51**, 1196–1206 (2004)
22. P. de Chazal, R. Reilly, A patient-adapting heartbeat classifier using ecg morphology and heartbeat interval features. IEEE Trans. Biomed. Eng. **53**(12), 2535–2543 (2006)

# Chapter 16
# Design for a System of Multimodal Interconnected ADL Recognition Services

**Theodoros Giannakopoulos, Stasinos Konstantopoulos, Georgios Siantikos, and Vangelis Karkaletsis**

## 16.1 Introduction

As smart, interconnected sensing devices are becoming ubiquitous, more applications are becoming possible by re-arranging and re-connecting sensing and sensor signal analysis in different pipelines. This appears to imply that the finer the grain of the distinct and referenceable services the better, since finer grain allows maximal flexibility; in reality, however, there are composite services that make more sense than their components as the finest grain that is exposed. The reasons vary, ranging from technical, to privacy, to business considerations, but the effect is that systems of services need to be properly designed to assume a satisfactory position between being too fine and too coarsely grained.

We shall focus here on detecting *activities of daily life (ADL)* in a smart home setting. Such a setting offers itself naturally to medical applications and especially independent living applications, where ADL logs can establish patterns and identify deviations. These can range from the time spent out of the house or carrying out a given activity in it, to sleeping patterns, to recognizing whether the user has changed clothes, washed, or other crucial indications required by the medical condition that necessitates monitoring.

To motivate our work, let us assume as an example a general-purpose acoustic feature extraction component that can provide input for different classification engines and models. For our application we have established that we will detect acoustic events using an SVN engine and that there is no further component that needs these features. In this situation, it makes sense to encapsulate the audio

T. Giannakopoulos • S. Konstantopoulos (✉) • G. Siantikos • V. Karkaletsis
Institute of Informatics and Telecommunications, NCSR 'Demokritos',
Aghia Paraskevi 15310, Greece
e-mail: tyianak@iit.demokritos.gr; konstant@iit.demokritos.gr; siantikosg@gmail.com; vangelis@iit.demokritos.gr



G. Keramidas et al. (eds.), *Components and Services for IoT Platforms*,
DOI 10.1007/978-3-319-42304-3_16

acquisition component, the feature extraction component, and the SVN engine under a single acoustic event detection service and not provide an interface for the acoustic features or, even more so, for the audio signal. We have based our decision on the observation that we re-analyse the lower-level feature extraction output into a higher-level acoustic event output that is more informative and more straightforward to be consumed by other services.

Let us now assume that we later decide to apply a fusion algorithm that combines audio and vision. The extracted audio-visual events represent the same real-world events as the acoustic and visual events that they fuse, except that they are associated with a higher confidence and/or carry more attributes than either. In such a situation our previous decision to bundle acoustic event extraction as a single service and not as an extraction–classification pipeline restricts us to event-level fusion and makes it impossible to apply feature-level fusion algorithms. On the other hand, if we were to use event-level fusion anyway, the finer-grained feature-level services would be unnecessary overheads.

Naturally, real-world scenarios are bound to be even more complex. To give another example, consider a system for our ADL monitoring application that comprises the following components:

- Waking up and getting out of bed is recognized in the depth modality
- Moving around the house is tracked in the depth modality
- Moving around the house is tracked and the person moving is identified in the image modality.

In this scenario, subsequent fused image/depth analysis both confirm and add attributes (the identity of the person) to the original getting-out-of-bed event. A system that measures the time it takes to transfer out of bed will need both the onset of the first event and the subsequent supporting information. However, this system of analysis components cannot be tried under a single service that log the *bed transfer* ADL, since movement tracking will also be useful for logging numerous other activities.

Further applications can also be envisaged, such as proactively offering automations that are relevant to the current ADL context. The design of this application as an extension of a medical monitoring application is straightforward, since it is based on analysing high-level ADL logs. Other applications, however, might require more concrete data and, thus, more flexibility regarding the components of the architecture that they will need to access. In general, the choice of how to best wrap the analysis components and pipelines of such components balances between micro-services that expose thin slices of functionality and heavier services that wrap larger sub-systems. Naturally, there can be no ideal solution and each application must use its own requirements to explore the design space.

ADL recognition is based on the recognition of events in (possibly multiple) audio-visual signals and on heuristics that characterize sequences or other compositions of events as more abstract ADL events.

In this article, we first present the audio (Sect. 16.2) and visual (Sect. 16.3) sensors and corresponding recognition methods that are typically used in our application domain. By studying the components that make up these methods and the kinds of information exchanges between them, we propose a conceptual architecture (Sect. 16.4). The purpose of our architecture is to integrate existing components developed in different contexts (robotics and the internet of things) and to make efficient use of processing and transmission resources. The key design points for this conceptual architecture are: (a) to establish appropriate articulation points for segmenting into components the pipelines commonly proposed in the multimodal processing literature; and (b) to specify the kind of information that is exchanged at the interfaces between these components.

## 16.2 Recognizing Events in Audio Content

Acoustic analysis pipelines include signal acquisition, acoustic feature extraction, and classification of the features into acoustic events. In particular, the following components are typically included in acoustic analysis pipelines:

- audio acquisition: uses the audio signal to produce a stream of short-term audio frames
- short-term feature extraction: uses an audio frame to produce a frame feature vector
- mid-term feature extraction: aggregates multiple frame feature vectors into a mid-term segment feature vector
- audio pattern analysis: uses frame and segment feature vectors to produce an event recognition.

In this section we will present these components and related design choices based on our ADL recognition application.

### *16.2.1 Feature Extraction*

*Audio acquisition* is based on the cross-platform, open-source PortAudio library [1]. Audio acquisition divides the audio signal into a stream of short-term windows (frames), and short-term features are extracted in an on-line mode from each frame.

*Short-term feature extraction* produces a stream of feature vectors of 34 elements (Table 16.1), where each vector is calculated from one frame. The time-domain features (features 1–3) are directly extracted from the raw signal samples. The frequency-domain features (features 4–34, apart from the MFCCs) are based on the magnitude of the discrete Fourier transform (DFT). Finally, the cepstral domain (e.g. used by the MFCCs) results after applying the inverse DFT on the logarithmic spectrum.

These features and the effect of short-term windowing in audio classification have been discussed in more detail by Kim et al. [6] and Giannakopoulos et al.

**Table 16.1** Audio features

| Index | Name | Description |
|---|---|---|
| 1 | Zero crossing rate | The rate of sign-changes of the signal during the duration of a particular frame |
| 2 | Energy | The sum of squares of the signal values, normalized by the respective frame length |
| 3 | Entropy of energy | The entropy of sub-frames' normalized energies. It can be interpreted as a measure of abrupt changes |
| 4 | Spectral centroid | The centre of gravity of the spectrum |
| 5 | Spectral spread | The second central moment of the spectrum |
| 6 | Spectral entropy | Entropy of the normalized spectral energies for a set of sub-frames |
| 7 | Spectral flux | The squared difference between the normalized magnitudes of the spectra of the two successive frames |
| 8 | Spectral rolloff | The frequency below which 90 % of the magnitude distribution of the spectrum is concentrated |
| 9–21 | MFCCs | Mel frequency Cepstral coefficients form a cepstral representation where the frequency bands are not linear but distributed according to the mel-scale |
| 22–33 | Chroma vector | A 12-element representation of the spectral energy where the bins represent the 12 equal-tempered pitch classes of western-type music (semitone spacing) |
| 34 | Chroma deviation | The standard deviation of the 12 chroma coefficients |

[2, 4]. What is important to note is that the frame size and the selected features depend on each other, so that a *different frame stream, using different frame size, would need to be produced for a different short-term feature extraction component.*

This observation suggests a very strong coupling between audio acquisition and short-term feature extraction. Furthermore, short-term feature extraction results in a drastically smaller stream than the original audio signal. This reduces the communications overhead of providing the short-term feature stream as a service, by comparison to providing the raw audio signal.

### 16.2.2 Feature Aggregation and Analysis

Another common technique in audio analysis is the processing of the feature sequence on a mid-term basis, according to which the audio signal is first divided into mid-term windows (segments). For each segment, *mid-term feature extraction* uses the short-term feature vector stream to produce a stream of feature statistics, such as the average value of the ZCR (Feature 1). Therefore, each mid-term segment is represented by a set of statistics.

The final stage is the analysis of patterns in the short- and mid-term features to infer event annotations. Two types of *audio pattern analysis* are performed:

- *Supervised:* A set of predefined classifiers is trained and used to extract respective labels regarding events [2, 11]. Apart from general audio events regarding activities (ADLs) mood extraction is achieved by applying regression and classification methods trained on speech-based emotion recognition data.
- *Unsupervised:* Apart from predefined taxonomies of audio events and activities, audio features are used in the context of a clustering procedure, according to which the extracted labels are not known a-priori. A typical example is *speaker diarization* or *speaker clustering*, the task of determining who spoke when [3].

For both types of audio pattern analysis, a decision is made for each short-term feature vector taking into account a combined short-term and mid-term feature vector. In a sense, the mid-term features provide a context within which the short-term features are interpreted. The signal energy, for example, is more informative when combined with the average energy in order to detect clangs and bangs even in overall noisy environments as sudden spikes above the average.

Typical values of the mid-term segment size can be from 200 ms to several seconds depending on the events that are being recognized. Segments can be overlapping, using a different sliding mid-term feature vector with each short-text feature vector, or non-overlapping, using the same mid-term feature vector throughout the duration of the segment.

Depending on the technical characteristics of the middleware used, some communication overheads can be reduced by keeping the same mid-term feature vector latched to the bus so that it can be read multiple times by the audio pattern analysis component. This potential gain is, however, relatively small as the information to be communicated has been reduced to 34 floating-point features, i.e. 136 bytes. Furthermore, different event recognizers often work better with different segment sizes, so that multiple (typically two or three) mid-term feature vectors need to be made available to audio pattern analysis. These observations suggest that there is little optimization value in non-overlapping segments, and that the (more accurate) rolling segments should be preferred.

This still leaves open the question of whether mid-term feature vectors should be bundled together with the acquisition/short-term feature extraction components as a complete audio feature extraction service, bundled together with the audio pattern analysis components, or provided as an independent service or services (for different segment sizes).

## 16.3 Recognizing Events in Visual Content

Visual information is recorded through a depth camera, therefore two types of information are adopted in this context: colour video and depth video. Both types of visual information are fed as input to a workflow of visual analytics (Fig. 16.1) whose purpose is to extract:

- activities of daily living (ADLs): a predefined set of events related to ADLs
- measures related to the user's ability to perform these ADLs.

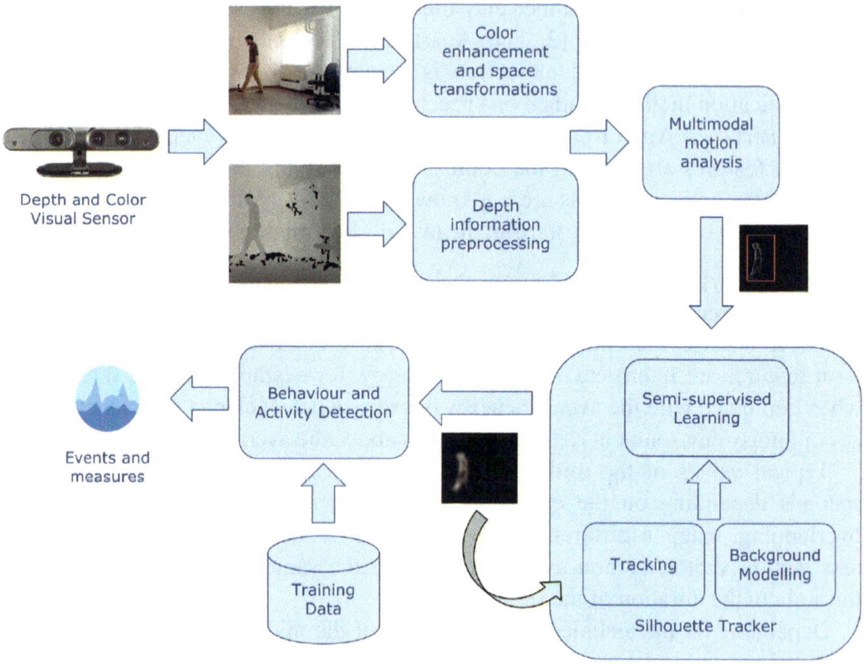


**Fig. 16.1** Depth and colour visual analytics workflow

## 16.3.1 Preprocessing

In our application area, stereoscopic or structured light visual sensors are used to provide both colour and 3D depth information channels. Both channels are preprocessed in order to provide better representations of the scene. Colour is transformed and normalized in order to achieve colour consistency and lighting conditions independence. In addition, de-noising and hole removal is applied on each depth frame in order to provide the next steps with smoother depth representations.

## 16.3.2 Motion Detection

This submodule aims to function not only as a triggering functionality for initializing the visual analytics workflow but also as a baseline estimator of the user's spatial information. The adaptation of motion detection as a separate submodule is twofold: (a) to enhance the performance of the next visual analytics steps (e.g. make the background estimation more robust) and (b) to minimize the computational complexity of the whole visual workflow by triggering the respective submodules only when necessary.

Motion detection is capable of checking the existence of motion within the visual information. For each processing time unit (i.e. video frame), a single thresholding rule is adopted for extra low computational complexity, in order to infer the existence of motion. If no motion is detected, then the background model is updated as described in the next paragraph. For better performance, separate channels are used in order to estimate the motion of each frame: depth-based distance metrics along with three different colour-space distance metrics. A simple weighted average fusion approach is used to generate the final two-dimensional distance metric that is used in the thresholding criterion that detects the motion area. The result of this procedure is a bounding box of the motion area.

Before combining the different modalities it is important to note that *visual registration* is applied, in order to align the different representations of the same scene into one common spatial representation. Especially between the depth and colour channels it is very important to apply such geometric transformations, since the two channels are heavily misaligned on most sensors.

### 16.3.3 Silhouette Tracking

The goal of this submodule is to detect and track the exact positions of the pixels associated with the person's silhouette. Towards this end the following algorithmic stages are adopted:

- *Background modelling* The goal here is to estimate a statistical model that describes the background of the visual information, so that it is subtracted in the main visual analysis steps. The background subtraction submodule is triggered from the motion detection service as described in the previous paragraph. As soon as the module is triggered, the following steps are executed:
  - the (detected as "motion") frames are captured to memory
  - contrast and light normalization is applied to remove unwanted dependencies
  - gamma correction is applied
  - the background model is extracted using the MOG operator [5]
  - the outline of foreground objects is finally estimated using a set of morphological operations [10]
- *Tracking* The goal of this step is to model the moving object's dynamics in order to track the exact position of the user.
- *Semisupervised learning* At each stage a clustering algorithm is applied based on the current estimated background, the feedback from the tracking algorithm as well as the raw depth info of the current frame, in order to decide on the final silhouette estimate.

### *16.3.4 Behaviour and Activity Detection*

Given the position and exact shape of the user's silhouette, a series of supervised and unsupervised machine learning approaches is applied in order to:

- extract a set of predefined activities (stand up, sitting, lying in bed, walking, running, eating, etc.). Towards this end, annotated data are used to train the respective classifiers.
- extract body keypoints using supervised models.
- detect faces and extract facial features. Also, if provided, the supervised database stores facial features of known users and the respective module also extract user ID (identification).
- extract clothing-related information (i.e. if the user has changed clothes since her last appearance on the sensor) [9].
- estimate metrics related to the user's ability to walk. Towards this end, unsupervised temporal modelling is adopted as the means to extract measures that quantify the gait: average speed, time required to walk four metres, etc.

### *16.3.5 Fused Audio-Visual Analysis*

Apart from the audio and visual workflows that extract respective high-level information and metadata regarding activities and measures, the *early* and *late fusion* approaches are used to extract information from the combined audio and visual modalities.

One example is combining facial features (cf. Sect. 16.3.4) with acoustic features (cf. Sect. 16.2.2) in the context of a speaker diarization method that extracts user labels based on speech and facial cues [8]. In this *early fusion* example, the acoustic and visual feature vectors are fused. By contrast, *late fusion* approaches as used for behaviour recognition. In particular, multimodal events can be recognized by combining events in each modality that represents the same physical event.

## 16.4 ADL Recognition Architecture

Based on our analysis of the ADL recognition methods above (Sects. 16.2 and 16.3), we will now proceed to bundle the relevant components into services and to specify these services' information outputs and requirements. These services will be used as the building blocks of our ADL recognition architecture in this section.

Our usage scenario is set in an assisted living environment with static sensors and a mobile robot which acts as a mobile sensing platform. In this setting, the clinical requirement is to monitor ADLs in order to report aggregated logs about the occurrence of specific ADLs, signs of physical activity, as well as performance measurements such as time needed to get off the bed or to walk a given distance [7].

In our design we foresee acoustic sensors that integrate a microphone with Raspberry Pi, a mobile TurtleBot2 robot[1] that integrates microphone, Xtion depth and colour camera and on-board computer, and a main computer that acts as the gateway to the home and the orchestrator of the overall monitoring and reporting. This physical infrastructure is used to deploy the sensing services and the ADL recognition services that use them.

There is a single acoustic features interface which publishes a stream of triplets of feature vectors. Each message in the stream contains the current short-term frame feature vector and two mid-term rolling averages of different numbers of frames, to accommodate analyses that require deeper or more shallow acoustic contexts.

This interface was chosen because at our 50 Hz frame rate volume of traffic generated by three floating-point feature vectors is insignificant and this interface lifts the requirement to have a middleware that can latch mid-term feature vectors or synchronize mid-term and short-term feature vectors. Exposing the complete acquisition-feature extraction pipeline as a single service also allows us to provide a unified acoustic feature service over two heterogeneous implementations [12]:

- The TurtleBot2 implementation comprises a microphone device driver and a feature extraction component that communicate using the ROS middleware.[2] The service end-point is a bridge that simultaneously connects to the robot-internal ROS middleware and to the home WiFi to access robot-external services.
- The Raspberry implementation comprises a microphone device driver and a feature extraction component that communicate using MQTT.[3] The service end-point is a bridge that simultaneously connects to MQTT and to the home WiFi to access external services.

All instances of the acoustic features service push their vector streams to the audio pattern analysis service. This service implements unsupervised and (previously trained) supervised machine learning methods that recognize ADL events from acoustic feature vectors. The audio pattern analysis service is also distributed, with instances executing at the Raspberry and the robot's computer.

The vision sensing components are analogously implemented as image acquisition, feature extraction, and pattern recognition services. One divergence from the acoustic analysis case is that the graph of dependencies between vision services is not a linear progression from the content to more abstract features and events: motion detection is the only service that constantly consumes features and it triggers more complex analyses as soon as motion is detected. Furthermore, there is no single feature set that is used by all visual analyses and analyses are occasionally stacked more deeply than the features/events/ADLs layers of acoustic ADL detection.

[1]Please cf. http://www.turtlebot.com for more details.

[2]The robot operating system (ROS) is a set of software libraries and tools for developing distributed applications in robotics; please cf. http://www.ros.org for more details.

[3]MQTT is an extremely lightweight publish/subscribe messaging middleware for the Internet of Things; please cf. http://mqtt.org for more details.

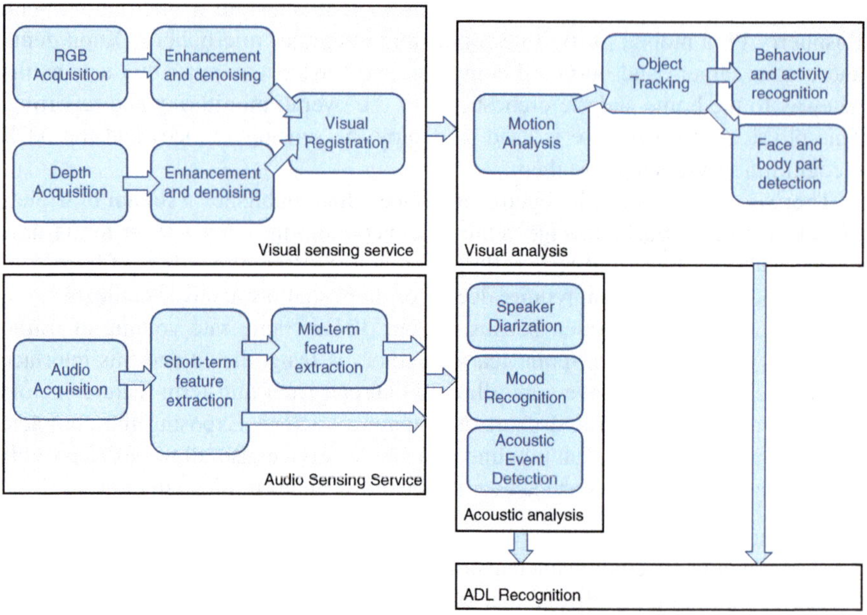


**Fig. 16.2** Conceptual architecture of audio-visual analysis

The services and interfaces design described here is also depicted in Fig. 16.2.

## 16.5 Conclusion

We presented a system of services that interact to recognize ADLs from audio-visual sensors. Our design integrates sub-systems which were originally integrated using heterogeneous middleware infrastructures. We have proposed articulation points for re-structuring these existing pipelines into a new set of services. In order to establish the right level of granularity for the functionality bundled under a single service, we used common patterns in the audio-visual analysis literature to identify services that would practically never need to be broken down into finer services.

The most prominent future research direction is the dynamic handling of early fusion methods. For such methods, the recognition component must have access to both the acoustic and the visual features. In our current design, this is not as issue as in the only situation where this is necessary (scenes captures by the robot) both sensors happen to be on the same ROS middleware and the audio-visual features can be easily consumed by the same component. As this will not be the case in general, our plan is to transfer concepts and technologies from the distributed processing literature. Although developed to address different problems (namely, processing large scale data), transferring such technologies to our application will allow us to develop distributed fusion components. In this manner, an early fusion component can perform calculations over distributed feature vectors without requiring that they are collected at a single computation node, applying the communication overhead optimizations developed by the distributed computation community to minimize the communication of intermediate results.

Another future research direction pertains to incorporating speech recognition in the system. Speech recognition uses radically different features than those computed for acoustic processing. Naturally, one can always concatenate or interleave the feature vectors produced for acoustic and speech processors in order to accommodate both. It is, however, worth investigating whether a study of the feature extraction methods proposed in the speech recognition literature can reveal opportunities for a more efficient integration of the acoustic and speech feature extraction components.

**Acknowledgements** This project has received funding from the European Union's Horizon 2020 research and innovation programme under grant agreement No 643892. For more details, please visit the RADIO Web site http://www.radio-project.eu

## References

1. R. Bencina, P. Burk, PortAudio-an open source cross platform audio API, in *Proceedings of the International Computer Music Conference*, Havana, pp. 263–266 (2001)
2. T. Giannakopoulos, pyAudioAnalysis: an open-source Python library for audio signal analysis. PloS One **10**(12) (2015)
3. T. Giannakopoulos, S. Petridis, Fisher linear semi-discriminant analysis for speaker diarization. IEEE Trans. Audio Speech Lang. Process. **20**(7), pp. 1913–1922 (2012)
4. T. Giannakopoulos, A. Pikrakis, *Introduction to Audio Analysis: A MATLAB Approach* (Academic Press, London, 2014)
5. P. KaewTraKulPong, R. Bowden, An improved adaptive background mixture model for real-time tracking with shadow detection, in *Video-Based Surveillance Systems* (Springer, New York, 2002), pp. 135–144
6. H.G. Kim, N. Moreau, T. Sikora, MPEG-7 *Audio and Beyond: Audio Content Indexing and Retrieval* (Wiley, New York, 2006)
7. RADIO Project: Deliverable 2.2: Early detection methods and relevant system requirements. Tech. rep. (2015), http://radio-project.eu/deliverables
8. N. Sarafianos, T. Giannakopoulos, S. Petridis, Audio-visual speaker diarization using Fisher LINEAR semi-discriminant analysis. J. Multimed. Tools Appl. **75**(1), 115–130 (2016)
9. D. Sgouropoulos, T. Giannakopoulos, G. Siantikos, E. Spyrou, S. Perantonis, Detection of clothes change fusing color, texture, edge and depth information, in *E-Business and Telecommunications* (Springer, Berlin, 2014), pp. 383–392
10. D. Sgouropoulos, E. Spyrou, G. Siantikos, T. Giannakopoulos, Counting and tracking people in a smart room: an IoT approach, in *Proceedings of the 10th International Workshop on Semantic and Social Media Adaptation and Personalization (SMAP 2015)* (IEEE, New York, 2015)
11. G. Siantikos, D. Sgouropoulos, T. Giannakopoulos, E. Spyrou, Fusing multiple audio sensors for acoustic event detection, in *Proceedings of the 9th International Symposium on Image and Signal Processing and Analysis (ISPA 2015)* (IEEE, New York, 2015), pp. 265–269
12. G. Siantikos, T. Giannakopoulos, S. Konstantopoulos, A low-cost approach for detecting activities of daily living using audio information: a use case on bathroom activity monitoring, in *Proceedings of the 2nd International Conference on Information and Communication Technologies for Ageing Well and e-Health (ICT4AWE 2016)*, Rome (2016)

# Chapter 17
# IoT Components for Secure Smart Building Environments

**Christos Koulamas, Spilios Giannoulis, and Apostolos Fournaris**

## 17.1 Introduction

As with almost any birth, evolution and market penetration of a new technology or combination of technologies wave, related acronyms and buzzwords are widely used in numerous research, development, and marketing contexts, often with slightly different or overlapping meanings. Typically, this inevitably introduces a form of blur, as this may be easily identified in relative existing literature, professional networks discussions and product descriptions, at least until maturity of concepts and of basic technology components is unambiguously captured by prevailing standards, products, and large scale installations, creating the basis of a common language agreement among the various stakeholders. Still, terms with wide meaning and technology coverage will continue to exist; think, for example, what could possibly anyone mean when mentioning "internet" or "web" technologies, in general. Focusing on the M2M/IoT/WoT terms, definition efforts can be found in numerous places, just noting a few most reputable sources or source collections in [1–6]. The generality of their main "seamless integration and interaction of physical and virtual objects" common ground makes evident that the concept is neither new nor it will probably cease shortly to include almost everything under this global connectivity vision. What is actually new is the available set of specific technologies as enabling components of this vision, as well as the real existing numbers today, and their short- and mid-term projection regarding the penetration and deployment of such objects and applications in everyday's business and people activities; a development which has its roots mainly on the radical cost and power

C. Koulamas (✉) • S. Giannoulis • A. Fournaris
Industrial Systems Institute, Research Centre Athena,
Patras Science Park Building,
Stadiou Str. Platani, 26504 Patras, Greece
e-mail: koulamas@isi.gr; sgiannoulis@isi.gr; fournaris@isi.gr



G. Keramidas et al. (eds.), *Components and Services for IoT Platforms*,
DOI 10.1007/978-3-319-42304-3_17

efficiency improvements in embedded sensing, processing and wireless connectivity components, and the consequent widespread adoption of supporting devices. It was these technologies from the embedded, cyber-physical and wireless (sensor) network, and mobile world, along with the worldwide increase of nodes, people, and business engagement with the conventional internet that triggered the evolution and practical applications of cloud and big data technologies that may today altogether consist the current snapshot of the Internet of Everything (IoE) definition and technology context.

Still, similar infrastructures, although neither global nor common, exist for decades in the industrial and building automation world. The inherent to these domains necessity to interconnect and manage hundreds and thousands of data sources (sensors) and sinks (controllers and actuators), configure automation loops, collect, store, and analyze their operational data, identify trends and events, and trigger higher level actions, carries a long history of technology evolution, debates, and market wars. Core quests that are also present in the IoT evolution today, such as semantic, information, data and network level interoperability, and device interoperation and self-configuration capabilities, along with other, domain specific requirements such as real-time operations, were addressed in one or another way through numerous standards or common industrial specifications, including specialized network protocols, device description languages, and configuration and management interfaces and tools. Moreover, many if not all of these silo technologies and applications included measures of internet connectivity, either by web-based access or by directly adopting IP, IP/TCP, and/or HTTP/Web Services as integral technology parts of their operation, including wireless extensions to their primary, typically wired infrastructure.

Such infrastructures may already lend past achievements and concepts to the IoT world—see, for example, the various sensor and actuator object descriptions of ZigBee profiles or IPSO, OIC, and AllSeen resource interfaces and their similarities with existing counterparts from various industrial and building automation standards—but mostly, it is expected to receive benefits from IoT relative developments, especially in relation to interoperability, complexity, scalability, and security improvements, as well as to components, installation, configuration, operation, and management costs reduction. In order to achieve this, and following the typical automation application patterns that implement interactions in a distributed way and not through a central repository or through the cloud, from the broad set of IoT technologies the focus is more on how to bring the web closer or into the edge device than how to bring or virtualize the device on the web or in a cloud. As a consequence, and taking into account the typical embedded resources in low end automation devices, the authors' view is that it is mostly the IETF and oneM2M suite of specifications, that may actually drive the concrete realization of the real IoT down or closer to the smallest possible "thing" as a web end point, ensuring security and interoperability at the lowest feasible architectural layer each time. A review of the most relative specifications, combined with widely used technologies and common practices in the smart buildings domain and the benefits from an integration shift, is presented in the next section. Section 3 identifies the elementary components of major importance to the realization of IoT architectures in the smart buildings application domain, while Sect. 4 provides comparative insights to the complexity and performance of specific implementations developed and exploited in a pilot demonstrator.

## 17.2 Relative Technology Review

The introduction and exploitation of new technological components in existing application domains and markets can be seldom and hardly achieved by ignoring current practices and commonly used technologies, especially in large installations such as commercial buildings, where the size of typical investments and the relative sizes of equipment and service providers in the domain define a consequent inertia of the market to changes. In the set of technologies of interest for smart building applications, we may safely first identify the most commonly used ones today, with very long history of existence and evolution, such as BACnet, KNX, and LonWorks, being technologies that cover the complete set of building functions, as well as other simpler specialized solutions for specific building parts, such as DALI for lighting systems and M-Bus for water and energy metering.

BACnet is a well established, widely used technology, especially in large building infrastructures, described in ANSI/ASHRAE [7] and ISO [8] standards. It defines the Network and Application layers over a number of transport media, with most popular the RS485-based MS/TP and the BACnet/IP, while ZigBee was added to the set in 2011, and more recently, IPv6. BACnet provides a sophisticated model for describing automation systems of all types and of complex hierarchies, ensuring internetworking and multivendor interoperability. Its application layer model is organized into Objects and Services for object access and parameter manipulation, initialization, discovery, time synchronization, configuration backup and restore, events, alarms, and file transfer. Regarding security, it is not mandatory but BACnet provides all needed mechanisms for peer entity, data origin, and operator authentication as well as data confidentiality and integrity.

KNX is the evolution and merge of the older EIB, EHS, and BatiBUS technologies, standardized today under the EN 50090 and ISO/IEC 14543-3 parts [9]. KNX supports various physical layers that can be used also in combination to form a heterogeneous media automation network. To provide flexibility and adaptation to different application domains, a wireless medium was also adopted, while to provide integration to existing IP networks, KNXnet/IP was introduced. Components that are used to create such an integrated network are KNXnet/IP routers used as segments couplers by tunneling KNX datagrams over IP, supporting both unacknowledged multicast and client/server communication patterns. The various abstractions of the domain are provided to the application as functional blocks, being objects with specified behavior and a set of inputs, outputs, and parameters, all called datapoints. Distributed applications consist of bindings between the inputs and output datapoints of functional blocks in different devices, typically mapped to group communications of the underlying stack. Mirroring the inherent complexities and the various scalability versus usability tradeoffs in the commissioning process of building automation systems, KNX offers 4 different configuration modes an installer might use in order to deploy a system, grouped in system (S) mode and easy (E) mode configuration profiles. The most versatile S-mode requires the usage of the ETS tool suite, in order to setup all advanced characteristics of each device profile, while the other configuration modes are based on simplified procedures (e.g., button push) or tools. Currently, KNX does not specify security mechanisms apart from a clear text password that may lock the configuration of a device.

The LonWorks technology is described by the ISO/IEC 14908, parts 1–4 and EN14908, parts 5 and 6 set of standards [10]. It is also a full-fledged control system, supporting various lower level networking stacks, including IP, with a network variable (NV) transfer and binding mechanism and an object oriented structure of the configuration and application layer service set definition on top of them. Application layer services include services for network variable propagation, generic message parsing, network management and diagnostics, as well as external protocol packet tunneling over the LonWorks network. Secure communications are supported, based on pre-configured shared keys, while configuration is achieved by specialized tools such as the Echelon's IzoT commissioning tool.

DALI is a lighting control system defined under the various parts of IEC 62386 standard [11]. DALI was designed to replace 1–10 V analog systems and proprietary digital systems such as digital serial interface, providing communication and control of multiple devices such as ballast, transformers, and other lighting equipment through a simple two wire physical interface. Its protocol architecture is simple master–slave commanding that supports besides basic switching, functions such as dimming, color control, and scene setting in unicast, broadcast, or group wise communications. Hierarchies can be built based on a 64-node addressing capability of subnets interconnected with hubs/routers, as well as through gateways interconnecting DALI systems to higher level control buses, such as KNX and BACnet. There is no security mechanism specified in the DALI standard, and the configuration of DALI systems and segments is achieved through specialized but simple tools that may query and set the parameters and the connections between master controller and slave device functions.

M-Bus is defined in the parts of EN13757 standard [12], and is being used for water, gas, electric and heat meters in buildings. It is also a simple master–slave protocol supporting low rate wired and sub-GHz wireless media in star topologies (WM-Bus), with primitive configuration space requirements and encryption support based on AES and pre-configured keys.

More recent but mature wireless counterparts used in the domain today, already considered as part of the wider IoT concept, are systems based on ZigBee, EnOcean, and ZWave.

ZigBee [13] builds on top of IEEE 802.15.4 wireless standard, by adding specifications for the network (NWK) and application (APL) layer structures. The ZigBee-PRO NWK supports multi-hop cluster tree and mesh topologies, with unicast, broadcast, groupcast, many to one, and source routing features in order to efficiently support large networks. The application layer model is defined on top of NWK, connected through the application support sub-layer (APS) and consists of the main ZigBee device object (ZDO) responsible for overall node management, and a series of individually addressable application end points/functional units that are grouped in various application profiles and are collections of functionality clusters specified from the ZigBee cluster library (ZCL). Application objects communicate directly through binding and exchange of commands and attribute values. ZigBee supports a distributed and a centralized security model for authentication and encryption, at network or application link level. ZigBee provides definitions for home and building automation profiles, and has been already accepted by ASHRAE as an official wireless BACnet transport.

EnOcean technology [83] penetrated the market by innovating mainly in extremely low resource devices with low duty cycle or infrequent and sporadic communication patterns, with its battery-less wireless design proposals that are powered by ambient energy through energy harvesting and smart energy management techniques. In its wireless medium it uses the ISO/IEC 14543-3-10 standard—part [15] of the KNX setup to the network layer definitions, including security provisions. At the application layer it introduces its own data model, structured in generic and equipment specific profiles, accessed by specialized configuration tools for remote commissioning and management. EnOcean wireless devices can be integrated to BACNet, KNX, and LonWorks backbone system infrastructures.

ZWave [14] is a wireless device ecosystem focused mainly on the home automation market. It uses the sub-GHz ITU G.9959 [84] for its radio layers, a mesh supporting networking structure, and a clearly defined simple command-based application layer for multivendor interoperability. It supports encryption based on AES128, but not as a mandatory feature.

Platform and technology independent information integration, interoperability and unified web access, besides simple IP level connectivity, has been a requirement in the domain for many years, too, addressed mainly through the BACnet/WS, oBIX, OPC-UA, and IEEE 1451 frameworks, as well as from the more recent initiative of the Haystack project.

BACnet/WS has been added in the ASHRAE standard initially to provide a service oriented architecture (SOA) based access and manipulation of data existing in servers through web services using XML formatted data, SOAP, and HTTP. Its data model is not bound to BACnet only and can theoretically be used to access an interface of any other control system. It was a simple model, based on tree structure of nodes and references, containing properties or other nodes which could be read or written, but not created or deleted through the WS API. Recently, a new series of addendums to the standard have been approved, introducing the BACnet Extended Data model (XD), deprecating SOAP, and providing RESTful web services and other capabilities including JSON encodings, resource creation and deletion, multiple resource access, CSML device descriptions, semantic information model and discovery and more, as well as adopting the security and authorization mechanisms to the state of the art, including TLS, OAuth 2.0, and PKI certificate management.

OPC was initially created to exploit Microsoft Windows OLE framework in automation systems, covering real-time and historical data access, as well as alarm and event related services, abstracting heterogeneous field level technologies. In its today unified architecture (OPC-UA) [16], published also under the parts of IEC62541 standard, it is a complete and complex automation level framework and information meta model based on nodes (objects, variables, methods, etc.) and their relationships (references). It is capable to support various concrete models from the industrial and building automation domain, such as BACNet, ISA95, and PLCopen, coupled with an extensive set of available services for discovery, management, monitoring, commanding, and querying under cross-platform SOA disciplines.

oBIX [17], standing for Open Building Information Exchange, is a specification managed today by the OASIS consortium. It is an extensible information model capturing the interactions between building control systems in a common abstraction layer integrating heterogeneous lower level automation technologies with enterprise systems. Its power is its simple modeling rules and its compositional capabilities. It provides bindings to SOAP web services, as well as to RESTful HTTP and CoAP, supporting JSON, XML/EXI, and custom binary encodings. It does not define though specific device or common function contracts. In a similar, simple but higher level approach, the Haystack project [18], an industrial open source initiative, addresses standardization of semantic data models and web services related to smart devices in building automation and control systems, through the provision of explicit but independent name/value pairs for all basic elements used to compose data points, equipment, and sites.

The IEEE 1451 is a family of standards which provide a set of common and network-independent interfaces for connecting transducers (sensors/actuators) to instrumentation and networking systems. Although it defines explicitly detailed protocols on how to connect transducer interface modules (TIM) to a Network Capable Application Processor (NCAP) entity, including wireless protocols such as WiFi, ZigBee, Bluetooth, and 6LoWPAN, the most relevant part of the standard from the back-end and web integration point of view is the NCAP information model described in IEEE21451-1 [19]. This includes an object model specifying the software component types which can be used by an application developer in order to compose specific application logic, a data model specifying the type and the form of the communicated information which corresponds to the interfaces provided by the objects, as well as a communication model, supporting client–server and publish–subscribe communication patterns among NCAPs.

From the side of the internet-enabled low resource devices, core specifications as main IoT building blocks come from the IETF working groups, as well as from industrial alliances and standardization organizations that integrate and complement them in certain ways. In all cases, the common foundations have been set by the definition of the IPv6 adaptation layer, widely known as 6lowpan [20–22] from the acronym of the relevant IETF Working Group (WG), which provided for an efficient applicability of the basic IPv6 connectivity mechanisms to constraint nodes and networks, over IEEE 802.15.4 links, and now over the TSCH mode of 802.15.4e [23]—in the 6tisch WG context—and over more link level technologies—in the 6lo WG context. Deeper multi-hop wireless IPv6 topologies and end to end connectivity was then made possible by the specification of the RPL routing protocol [24], addressing explicitly identified routing requirements from specific application domains, including home [25] and building automation [26], while multicast capabilities are specified by MPL [27], as an outcome of the same WG. With the accomplishment of the very first step towards IoT, i.e., this of direct internet connectivity, the next move, already conceived being towards the Web of Things (WoT), was to incorporate into the higher layers of constraint nodes, the known HTTP and REST architectural elements, in a seamless way, leading thus to the creation of the Constrained Application Protocol (CoAP) [28] from the IETF CoRE WG. CoAP provides efficient definitions for all CRUD mechanisms for handling resource hierarchies, for distributed resource discovery and multicast requests, for URI schemes, option mappings, and cross-proxying with HTTP, as well as for a lightweight reliable transport layer based on UDP, in avoidance of the complexity and the pitfalls of TCP in this kind of systems. In support of secure web transfers, the "coaps" URI scheme is registered as the equivalent of "https," normatively requiring the usage of DTLS [29], and more precisely, designating as mandatory the support of the RawPublicKey mode. Important specifications accompanying the core CoAP are this for observation streams management [30]—providing for the final "N" in the CRUDN method set—these for large data block transfers [31], group communications [32], and web linking [33], as well as for resource directories (CoAP-RD) [34] that may provide a fine grained network resource registry, either independent or in combination with standard multicast DNS (mDNS) [35] or DNS Service Discovery (DNS-SD) [36] deployments, assisting also to the "sleepy nodes" problem solutions [37]. Following the established trend to efficiently port existing protocols, disciplines, and tools from the wider internet to its constrained subset, noticeable efforts that are gaining fast acceptance from the research community and the industry are the creation of binary encoding formats such as CBOR [38] which is seamlessly coupled with JSON [39] as it is based on, and extending the latter's data model, as well as current proposals [40, 41] of using CoAP, CBOR, and hashes for the manipulation of Management Information Base (MIB) modules specified in YANG [42], transferring the access model of RESTCONF [43]. Finally, in the front of establishing secure architectures with efficient mechanism and protocol choices for the system provisioning and operation [44–46], besides the exploitation of transport level security with DTLS, application level security mechanisms, encodings and frameworks are currently proposed, such as OAuth 2.0 [47], DCAF [48], and OSCOAP [49], using the encodings specified in [50, 51], being roughly the CBOR-based equivalent of the JavaScript Object Signing and Encryption (JOSE) encodings.

Leveraging on the above IETF infrastructure, various industrial alliances incorporate selected components into their existing architectures or define new frameworks. ZigBee alliance introduced ZigBee/IP [52] in support of the IEEE 2030.5-2013/Smart Energy profile v2.0 [53], utilizing TLS, mDNS/DNS-SD, PANA, and MLE frameworks on a TCP/UDP over RPL/6LoWPAN stack. Open Mobile Alliance Lightweight M2M (OMA LWM2M) [54] is the lightweight variance of

OMA's device management and application data framework, focused on constrained cellular and other devices. It defines an extensible object model for device objects and resources, accessible through CoAP/DTLS over UDP and SMS bindings, using simple URIs of the form object-id/instance/resource-id. The specification includes normative objects for the server, access control, device, connectivity monitoring, firmware and location parameter management, as well as a set of interfaces for bootstrapping, registration, management, and information reporting. IPSO Alliance uses the OMA LWM2M object model to specify additional registered smart objects for functionalities of generic analog and digital I/Os and specific sensors, actuators, and control switches [55, 56]. OCF/OIC specifications [57] adopt also a RESTful architecture supporting a CoAP over UDP/DTLS and TCP/TLS transport, with CBOR as main, and JSON or XML/EXI as alternative encodings, focusing on the provision of a common resource model, for core and device/domain specific resources, supporting session (DTLS) and resource level (ACLs) protection in its security model. Currently, a set of device type specifications has been defined for home automation, based on the available resource type definitions, but OIC accepts at its core the existence and possible negotiation of different or even multiple resource model profiles, providing mapping examples with the IPSO object model. The Thread group [58] introduces its own mesh network layer structure over 6LowPAN, along with its security and commissioning model. Its management framework is based on CoAP and the focus is on smart homes but it does not define application level models, expecting and accepting any existing in the domain, such as IPSO, ZigBee, EchoNet, AllJoin, etc.

Finally, looking at the wider standardization organizations activities in the context of oneM2M and the current status of specifications [59], it is important to mark the acceptance and mapping definitions of specific framework components, particularly for these applicable to the low resource, IP enabled device type targets relevant to the smart home and building domains. Among these, mappings for OMA LWM2M provisions for device management should be noted [60, 85], paired with explicit bindings of the CoAP usage [61] and their common minimum support requirements for DTLS and the underlying cipher suites [62]. Furthermore, in the forthcoming version 2.0, oneM2M specifications proceed in higher level semantic definitions of application layer entity models [63], providing also model definitions for home appliances [64] and their mappings and interworking with OIC [65] and AllJoyn [66] specifications.

## 17.3 Architectural Considerations and Principal Components

During the past, the main point under consideration in industrial and building automation applications and systems (BAS) design, installation, and maintenance was device interoperability and interoperation, served by the development of open standards and driven by the need to avoid dependency on a single equipment provider. This standardization activity though did not actually lead to a true satisfaction of the end user's expectation, as there are just plenty of relative standards active in the market today, and current typical practices inherently assume having total ownership and control over the whole infrastructure, configured and managed through a specific technology's disciplines and tools. Most technologies though, do define high level objects which deal with the management of the whole system's parameter space and "virtualize" the application level "wiring" of input and output parameter points, supported by the relevant tools in order to assist the various stakeholders in the design, installation, commissioning, and operation of such systems [67, 68, 86]. The new paradigm involves development of a high level smart automation function based on logical input values and logical/virtual communication channels mapped to an underlying unified REST infrastructure. This virtual wiring and the associated end point resource binding rules can be eventually combined in the formation of a new single configuration space of application and networking system variables and parameters, to handle the system-wide control, clearly separating decisions that can be made statically during design time from these that can be made dynamically either during installation/commissioning or during normal operation under a well-defined bootstrapping/configuration/adaptation process based on the common CoAP mechanisms, and automated through a common resource directory structure (CoAP-RD).

Moreover, in typical existing installations in large commercial buildings, there is usually a large number of heterogeneous network and device technologies, which, although interoperable, lead to complex hierarchies and topologies with extensive usage of gateways, even when modern commercial mixed wired/wireless structures are used. The benefits of a flat networking architecture providing end-to-end connectivity between all devices, leveraging a horizontal IP-based structure, have been already widely acknowledged and we already see large companies, industry alliances, and standardization bodies switching towards an all-IP based solution, seeking the removal of hardware redundancy, the exploitation of a flat, possibly dynamically configured global addressing scheme and the seamless integration with traditional backbone infrastructures, utilizing as much as possible from the well-known existing internet protocols and engineering disciplines and tools.

The various possible topological and architectural choices for the interconnection to the web or in-between different network segments, heterogeneous or not, follow one of a small set of patterns, depending on the layer that heterogeneity may appear and the application's main communication pattern. Considering low resource IP enabled devices, the most widely used pattern from the set described in [5], is so far the device to cloud interaction, a rather not first order requirement for automation installations, although used for small residential cases. Instead, the domain requires rather a mix of patterns, with device to device direct bindings for the operational control, open back-end sharing for high level management and analytics, and local interconnecting devices to cope with highly heterogeneous installations, ranging in device architecture from simple OSI L2 bridges to application layer protocol translators and complex gateways supporting also server back-end sharing, as well

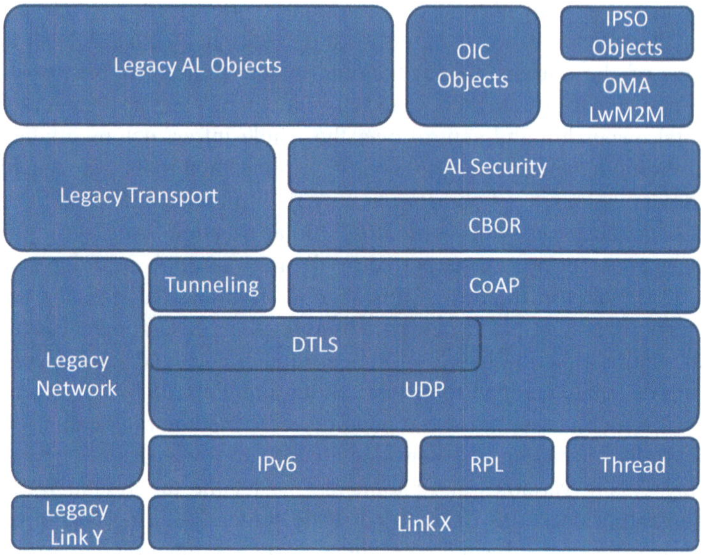


**Fig. 17.1** Peripheral interface components

as local configuration, security and commissioning functions [69–72]. This high diversity and complexity of the various interconnecting device architectures can be largely reduced taking into account the acceptance and rapid introduction of common components into different standards and industrial specifications. Today, it is already evident that the CoAP/DTLS/UDP/IPv6 over various links is the most common transport binding of different application level REST models for resource constrained devices. Along with the relative common acceptance of JSON and its direct binary CBOR encodings, it provides a much better ground for efficient device to device communication patterns, unified management, and simpler higher level model mapping implementations that can be pushed from the back-end or the gateways to the periphery.

The generic structure of a peripheral interconnection unit that is being used in the context of a heterogeneous building automation test-bed and demonstrator is presented in Fig. 17.1. The test-bed supports wired BACNet and KNX segments with off-the-shelf various I/O devices and BACNet/IP and KNX/IP routers, as well as a wireless 6LoWPAN network of programmable devices from different manufacturers. The objective is to push to the peripheral interfaces all components that are common to the majority of current IoT specifications, instead of overwhelming a multi-protocol central device—be it a gateway or application server/controller—allowing thus direct distributed flows and a unified remote configuration and management over a secure common transport. The relative generic structure of centralized gateway or server devices appears in Fig. 17.2.

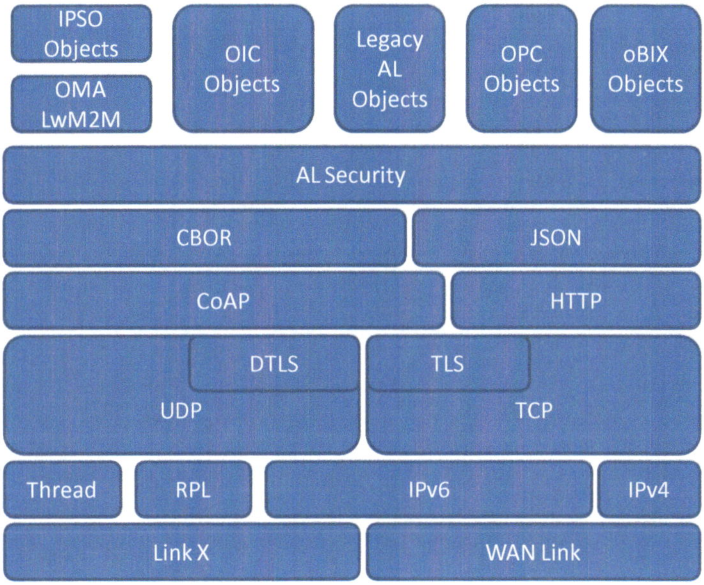


**Fig. 17.2** Gateway and server components

As long as the lower protocol components are agreed, the remaining major points of required attention are the realization of mappings between the various possible application level object models, as well as performance issues of specific component implementations.

## 17.4 Implementation Insights

The choice of which application layer models to implement in a peripheral concentrator or central gateway is not straightforward. Relevant approaches that base interoperability on mappings to a common schema are again design choices of that schema at the end, and can never guarantee that can capture all aspects of different underlying models. The experience based on decades of history of market wars on these matters indicates that it will be too optimistic to expect one single global agreement, even if international standards exist. Application layer models of industrial and building automation systems can be sometimes complex designs of objects and property collections that may capture many different details or configuration options of an object's behavior. Various specifications in their effort to establish generality or simple interoperability with other competing or complementary ones include mappings between models. BACNet already provides such mappings with KNX in its specifications. OIC suggests also mappings with IPSO objects. oneM2M, specifies integration mappings with OIC, AllJoyn, and

| BACnet | | IPSO | |
|---|---|---|---|
| Property | Type | Property | Type |
| Object_Identifier | BACnetObjectIdentifier | <instance id> | URI path |
| Object_Name | CharacterString | - | - |
| Object_Type | BACnetObjectType::ANALOG_INPUT | GenericSensor | URI Path |
| Present_Value | REAL | SensorValue | Float |
| Description | CharacterString | Application Type | String |
| DeviceType | CharacterString | Sensor Type | String |
| Status_Flags | BACnetStatusFlags | - | - |
| Event_State | BACnetEventState | - | - |
| Reliability | BACnetReliability | - | - |
| Update_Interval | Unsigned | - | - |
| Out_Of_Service | BOOLEAN | - | - |
| Units | BACnetEngineeringUnits | Units | String |
| Min_Pres_Value | REAL | Min Range Val | Float |
| Max_Pres_Value | REAL | Max Range Val | Float |
| Resolution | REAL | - | - |
| COV_Increment | REAL | - | - |
| Time_Delay | Unsigned | - | - |
| High_Limit | REAL | - | - |
| Low_Limit | REAL | - | - |
| Deadband | REAL | - | - |
| Limit_Enable | BACNetLimitEnable | - | - |
| Event_Enable | BACNetEventTransitionsBits | - | - |
| Acked_Transitions | BACNetEventTransitionsBits | - | - |
| Notify_Type | BACNetNotifyType | - | - |
| Event_Time_Stamps | BACNetTimeStamp[3] | - | - |
| Profile_Name | CharacterString | - | - |
| - | - | Min Meas Val | Float |
| - | - | Max Meas Val | Float |
| - | - | Reset MinMax Meas | Event |

**Fig. 17.3** Mappings of different sensor models

OMA LwM2M—thus implicitly with IPSO objects, too. However, in Fig. 17.3, an example of a mapping between a BACNet Analog Input and an IPSO generic sensor objects is presented, where it is evident that each model may include additional aspects of behavior of the modeled entity, which can be never mapped.

Nevertheless, having a common access, observation and encoding agreement such as CoAP with JSON/CBOR beneath different application layer structures, the cost of implementing more than one model, simply as different, complementary, or merged views of the device's or segment's functionality is, at least, not prohibitive in a first place.

On a different axis, execution loads and energy consumption performance [73, 74] of resource constrained embedded devices are important aspects in determining the minimum acceptable cost. During the development and integration of components in the various architectural or operational options, the bottleneck in performance was always introduced by the enablement and execution of security supporting operations. In a 16 MHz MSP430x-based platform, the activation of

DTLS imposed an increase of about 250 % in the roundtrip delay of a single CoAP transaction and a relative decrease in the maximum throughput in a point to point link. More importantly, the time it took the two nodes to establish a session key was no less than 9.68 s using the PSK (Pre-Shared Key) mode of DTLS with the TLS_PSK_WITH_AES_128_CCM_8 cipher suite, while in RawPublicKey mode with TLS_ECDHE_ECDSA_WITH_AES_128_CCM_8, it topped at more than 38 s, delays that are considerably large resulting in nodes being unresponsive until the key establishment phase is concluded, with amplifying effects in case that the DTLS session needs to be re-handshaked.

The software delay problem is more evident in ECs defined over binary extension fields GF($2^k$) since operations in such fields do not fit optimally in most of existing processor instruction sets. Implementing in software Elliptic curve cryptographic operations, apart from the work presented here, has shown that in low end embedded system μC (ATMega128, MSP430) scalar multiplication operation is performed in around 1 s (e.g., in tinyECC, nanoECC libraries) [75, 76] while in high end embedded system processors like the ARM Cortex M series [77, 78] scalar multiplication is performed with a few tenths of msec. In an effort to find a solution in the above problem, the hardware/software co-design of EC cryptographic primitives is also explored. So, apart from software implementation of an ECC protocol, a hardware IP core accelerator for the computationally intensive EC operation of scalar multiplication is introduced in the embedded system design flow. The goal of this approach is to achieve an increase in ECC operations speed, reduction in power consumption, and stronger security compared to purely software solutions (better protection against side channel attacks [79]).

There exist a wide range of scalar multiplication algorithm variations, however, one of the most promising one, that provides protection against simple power analysis attacks (a category of side channel attack that exploit hardware power consumption leakage), is the Montgomery Power Ladder (MPL) Algorithm. The algorithm performs scalar multiplication $Q = sP$, where s is an element of a GF($2^k$) field and P is the base point of an Elliptic Curve, in a balanced way and supports parallelism of point operations. To achieve that, it uses point addition $R_0 + R_1$ and point doubling $2R_0$ or $2R_1$. Taking into account that each point of the EC is represented by its coordinates [$P$:($x$,$y$) in the affine coordinate plane or $P$:($x$,$y$,$z$) in the projective coordinate space] belonging to GF($2^k$), each point operation is analyzed in a series of GF($2^k$) operations (GF($2^k$) multiplication, addition, and squaring) following the group law [80]. By using the Lopez–Dahab projective coordinate space, we analyzed the interdependences between the needed GF($2^k$) operations for point addition or doubling and estimated which of those operations can be performed in parallel and in what execution order. Our goal is to group the GF($2^k$) operations needed for a single MPL round in parallelism stages regardless of the point operation that they implement. Each parallelism stage is concluded in one clock cycle and all GF($2^k$) operations of the stage are performed in parallel. All parallelism stage outcomes must be calculated by reusing (for each parallelism stage) a parallelism stage module implemented in hardware.

An important factor that determined the architecture of the parallelism stage module can be the design concept behind GF($2^k$) multiplication. GF($2^k$) multipliers can be realized in various ways (i.e., bit serial, digit serial, and bit parallel designs) [81]. Among the existing design GF($2^k$) multiplication approaches, the most efficient in terms of speed implementations are provided for GF($2^k$) bit parallel designs. However, bit parallel GF($2^k$) multipliers require considerable number of gates (so only one or two of them can be used in a design). On the other hand, the smallest, in terms of gate number, designs can be realized when using the bit serial design approach. This approach also offers flexibility (the same design can be reused for various GF($2^k$) with arbitrary $k$ values) but, however, leads to slow implementations in term of computation speed. From the above remarks it can be concluded that if the bit parallel GF($2^k$) multiplication design approach is adopted in the parallelism stage module architecture then only one or two GF($2^k$) multiplications can be performed in parallel per stage (due to the high gate number of the implemented GF($2^k$) multiplier) while if the bit serial GF($2^k$) multiplication design approach is adopted in the parallelism stage module architecture, then no non-functional restriction is placed in the GF($2^k$) multiplications to be performed per stage (the multiplication number is only restricted by data dependencies and not by exceeding gate number). A study on the restrictions posed by the GF($2^k$) multiplication design approaches on the GF($2^k$) operation distribution per parallelism stage for a single MPL round is presented in [81], while in [82] the above described technique based on parallelism stages categorization is applied in Weierstrass ECs using the bit serial GF($2^k$) multiplication approach.

To prove that a hardware–software co-design approach can considerably improve the behavior of the DTLS-ECC protocol, an FPGA-based system-on-chip realization of the above described methodology was conceived. A 32 bit Microblaze softcore processor along with necessary peripherals is used in order to emulate an actual embedded system design where software code can be executed. On the main Processor Local Bus (PLB), a scalar multiplication accelerator module is added that communicates with the bus and the processor through an appropriate Input/Output bus Scalar multiplication interface. The test architecture is described in Fig. 17.4.

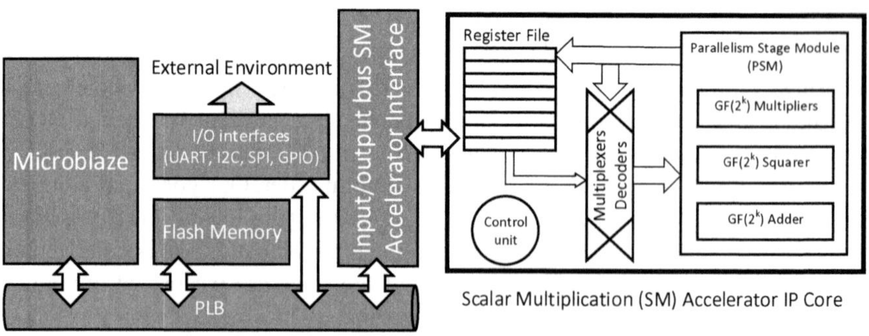


**Fig. 17.4** Case study hardware–software co-designed test architecture

The scalar multiplication (SM) accelerator architecture relies on the parallelism stage module (PSM) to perform all parallelism stage GF($2^k$) operations. The PSM takes as inputs values from the register file that are chosen using a multiplexer–decoder module. The PSM then perform a parallelism stage operation and stores the results back to the Register file. This procedure is done repeatedly for all parallelism stages of a single round by reusing the same PSM and consecutively for all rounds of the MPL algorithm. The whole process is managed through a control unit that also takes as input the bits of the scalar $s$ (from the Input/Output bus SM accelerator interface).

The whole design is supported by an appropriate software program that is executed in the Microblaze processor and is responsible for storage/retrieval of EC domain parameters, public–private keys and messages to be authenticated—digitally signed as well as transforming those data in an SM accelerator friendly form, setting up the bus control for input/output and eventually sending or retrieving data to or from the SM accelerator IP core. This program can also be used for constructing more complex EC related structures like ECC protocols.

To provide actual results from the case study architecture, Fig. 17.4 design was implemented in FPGA Xilinx Virtex 5 technology in [82] for a GF($2^{233}$)-based Elliptic Curve. The EC Diffie Helman protocol (ECDH) was implemented in software on the Microblaze in order to test the Hardware SM accelerator overall performance and estimate the hardware–software co-design concept. The Microblaze processor works in 125 MHz maximum frequency and is striped from any performance enhancing technology like multiplier units or cache memory. The results presented in [82] show that an ECDH key agreement operation can be concluded in 8.8 ms using the Hardware SM accelerator realizing the adopted design approach. The SM accelerator itself can perform one SM in approximately 800 μs. Taking into account that the MSP430x processor executing a DTLS purely software solution was operating at 16 MHz, a normalization can be made on the results from [82] showing that SM can be concluded in 6.25 ms and an ECDH in 68.8 ms, showing an improvement of roughly one order of magnitude.

## 17.5 Summary and Conclusions

Building automation systems and smart applications built on top of them has been already a commodity, with a long history of evolution. General principles of today's IoT and WoT nomenclature were already present through a vast set of available competing or complementary technologies with which object level interactions and web presence of device data were possible. The significant difference and step forward comes from the advancement of wireless technologies and of specifications and standards that enable resource efficient implementations of IP and web technologies down to the resource constrained peripheral devices. This latter step provides already the basis for feasible agreement on common stacks, acceptable to this special range of devices, at least up to the application layer protocols and object models. Although the efforts now concentrate on interoperability aspects at the highest layer of the semantics of these models, efficient coexistence of more than one of an already small prevailing set of them in low resource device and interconnecting unit implementations may be a feasible intermediate step towards a unified management of such complex systems. In these implementations' efficiency, most of the loads may attributed to the support of secure bootstrapping procedures, security protocols, and cryptographic operations. In view of this, it has been shown that hardware–software co-designed components can be a viable alternative for significant improvements in responsiveness for a low end microprocessor-based system in support of secure smart building environments.

## References


1. K. Ashton, That 'Internet of Things' thing. RFID J. 2009; 1999
2. O. Vermesan, P. Friess, P. Guillemin, S. Gusmeroli et al., *Internet of Things Strategic Research Agenda* (River Publishers, Aalborg, 2011)
3. International Telecommunication Union, ITU-T Y.2060, Next generation networks—frameworks and functional architecture models—overview of the Internet of things, June 2012
4. A. Bassi, M. Bauer, M. Fiedler, T. Kramp, R. van Kranenburg, S. Lange, S. Meissner, *Enabling Things to Talk: Designing IoT Solutions with the IoT Architectural Reference Model* (Springer, Heidelberg, 2013). ISBN 978-3-642-40402-3
5. H. Tschofenig, J. Arkko, D. Thaler, D. McPherson, Architectural considerations in smart object networking (IETF RFC 7452, Mar 2015)
6. IoT Definitions, How the IoT has been used and defined over the years, http://postscapes.com/internet-of-things-definition
7. ASHRAE Standard 135, BACnet—A data communication protocol for building automation and control networks (ANSI Approved) (2012)
8. ISO 16484–5, Building Automation and Control Systems (BACS)—Part 5: Data Communication Protocols (2014)
9. KNX Association, *KNX System Specifications—Architecture* (2013), http://www.knx.org/media/docs/downloads/KNX-Standard/Architecture.pdf
10. ISO/IEC 14908 LonWorks Family of Standards, *LonMark International*, http://www.lonmark.org/technical_resources/standards
11. IEC 62386, *Digital Addressable Lighting Interface*, http://www.dali-ag.org/discover-dali/dali-2-the-new-version.html
12. EN 13757-x, Communication System for Meters and Remote Reading of Meters—Parts 1–6
13. ZigBee 3.0 Specification, http://www.zigbee.org/
14. Z-Wave Alliance, http://z-wavealliance.org/
15. ISO/IEC 14543-3-10 Information technology—Home Electronic Systems (HES)—Part 3–10: Wireless Short-Packet (WSP) protocol optimized for energy harvesting—Architecture and lower layer protocols
16. OPC Unified Architecture Specifications, OPC-UA, https://opcfoundation.org/developer-tools/specifications-unified-architecture
17. oBIX v1.1, OASIS Committee Specification, http://docs.oasis-open.org/obix/obix/v1.1/cs01/obix-v1.1-cs01.pdf
18. Project Haystack, http://project-haystack.org/
19. ISO/IEC/IEEE 21451-1—Information technology—Smart transducer interface for sensors and actuators—Part 1: Network Capable Application Processor (NCAP) information model

20. G. Montenegro, N. Kushalnagar, J. Hui, D. Culler, Transmission of IPv6 packets over IEEE 802.15.4 networks (IETF RFC 4944, Sept 2007)
21. J. Hui, P. Thubert, Compression format for IPv6 datagrams over IEEE 802.15.4-based networks (IETF RFC 6282, Sept 2011)
22. Z. Shelby, S. Chakrabarti, E. Nordmark, C. Bormann, Neighbor discovery optimization for IPv6 over low-power wireless personal area networks (6LoWPANs) (IETF RFC 6775, Sept 2012)
23. P. Thubert, An architecture for IPv6 over the TSCH Mode of IEEE 802.15.4e (6TiSCH WG Internet Draft draft-ietf-6tisch-architecture-09, Nov 2015)
24. T. Winter, P. Thubert, A. Brandt, J. Hui, R. Kelsey, P. Levis, K. Pister, R. Struik, J.P. Vasseur, R. Alexander, RPL: IPv6 routing protocol for low-power and lossy networks (IETF RFC 6550, Mar 2012)
25. A. Brandt, J. Buron, G. Porcu, Home automation routing requirements in low-power and lossy networks (IETF RFC 5826, Apr 2010)
26. J. Martocci, P. De Mil, N. Riou, W. Vermeylen, Building automation routing requirements in low-power and lossy networks (IETF RFC 5867, June 2010)
27. J. Hui, R. Kelsey, Multicast protocol for low-power and lossy networks (MPL) (IETF RFC 7731, Feb 2016)
28. Z. Shelby, K. Hartke, C. Bormann, The constrained application protocol (CoAP) (IETF RFC 7252, June 2014)
29. E. Rescorla, N. Modadugu, Datagram transport layer security version 1.2 (IETF RFC 6347, Jan 2012)
30. K. Hartke, Observing resources in the constrained application protocol (CoAP) (IETF RFC 7641, Sept 2015)
31. C. Bormann, Z. Shelby, Block-wise transfers in CoAP (CoRE WG Internet Draft, draft-ietf-core-block-18, Sept 2015)
32. A. Rahman, E. Dijk, Group communication for the constrained application protocol (CoAP) (IETF RFC 7390, Oct 2014)
33. Z. Shelby, Constrained RESTful environments (CoRE) link format (IETF RFC 6690, Aug 2012)
34. Z. Shelby, M. Koster, C. Bormann, P. van der Stok, CoRE resource directory (CoRE Internet Draft, draft-ietf-core-resource-directory-05, Oct 2015)
35. S. Cheshire, M. Krochmal, Multicast DNS (IETF RFC 6762, Feb 2013)
36. S. Cheshire, M. Krochmal, DNS-based service discovery (IETF RFC 6763, Feb 2013)
37. T. Zotti, P. van der Stok, E. Dijk, Sleepy CoAP nodes (CoRE WG Internet Draft, draft-zotti-core-sleepy-nodes-04, Sept 2015)
38. C. Bormann, Concise binary object representation (CBOR) (IETF RFC 7049, Oct 2013)
39. D. Crockford, The application/JSON media type for JavaScript object notation (JSON) (IETF RFC 4627, July 2006)
40. P. van der Stok, A. Bierman, CoAP management interface (CoRE WG Internet Draft, draft-vanderstok-core-comi-09, Mar 2016)
41. M. Veillette, A. Pelov, S. Somaraju, R. Somaraju, Constrained objects language (draft-veillette-core-cool-00, Nov 2015)
42. M. Bjorklund, YANG—A data modeling language for the network configuration protocol (NETCONF) (IETF RFC 6020, Oct 2010)
43. A. Bierman, M. Bjorklund, K. Watsen, RESTCONF protocol (Network WG Internet Draft draft-ietf-netconf-restconf-09, Dec 2015)
44. L. Seitz, S. Gerdes, G. Selander, M. Mani, S. Kumar, Use cases for authentication and authorization in constrained environments (IETF RFC 7744, 29 Jan 2016)
45. S. Gerdes, L. Seitz, G. Selander, C. Bormann, An architecture for authorization in constrained environments (IETF draft-ietf-ace-actors-02, 19 Oct 2015)
46. S. Gerdes, Managing the authorization to authorize in the lifecycle of a constrained device (ACE WG Internet Draft, draft-gerdes-ace-a2a-01, Sept 2015)

47. L. Seitz, G. Selander, E. Wahlstroem, S. Erdtman, H. Tschofenig, Authorization for the internet of things using OAuth 2.0 (ACE WG Internet Draft draft-ietf-ace-oauth-authz-01, Feb 2016)
48. S. Gerdes, O. Bergmann, C. Bormann, Delegated CoAP authentication and authorization framework (DCAF) (ACE WG Internet Draft, Oct 2015)
49. G. Selander, J. Mattsson, F. Palombini, L. Seitz, Object security of CoAP (OSCOAP) (ACE WG Internet Draft, draft-selander-ace-object-security-03, Oct 2015)
50. J. Schaad, CBOR encoded message syntax (COSE WG Internet Draft, draft-ietf-cose-msg-10, Feb 2016)
51. E. Wahlstroem, M. Jones, H. Tschofenig, CBOR web token (CWT) (ACE WG Internet draft, Dec 2015)
52. ZigBee IP/920IP, http://www.zigbee.org/zigbee-for-developers/network-specifications/zigbeeip
53. IEEE 2030.5-2013, IEEE adoption of smart energy profile 2.0 application protocol standard, http://standards.ieee.org/findstds/standard/2030.5-2013.html
54. Open Mobile Alliance (OMA), Lightweight M2M specification v1.0 (2014), http://openmobilealliance.hs-sites.com/lightweight-m2m-specification-from-oma
55. IPSO Smart Object Guideline, Smart objects starter pack, v1.0 (Sept 2014), http://www.ipso-alliance.org/
56. IPSO Smart Object Guideline, Smart objects expansion pack, v1.0 (Oct 2015), http://www.ipso-alliance.org/
57. OCF/Open Interconnect Consortium (OIC) Specifications V1.0, http://openconnectivity.org/resources/specifications
58. Thread, http://threadgroup.org/
59. OneM2M standards for M2M and the internet of things, http://www.onem2m.org/technical/latest-drafts
60. Management Enablement (OMA), TS-0005-V1.0.1, OneM2M technical specification
61. CoAP Protocol Binding, TS-0008-V1.0.1, OneM2M technical specification
62. Security Solutions, TS-0003-V1.0.1, OneM2M technical specification
63. Base Ontology, TS-0012-V-0.7.0, OneM2M technical specification
64. Home Appliances Information Model and Mapping, oneM2M-TS-0023-v0.1.0, OneM2M technical specification
65. oneM2M and OIC Interworking, oneM2M-TS-0024-v0.1.0, OneM2M technical specification
66. oneM2M and AllJoyn Interworking, oneM2M-TS-0021-V-0.0.1, OneM2M technical specification
67. F. Kerasiotis, C. Koulamas, G. Papadopoulos, Developing wireless sensor network applications based on a function block programming abstraction. 2012 IEEE International Conference on Industrial Technology (ICIT), Athens, Mar 2012, pp. 372–377
68. F. Kerasiotis, C. Koulamas, C. Antonopoulos, G. Papadopoulos, Middleware approaches for wireless sensor networks based on current trends. 4th Mediterranean Conference on Embedded Computing (MECO), 2015
69. C. Koulamas, S. Koubias, G. Papadopoulos, Using cut-through forwarding to retain the real-time properties of profibus over hybrid wired/wireless architectures. IEEE Trans. Ind. Electron. **51**(6), 1208–1217 (2004)
70. G. Bovet, J. Hennebert, A web-of-things gateway for KNX networks. Smart SysTech 2013 European Conference on Smart Objects, Systems and Technologies, June 2013, Erlangen, Germany
71. M. Jung, J. Weidinger, C. Reinisch, W. Kastner, C. Crettaz, A. Olivieri, Y. Bocchi, A transparent IPv6 multi-protocol gateway to integrate Building Automation Systems in the Internet of Things. Proceedings of the IEEE International Conference on Internet of Things (iThings 2012), Besancon, France, Nov 2012
72. S.C. Park, W.S. Lee, S.H. Kim, S.H. Hong, P. Palensky, Implementation of a BACnet-ZigBee gateway. 8th IEEE International Conference on Industrial Informatics (INDIN), 2010

73. C. Antonopoulos, A. Prayati, T. Stoyanova, C. Koulamas, G. Papadopoulos, A modeling approach on the TelosB WSN platform power consumption. J. Syst. Softw. **83**, 1355–1363 (2010)
74. C. Antonopoulos, F. Kerasiotis, C. Koulamas, G. Papadopoulos, Experimental evaluation of the waspmote platform power consumption targeting ambient sensing. 4th Mediterranean Conference on Embedded Computing (MECO), 2015
75. A. Liu, P. Ning, TinyECC: A configurable library for elliptic curve cryptography in wireless sensor networks. 2008 International Conference on Information Processing in Sensor Networks (IPSN 2008), 2008, pp. 245–256
76. P. Szczechowiak, L.B. Oliveira, M. Scott, M. Collier, R. Dahab, NanoECC: Testing the limits of elliptic curve cryptography in sensor networks. Jan 2008, pp. 305–320
77. E. Wenger, T. Unterluggauer, M. Werner, 8/16/32 shades of elliptic curve cryptography on embedded processors. Progress in Cryptology—INDOCRYPT 2013, 2013, pp. 244–261
78. C. Pendl, M. Pelnar, M. Hutter, Elliptic curve cryptography on the WISP UHF RFID tag. RFID. Security and Privacy, 2012, pp. 32–47
79. J. Fan, I. Verbauwhede, An updated survey on secure ECC implementations: Attacks, countermeasures and cost, in *Cryptography and Security: From Theory to Application*, ed. by D. Naccache, vol. 6805 (Springer, Berlin and Heidelberg, 2012), pp. 265–282
80. A.P. Fournaris, O.G. Koufopavlou, *Hardware Design Issues in Elliptic Curve Cryptography for Wireless Systems* (CRC Press, 2007)
81. A.P. Fournaris, I. Zafeirakis, P. Kitsos, O. Koufopavlou, Comparing design approaches for elliptic curve point multiplication over GF($2^k$) with polynomial basis representation. Microprocess. Microsyst. **39**(8), 1139–1155 (2015)
82. A.P. Fournaris, J. Zafeirakis, C. Koulamas, N. Sklavos, O. Koufopavlou, Designing efficient elliptic curve Diffie-Hellman accelerators for embedded systems. Proceedings of 2015 IEEE International Symposium on Circuits and Systems (ISCAS 15), Lisbon, Portugal, 24–27 May 2015
83. EnOcean Alliance, https://www.enocean-alliance.org/en/enocean_standard/
84. International Telecommunication Union (ITU) G.9959, *Short Range Narrow-Band Digital Radio Communication Transceivers—PHY and MAC Layer Specifications* (2012)
85. LWM2M interworking, TS-0014-V0.9.0, OneM2M technical specification
86. IEC 61499–1, Function Blocks—Part 1: Architecture

# Chapter 18
# Building Automation Systems in the World of Internet of Things

**Konstantinos Christopoulos, Christos Antonopoulos, Nikolaos Voros, and Theofanis Orfanoudakis**

## 18.1 Introduction

The idea of a fully functional smart home has been a dream since the mid of the twentieth century. The first trials started in the 1960s with the use of domestic smart devices, but this has become a reality only during the last decades. Smart home, as a term, was also introduced in 1984 by the American Association of Housebuilders. Since then, things have dramatically changed with significant advances in the domain of intelligent building control. More specifically, houses, schools, and offices security can be drastically enhanced through electrical and electronic equipment while their energy footprint minimized with inner climate control which maximizes the comfort and safety of the occupant. This can be

K. Christopoulos
Department of Computer and Informatics Engineering, Embedded System Design and Application Laboratory, Technological Educational Institute of Western Greece, Patras, Greece

School of Science and Technology, Digital Systems and Media Computing Laboratory, Hellenic Open University, Patras, Greece
e-mail: kchristopoulos77@gmail.com

C. Antonopoulos (✉)
Department of Computer and Informatics Engineering, Embedded System Design and Application Laboratory, Technological Educational Institute of Western Greece, Patras, Greece

N. Voros
Technological Educational Institute of Western Greece, Embedded System Design and Application Laboratory (ESDA), Antirio, Greece

T. Orfanoudakis
School of Science and Technology, Digital Systems and Media Computing Laboratory, Hellenic Open University, Patras, Greece



G. Keramidas et al. (eds.), *Components and Services for IoT Platforms*,
DOI 10.1007/978-3-319-42304-3_18

achieved, for example, by remote mobile control, automated lights, automated thermostat adjustment, scheduling appliances, mobile/email/text notifications, and remote video surveillance.

This chapter provides an overview of the latest technologies for smart home/building related technologies and introduces KNX, as an emerging standard for interconnecting KNX of Things and combining them with the world of Internet of Things (IoT). The rest of the chapter is organized as follows: Sects. 18.2, 18.3 and 18.4 provides an overview of the technology evolution of building automation systems, while Sect. 18.5 presents the major requirements of modern home/building automation. Sections 18.6 and 18.7 detail how building automation systems can be combined with IoT and KNX of Things, respectively. Section 18.8 outlines how KNX gateways can be used for developing intelligent home and building infrastructure, while Sect. 18.9 presents a real world implementation that combines KNX of Things with IoT prominent technologies. Finally, Sect. 18.10 concludes and presents the future work.

## 18.2 Evolution of Building Automation Systems

Most *Building Automation Systems* (*BAS*) consist of wired buses or wireless broadcast networks, which connect high-level controllers. Initially, bus systems were developed in the field of computers and protocols for BAS such as the X10 protocol which offered a limited number of instructions, low data rate while characterized by compatibility problems. The second stage of development with bus systems such as EIB, LON, and other proprietary systems with devices from different manufacturers aimed for better compatibility. Currently, in the third development stage, there is further market penetration for bus systems, increased degree of system integration and an increasing use of the radio frequency (RF) bus systems (Fig. 18.1).

In BAS, there is a wide range of technology platforms and protocols a user/designer/developer can select from. Each of them has pros and cons. Amongst them,

**Fig. 18.1** Bus systems development stages

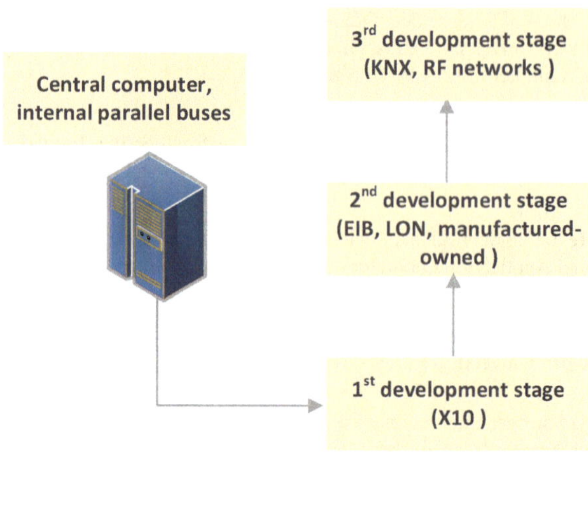

most popular are those offering *open protocols* and *support a large number of devices*, as well as those that offer *device interoperability*. Cost, energy consumption, and bandwidth are also essential parameters, as well as the programming language(s) and the maximum number of connected devices supported. The most popular home automation wired and wireless platforms are detailed in the following sections.

## 18.3 Wired Protocols

### *18.3.1 BACnet*

BACnet (http://www.bacnet.org/Tutorial/HMN-Overview/sld039.htm), [9, 11] is one of the most popular communication protocols with many implementations in building automation systems (HVAC—Heating, Ventilating and Air-conditioning Control, Lighting Control, Security, Smart elevators) and control networks. The BACnet protocol was initially developed by the American Society of Heating, Refrigerating and Air-conditioning Engineers (ASHRAE) project committee. Today, BACnet is covered by two standards: international ISO 16484–6 and ANSI/ASHRAE STANDARD 135 (http://www.bacnet.org/Overview/). BACnet supports many different communication protocols (ARCNET, Ethernet, BACnet/IP, Point-To-Point over RS-232, Master–slave/Token-Passing over RS-485, Lon Talk, or KNX-IP). The most common serial version is called BACnet MS/TP (master–slave/token-passing) which is a twisted pair wiring and is intended for devices with lower requirements in terms of speed. It runs at speeds of up to 1 Mbps while the dominant Ethernet version is BACnet/IP and supports several types of media UTP, fiber, or even wireless with data rate of up to 10 Gbps. BACnet communication is based on "objects" (analog input/output/value, binary input/output/value, loop, calendar, program . . . ). An object is effectively a collection of information related to a particular function that can be uniquely identified and accessed over a network in a standardized way.

One of the disadvantages of the BACnet is that there is no common configuration tool (in contrast to KNX). Therefore, manufacturers have to develop their own new (and typically non open-source) tools for their devices. Secondly, the device interoperability is limited, since some BACnet products are not fully compliant to each other.

### *18.3.2 LON*

LonWorks [1, 2, 9, 11] (http://www.rtaautomation.com/technologies/lonworks/, http://www.lonmark.org/technical_resources/standards) was created by Echelon Corporation in 1988. LonWorks is a Building Automation Platform based on LonTalk communication protocol. The LonWorks platform is a distributed control system that enables peer-to-peer as well as master–slave communication among intelligent devices. The LonWorks platform supports a wide range of different physical media including free topology twisted pair transceiver (FFT-10A) [3], power line, radio frequency, infrared, fiber optic, and coaxial cable.

One of the main features of Lon is IP-tunneling based on the ISO/IEC 14908–4 (ANSI/CEA-852) standard (http://www.lonmark.org/technical_resources/standards). There is a special transceiver for every physical layer which is transforming its specific signals to standardized input signals for the connected Neuron chip. The LonWorks platform can manage almost every task in BAS from room temperature control to security and access control systems.

LonTalk, similarly to BACnet or KNX, supports several international and national standards which are shown in Table 18.1. All LonWorks-based devices comprise of a Neuron chip which contains a complete system. The most important part of the neuron chip is the 48-bit serial number, which is unique for every device. This is important for the identification in an installation, while the main disadvantage of all Lon talk devices is their dependency on the neuron chip itself. The Neurons contain the entire LonTalk protocol stack and are comprised of three 8-bit CPUs (two for the protocol and one for the application). There are two basic types of Neuron chips, the 3120 and the 3150 which are functionally equivalent except from memory configuration and packaging.

In order to allow system interoperability, the standardization of the variables used to describe physical things in LonWorks is of paramount importance. This list of standards is maintained by LonMark International and each standard is known as Standard Network Variable Types (SNVTs). One drawback of Lon is the fact that the devices made by different manufacturers have different development software. Most of the tools are based on the additional programs of the unified Echelon database platform (LNS) (plug-in) which enables the parameterization via Windows dialog boxes.

### *18.3.3 KNX*

KNX [4–9, 11] is an open, worldwide standard used for more than 25 years. The KNX building automation system was developed at the beginning of 1990s by the European Installation Bus (EIB). In 1999, EIB Association, Batibus Club International (BCI, France), and the European Home Systems Association (EHSA, Netherlands) have been merged. The result of this merging is a new association called “KNX Association.”The supported standards for KNX are shown in Table 18.1. The KNX association nowadays consists of more than:

- 400 manufacturers in 37 countries
- 48200 KNX Partners in 139 countries
- 360 KNX Training Centers in 59 countries
- 120 Scientific Partners in 30 countries comprised of Universities all over the World

**Table 18.1** Features of wired protocols employed in BAS

| Standard | KNX | LonWorks | BACnet |
|---|---|---|---|
| Number of bus devices | 58384 addressable devices, whereby many devices have realised several input/output points | 32385 devices per domain. Several input/output points are often implemented in a device | 127 MS/TP masters |
| Expansion | 700 meters per galvanic unit, possible to extend complete system by many kilometers | | 1200m MS/TP Twisted Pair cable |
| Transmission medium | Twisted twin core cable, Powerline, Radio, IP network individual sections of Optical fibre | Various types with twisted twin core cable, Powerline, Radio, Infrared, Fibre glass | BACnet IP, BACnet LonTalk, or BACnet MS/TP |
| Topology | With twisted twin core cable: tree structure | Very varied depending on the selected transmission type | Line or star topology (Standard Ethernet topology) |
| Applications | Building Management Automation, Lighting, blind control, heating, ventilation access control, media, security, monitoring, visualisation and load management | Lighting, blind control, heating, ventilation access control, monitoring, visualisation and load management | Building Managment Automation, HVAC plants, fire control panels, smart elevators, intrusion detection and access control systems |
| Maximum Data Rate | 9,6 kbit/s | 78 kbit/s | 10/100 Mbit/s full duplex |
| Standards | International Standard (ISO/IEC14543-3)<br>European Standard (CENELEC EN50090 and CEN EN 13321-1 and 13321-2)<br>Chinese Standard (GB/T 20965)<br>ANSI/ASHRAE Standard (ANSI/ASHRAE 135) | ANSI/CEA-709.1-B<br>EN 14908-1:2005<br>IEEE 1473-L<br>GB/Z 20177.1-2006 | international ISO 16484-6<br>ANSI/ASHRAE STANDARD 135 |

The technology used in modern KNX devices is compatible with that of the old EIB system, so all devices bearing either the KNX or the EIB logo are mutually compatible. The KNX certification process ensures seamless cooperation and communication amongst different products of different manufacturers used in different applications. This ensures a high degree of flexibility concerning the extension and the modification of installations and wide scale deployments.

Similarly, to other typical complex standardized protocols, KNX uses more than one physical layers:

- KNX TP (separate bus cable)
- KNX PL (existing mains network)
- KNX RF (via radio signals)
- KNX IP (via Ethernet or Wi-Fi)

The KNX installation can be configured with a manufacturer-independent tool called Engineering Tool Software (ETS) [4]. Thus, the KNX system avoids typical problems of other standardized systems (e.g., BACnet, LonWorks) where each manufacturer has a special configuration tool, while it is often not possible to use one manufacturer's tool for devices (supporting the same protocol) made by another vendor. The ETS tool is managed by the KNX association and the manufacturers only have to add a plug-in to it.

In contrast to other protocols, KNX is independent of any hardware or software technology meaning that it can be realized on any microprocessor platform.

## 18.4 Wireless Protocols

### 18.4.1 Z-Wave

Z-Wave (http://z-wavealliance.org/) uses ITU-T G.9959 PHY/MAC with protocol stack from Sigma Designs. The international standard is maintained by Z-Wave Alliance. Z-Wave uses low power sub 1 GHz RF and works within a mesh topology. Z-Wave is one of the most popular wireless protocols for home automation. It transmits on 868.42 MHz (Europe) frequency band, while it is not affected by the interference generated by other household wireless products that usually work on 2.4 GHz. A significant advantage of Z-Wave is its interoperability; all Z-Wave devices talk to each other, regardless of their type, version, or brand, while Z-Wave devices are backward and forward interoperable. Z-Wave provides access to a wide range of devices with low power consumption. Z-Wave's mesh network makes all devices double as repeaters, which means that each Z-Wave device will pass the signal along to another until the final destination is reached.

### 18.4.2 ZIGBEE

ZigBee PRO (http://www.zigbee.org/) [11] offers full wireless mesh, low power networking capable of supporting more than 64,000 devices on a single network. It provides standardized networking designed to connect an extended range of devices from different manufacturers into a single control network, transcending interoperability issues with older versions. The latest version of ZigBee Home Automation standard, which is fully interoperable with the previous versions, adds several important new features that improve the battery lifetime for security sensors (with sensor lifetime over 7 years), standardize device pairing and simplify installation and maintenance. These features have significant impact on operational and equipment costs for the service providers, and on quality of service for the consumers. The Green Power feature of ZigBee PRO allows battery-less devices to securely join ZigBee PRO networks. It is the eco-friendliest way to power ZigBee products such as sensors, switches, dimmers, and many other devices.

### 18.4.3 Bluetooth Low Energy

Bluetooth (https://www.bluetooth.com/what-is-bluetooth-technology/bluetooth) is contained in many devices. It has higher data bandwidth than ZigBee and Z-wave. In Bluetooth 4 version (Bluetooth Low Energy or BLE) there is a significant decrease in power consumption. The latest version of the protocol, Bluetooth 4.2, makes Bluetooth a promising wireless technology for the IoT.

It is one of the most reliable and secure wireless standards. BLE Secure Connection includes Federal Information Processing Standards (FIPS) (https://en.wikipedia.org/wiki/Federal_Information_Processing_Standards) approved encryption, granting the ability to encrypt connections between devices under P-256 elliptic curve public-key cryptography. According to the ABI Research (https://www.abiresearch.com/) the market share for smart home applications, such as lighting, security, and energy management, based on Bluetooth will increase at a faster rate than any wireless technology over the next 5 years.

Bluetooth 4.2 uses the Internet Protocol Support Profile (IPSP) to connect Bluetooth Smart sensors to the Internet in order to send and receive transmissions through gateways via IPv6 (https://el.wikipedia.org/wiki/IPv6) and 6LoWPAN (https://en.wikipedia.org/wiki/6LoWPAN, http://www.ti.com/lit/wp/swry013/swry013.pdf).

The basic topology used within a Bluetooth ecosystem is the *piconet*, in which a master device can be interconnected with up to seven slaves. When there are more than two piconets interconnected to each other, the topology is referred to as *scatternet* (Fig. 18.3). A critical advantage compared to any competitive technology stems from the fact that BLE capitalizes on the wide acceptance and wide spread of previous versions of Bluetooth protocol. Therefore, even nowadays BLE interfaces are integrated in essentially any mobile phone, tablet, wearable device, etc., offering a huge precedence in dominating relative markets.

### 18.4.4 EnOcean

EnOcean (https://www.enocean.com/en/) technology is used primarily in building automation systems. The EnOcean standard is based on the international wireless standard ISO/IEC 14543-3-1X, which is optimized for ultra-low power wireless applications and energy harvesting. EnOcean energy harvesting wireless sensor solution is able to generate a signal of astonishing range from an extremely small amount of energy. Consuming just 50 $\mu$W a standard EnOcean energy harvesting wireless module can easily transmit a signal 30 m away (in-doors). The technology behind this is based on the signal duration since the entire process is started, executed, and completed in no more than a thousandth of a second. The latest generation of EnOcean energy harvesting wireless sensors requires standby currents of only 100 nA or less via gateways. There are also products which are self-powered (without battery) such as switches and sensors. EnOcean wireless solutions communicate with all major wired bus systems such as KNX, LON, DALI, BACnet, or TCP/IP.

### *18.4.5 Wi-Fi*

Wi-Fi (http://www.wi-fi.org/) is supported by many applications and devices. The Wi-Fi Alliance owns and controls the "Wi-Fi Certified" which certifies all 802.11-based products for interoperability. It offers significantly higher data rates than any other wireless standard but its high power consumption comprises a notorious problem for the battery life in mobile devices. The Wi-Fi Alliance introduced Wi-Fi HaLow (http://www.wi-fi.org/) in January 2016 as an extension of the upcoming 802.11ah standard which is intended to be competitive with low power Bluetooth, but with a wider coverage range. The new wireless protocol operates on the 900 MHz band but retaining support for existing 2.5 and 5 GHz access points (http://www.wi-fi.org/).

### *18.4.6 THREAD*

Thread (https://www.threadgroup.org/) is a new open wireless protocol for home automation, which was founded in July 2014 by vendors including Google's Nest Labs, Samsung Electronics, ARM Holdings, Silicon Labs, and other key players. Thread is based on the 802.15.4 radio standard. Contrary to the Wi-Fi paradigm, it is an IPv6-based mesh network in which product developers and consumers can easily and securely connect more than 250 devices into a low power, wireless network that also includes direct Internet, and cloud access for every device (Table 18.2).

## 18.5 Requirements for Building Automation System

When it comes to making a decision on choosing the appropriate BAS for a house or a building, there is a significant number of parameters that should be considered. The major requirements mainly needed are outlined in the sequel:

- *Basic BAS operation support.* Tasks such as scheduling occupancy, adjusting setpoints, troubleshooting complaints, responding to alarms, or checking of the system should be as simple and as intuitive as possible. It is essential to accomplish them in a fast and easy way and from any available terminal or remote device.
- *Low energy consumption.* The current demands must be constantly, accurately measured and analyzed, in order to have an effective and environment-friendly use of energy. BAS must provide facilities and services that will allow both monitoring and decision making in an efficient way.

**Table 18.2** Features of wireless protocols employed in BAS

| Standard | Z-WAVE | ZigBee | enocean | Bluetooth | WiFi | THREAD |
|---|---|---|---|---|---|---|
| Network Topology | Mesh | Star, Tree, Mesh | Mesh | Star/Point-toPoint | Star, Mesh | Star, Mesh |
| Power Usage | Low | Low | Low | Low | | Low |
| PHY/MAC Standard | ITU-T G.9959 | IEEE 802.15.4 | IEEE 802.15.4 | IEEE 802.15.1 | IEEE 802.11 | IEEE 802.15.4 |
| ISM Radio bands | 868.42 MHz (Europe) 908.42 MHz (United States) | 2408 to 2480 MHz | 868 MHz (Europe) 902 MHz(United States) | 2.4 to 2.485 GHz | 2.4 GHz | 2.4 GHz |
| Number of bus devices | up to 232 | up to 64000 | | 7 devices in one piconet region | | up to 250 |
| RF range (indoor - line of sight) | 30 - 100m | 70 - 400m | 30 - 300m | 30 - 100m | 200m | 30m |
| Maximum Data Rate | 9,6 - 100 kbit/s | 20 - 250 kbit/s | 125 kbit/s | 1 - 24 Mbit/s | 250 Mbit/s | 250 kbit/s |
| Manufacturers | up to 330 | up to 76 | up to 100 | | | |
| Applications | Home Automation | Home Automation, Building Management Automation | Home Automation, Building Management Automation | | | Home Automation |
| Power Efficiency | √ | √ | √ | √ | x | √ |
| Certified products | up to 1400 | up to 939 | up to 1200 | | up to 25000 | up to 30 |

- *Assurance of high quality indoor living*. Comfort management is one of the major requirements that must also be taken into account. Depending on the type of the inhabitant, it can range from simple environment monitoring to environment adjustment according to the need of people having chronic diseases, e.g., people having asthma.

In order to support the aforementioned requirements, there are several features that need to be supported and must be identified beforehand, since there is a wide variety of technology platforms or protocols on which a BAS can be build on. It is also important for the system integrator to be informed about the type of the building, i.e., whether it is a commercial or a residential one or any other type of different use. Furthermore, the region where the building is located, as well as the characteristics of the people who might be living or working there, should be taken into consideration.

As far as building automation is concerned, several of the protocols already described can cover one or more control levels. There is a three-level hierarchical model for automation and control systems [7], upon which building automation protocols can be employed. The first one is the *field level* which contains the connected sensors and actuators. The second one is the *automation level,* which includes all the applications that can be automated in a building. The last is the *management level* where control, operation, and monitoring take place. Figure 18.2 shows the three-level hierarchical model and indicates the level in which each standard can be used.

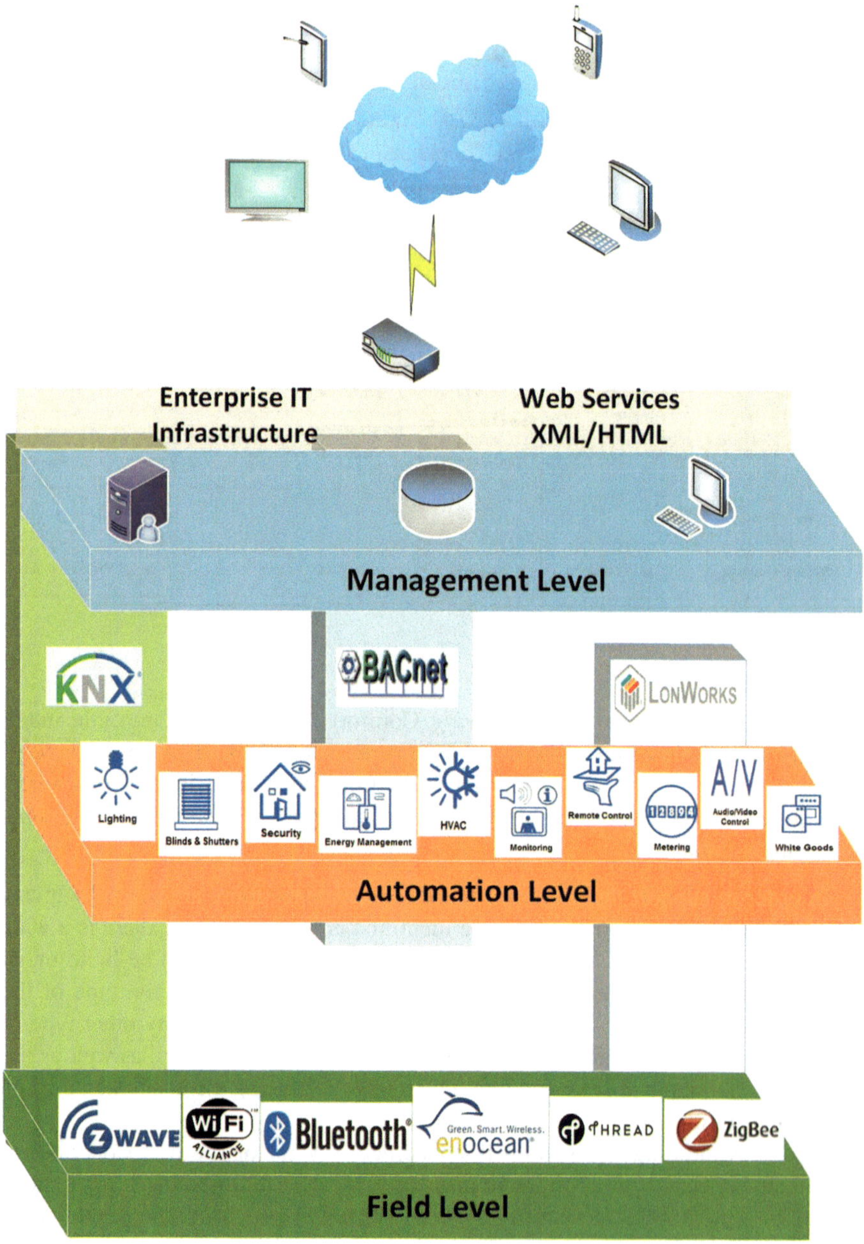


**Fig. 18.2** Three-level hierarchical model of automation and control system

## 18.6 Building Automation Systems and "Internet of Things"

Nowadays, there are about 25 billion connected devices [10] on the Internet and it is estimated that by 2020 that number will be up to 50 billion. In a BAS, sub-networks employing different communication protocols have the ability to connect to the Internet either directly or through gateways. The same applies when we have to connect different sub-networks with each other in a single installation. The goal, in this case, is to use a protocol suite, which covers the majority of the installed devices and their corresponding applications within the specific context. Furthermore, a protocol suite must cover all or most of the communication media (twisted pair, power line, and radio frequency) and internet protocol (IP)-based networking. The interconnection of all BAS is realized through a main router with the appropriate IP gateway of each communication protocol as shown in Fig. 18.3.

In a *new building*, the use of a *wired decentralization protocol* like a bus system is strongly recommended. This approach offers critical advantages such as reliability, connectivity, and facilities for future improvements. For instance, if one unit of a bus system fails, the remaining system continues to operate without any problem in contrast to a centralized one. An Ethernet cable CAT 6 should also be placed near TVs, computers, IP cameras, security systems, and anywhere else where it may be useful to connect via a high-speed interface appropriate for media applications and

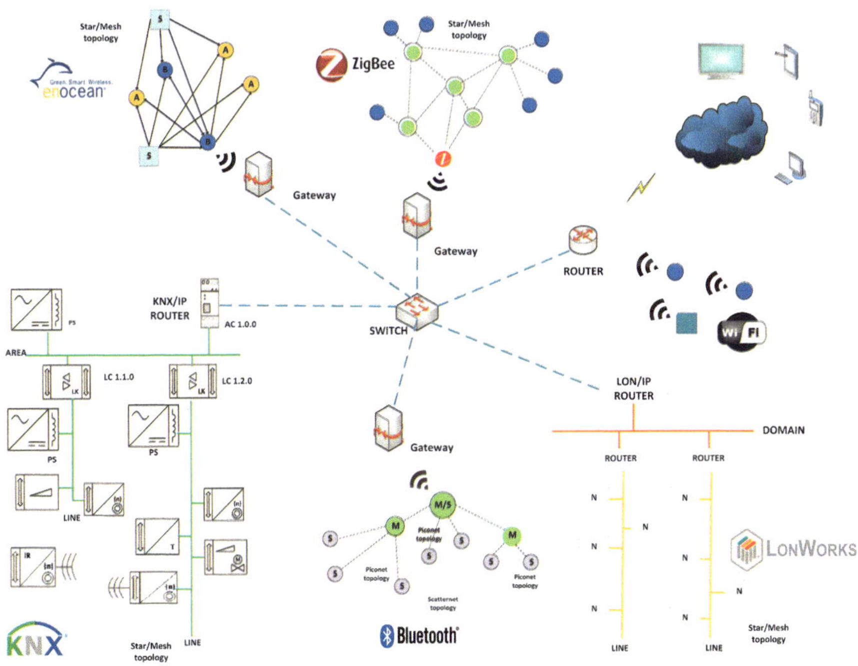


**Fig. 18.3** A typical topology allowing BAS to participate in Internet of Things

communication to application servers. This makes possible for remote access and control of the installation covering any future needs.

On the other hand, in cases such as a *reconstruction* or a *renovation*, there are different approaches depending on the variety of uses. In such circumstances, the use of *wireless home automation protocols* offers significant practical advantages. However, it would be essential to have a *combination of wireless and wired communication* media with the same or different protocols.

During the last years, wireless protocols have been improved and they are leading the IoT paradigm. They have an extended use and one of the most pronounced objectives is energy efficiency. Ultra-low power devices with current demands as low as few μA offer a gain of 10–15 years battery life with a small coin cell battery. They can be even more efficient by incorporating intelligence enabling to turn into sleep mode power hungry components (e.g., radio interface, processing unit, and mechanical parts) yielding even less power consumption. Furthermore, each device tends to be more and more complex since they are composed of multiple sensors. Finally, there is no need to have a wiring infrastructure running from a central controller to switches and keypads throughout the house in order to enjoy a more connected home.

The most recent trend is towards a direct IP addressability of every single node of a wireless network, as in the case of THREAD (described in Sect. 18.4.6) for example, which allows a more immediate communication between the end nodes. Apart from THREAD, IPv6 communications are now also used in newer Bluetooth versions as well as Wi-Fi. IPv6 is able to offer a complete cover across all different networks in a home/building automation. However, native IPv6 implementations often prove to be too complex and resource demanding for typical hardware encountered in IoT platforms coming from the words of embedded systems and WSNs characterized by extremely limited resources in all aspects. A prominent and widely tested solution is the 6LoWPAN (https://en.wikipedia.org/wiki/6LoWPAN, http://www.ti.com/lit/wp/swry013/swry013.pdf), effectively comprising an implementation for IPv6 specifically targeting platform with scares resource availability (e.g., IEEE 802.15.4, ZigBEE, etc.) allowing each sensor to have its own IP address and in this way realize the ultimate objective of IoT approach.

## 18.7 Building Automation Systems and the "KNX of Things"

IoT comprises by devices of any standard that is already or will be able to offer Internet connectivity in the near future. There is an interconnection between several decentralized intelligent components, which communicate directly and autonomously via the Internet. KNX with all these applications/devices and interfaces to other systems is already used similarly to the "Internet of Things" and it provides the possibility of central controlling and monitoring through the KNX/IP gateway as shown in Fig. 18.4.

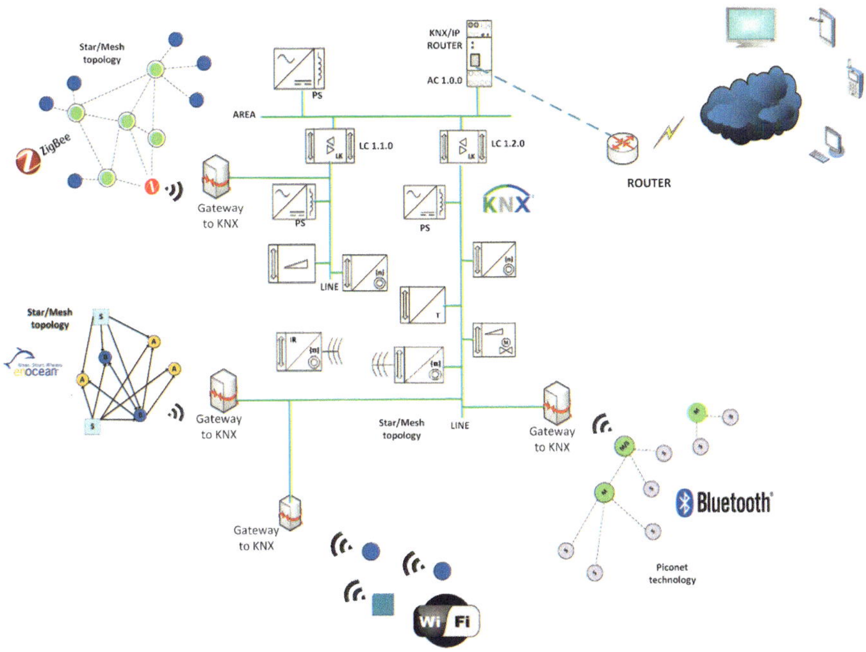


**Fig. 18.4** Topology for KNX of Things

Many companies provide solutions for home/building control such as Google's nest (https://nest.com/) and Apple's Homekit (http://www.apple.com/ios/homekit/), through applications which unify different standards under a single interface. However, this trend has many drawbacks such as increased complexity and sometimes reliability problems. Such solutions require a central server or controller for the communication of different protocols. This is the reason why systems should be chosen to enable a more sophisticated level of customization, including the ability to work with a broader field of products and systems. For instance, typically there is no strict need to access every single data object in a home/building.

KNX is one of the most complete communication protocols designed as a decentralized bus system supporting all kinds of physical layer alternatives such as twisted pair, radio frequency, power line, IP/Ethernet, and Wi-Fi. Bus devices can be either sensors or actuators required for controlling the building management equipment. KNX can be used for all application areas in home and building control and for a variety of functions, ranging from lighting and shutter control to security, heating, ventilation, air-conditioning, monitoring, water control, energy management, as well as household appliances, and audio systems (Fig. 18.5). Another advantage of KNX is that it is easily adapted for use in different kind of buildings both new and existing ones.

The KNX standard has many different configuration modes which can be used according to the target market and typical applications as shown in Fig. 18.6. One

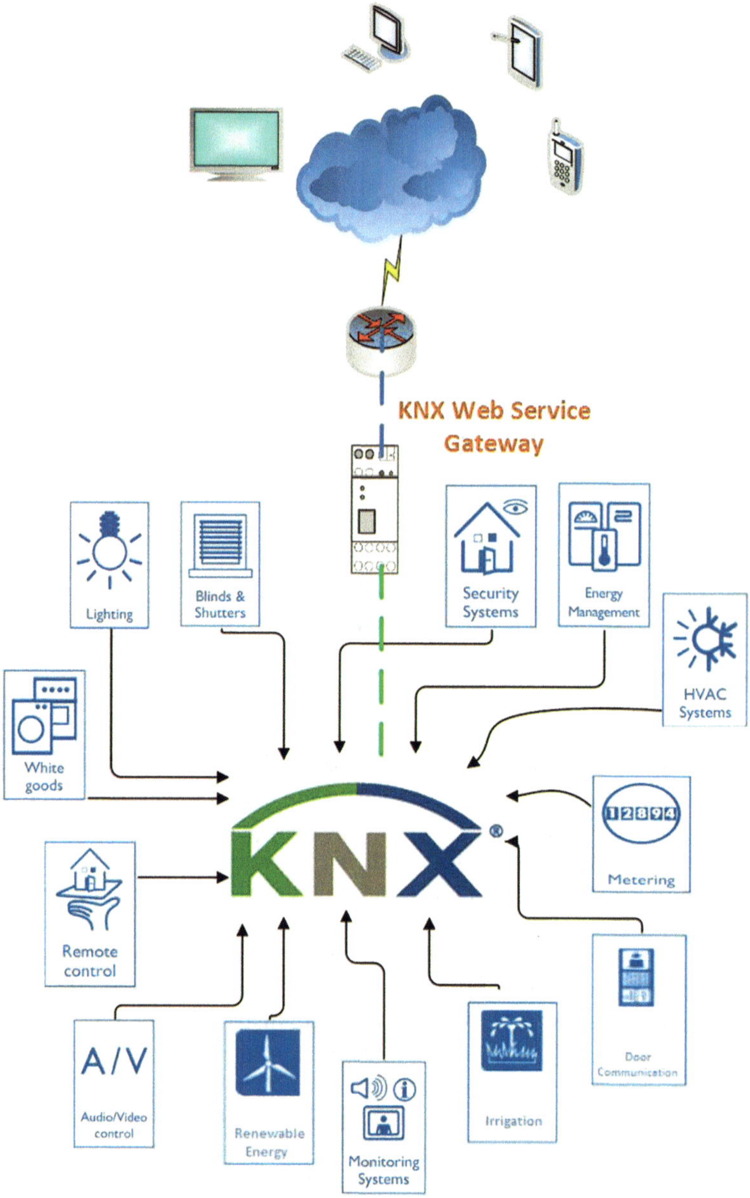


**Fig. 18.5** KNX Web Services

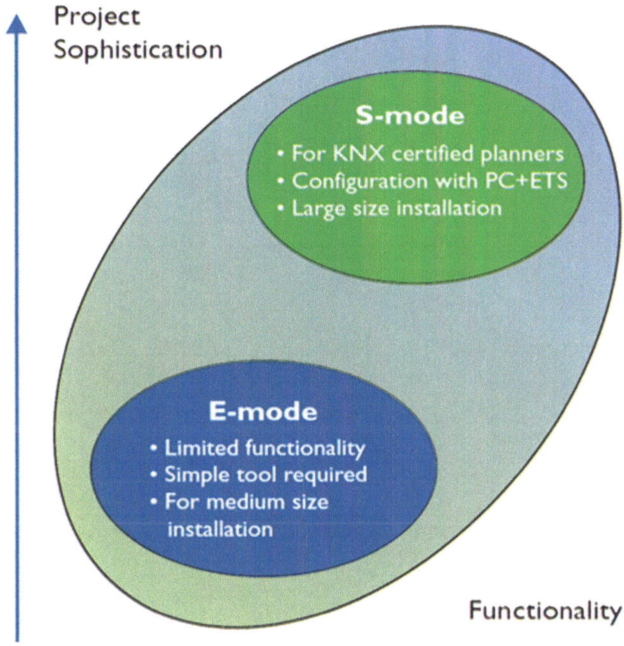


**Fig. 18.6** KNX operation modes "Konnex Association"

of its tools is ETS Professional for the "S-Mode" components which are the same for all the KNX products, regardless the manufacturer.

## 18.8 KNX Web Service Gateway

The communication of KNX applications via an IP-based network is available for more than 10 years. The IP router uses the KNXnet/IP standard defined by KNX Association and ensuring two functionalities. The first is *KNX/IP Routing* which allows the interconnection of any remote KNX installation, especially for extended installations where considerably higher bandwidth may be required in the backbone. Here, the high bandwidth of a LAN network offers an optimal solution. While with KNX twisted pair a maximum of only 25–50 telegrams (data packets) can be transmitted per second, according to their length, transmission via LAN exceeds 10,000 telegrams at 10 Mb/s. To process this traffic without loss or excessive delays, a high computational power is required in addition to an adequate telegram buffer from IP to KNX TP. An IP router is used as line or backbone couplers in a KNX installation. Figure 18.7 depicts an example of a KNX/IP Routing.

The second is *KNX/IP Tunneling* which enables the IP-based access of a terminal device to a KNX installation. KNX tunneling is the technique used by web clients,

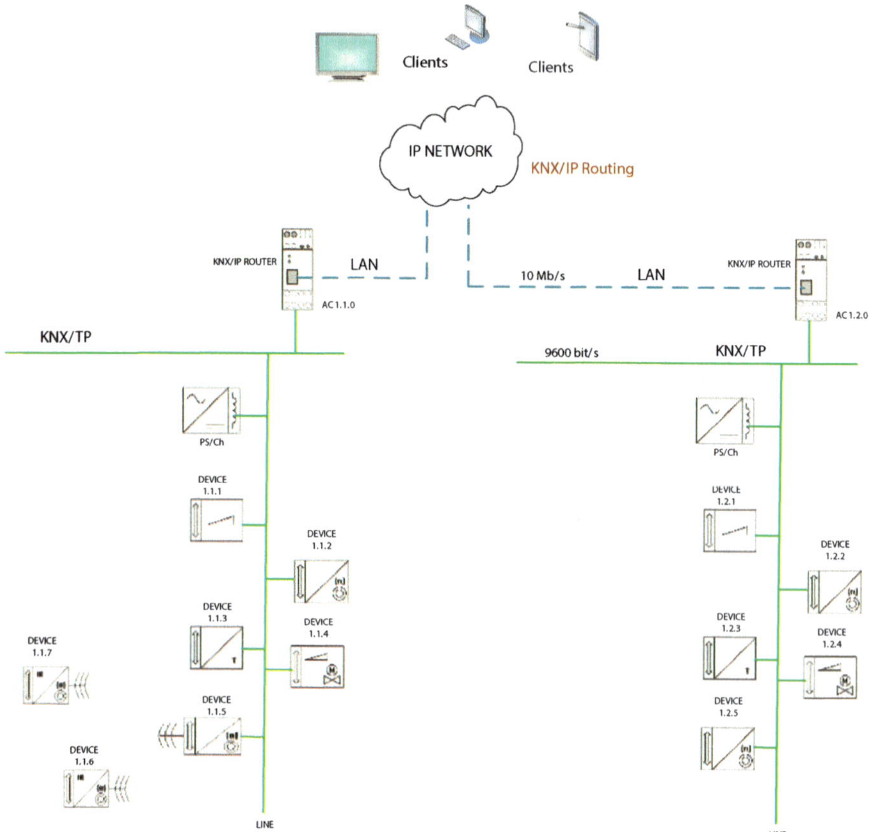


**Fig. 18.7** KNX/IP Routing

computers, and smart phones to communicate with KNX devices in order to provide services to the end user. A typical example of KNX/IP Tunneling is shown in Fig. 18.8.

Using of a simple IP/KNX Router it is not easy to have access over the Internet to a KNX installation without support from an Information Technology (IT) expert. However, typically building automation is an unknown field for IT experts. A solution for this sector would be a translator connecting both worlds without the need for each party to have insight of other side. KNX Association has recognized this trend and developed the corresponding solution "KNX Web Services" (KNX WS) [8], which is realized using web services like oBIX, OPC UA, and BACnet-WS. Web services are self-contained modular software components that can be described, published, and activated via the web. Usually, they are employed by applications and not by persons. Thus, a simple and multi-faceted communication between web services and systems of building automation is possible.

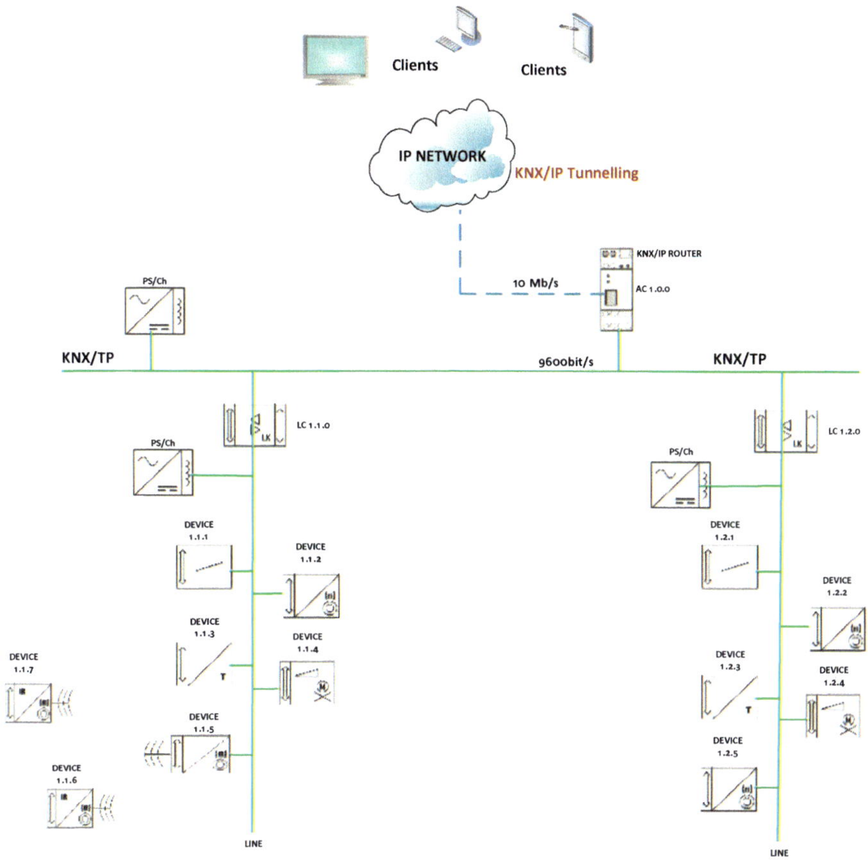


**Fig. 18.8** KNX/IP Tunneling

## 18.9 KNX of Things in Practice: Design and Implementation of an Ambient Assisted Living Residence

Using the components described in the previous sections, an Ambient Assisted Living Environment that combines the world of the IoT with the world of KNX of Things has been designed and developed. It is called Ambient Assisted Living House (AAL House) [13] and it is a fully functional 60 m$^2$ residence for *applying*, *experimenting*, and *evaluating* state-of-the-art ambient assisted living technologies. One of the main design goals of AAL House is to provide an ambient environment with services for elders and people with chronic diseases.

The architecture of the AAL House is designed in accordance with the three-level hierarchical model for automation and control systems already described in Sect. 18.5, where the wireless sensors (ZigBee, BLE, and EnOcean) are located in the *field level*. Their task is to record human physiologic parameters, such as

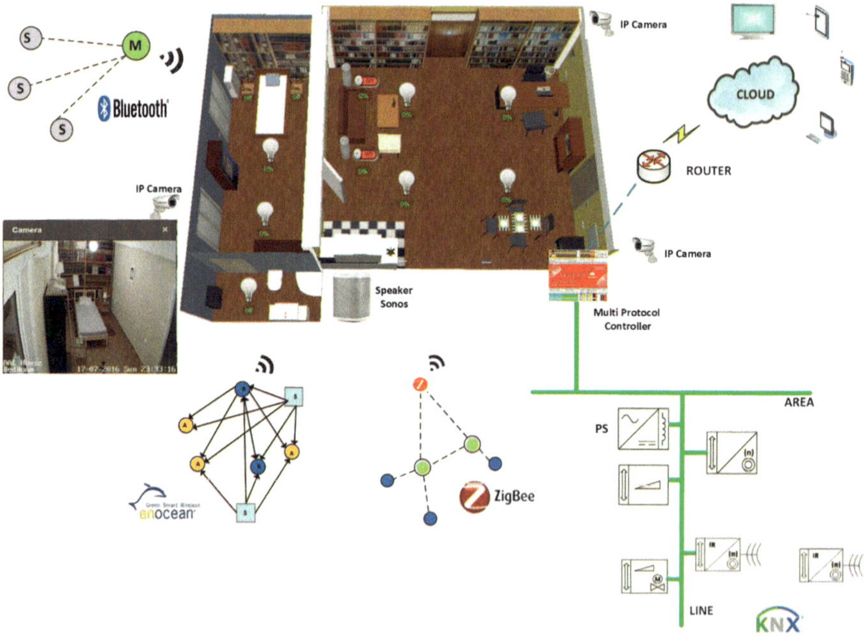


**Fig. 18.9** Internet/KNX of Things infrastructure of AAL House

temperature and heart rate, as well as room air quality ($CO_2$) and humidity. A wired protocol is also available in the field level which is supported by KNX and it is responsible for the functionality of home automation. It basically triggers the appropriate functions according to the signals received from the sensors.

The programming of the automation functions is located at the *automation level*. Last but not least is the *management level*, where the monitoring and control of AAL House is realized. This is where a controller, LogicMachine Re:Actor V3 (http://openrb.com/logicmachine-reactor-v3/), is located, which contains all the interfaces for the heterogeneous communication protocols of the AAL House and is also in charge of the interaction between the different protocols. The controller is also connected to the Internet and it offers the possibility of remote control and visualization of the AAL House. An overview of the AAL House infrastructure is presented in Fig. 18.9.

As already mentioned above, special groups of population can benefit from living in a smart home environment. Currently, there are protocols and sensors (mainly wireless coming from the CyberPhysical and Wireless Sensor Network domains) which are used for monitoring individuals through special parameters of their body. These could be adopted and integrated into the KNX installation in such a way that the greatest comfort and safety is achieved. So far, KNX has the ability to regulate heating and air-conditioning systems, so as to reach the desired room temperature

**Fig. 18.10** AAL House project

with energy efficiency. But this could also be done automatically, according to the special needs of a person which are determined by biological parameters using Z-Wave, ZigBee, and BLE sensors.

Remote control of room functions or making phone calls through the room's sound system and voice demand are also typical examples how inhabitants with kinetic disorders can benefit from the infrastructure and services offered by AAL House. In addition, alert generation is also supported in case of a fall is identified or in cases of rapid change in the body's physiologic parameters (e.g., in body temperature or in electrocardiography parameters).

AAL House smart home infrastructure allows also the monitoring of a wide range of parameters related to environment, such as humidity and $CO_2$ monitoring. Especially for the case of emergency, like a fire, AAL House is equipped with devices and services that allow automatic door unlock, lighting of safety path, and automatic call of fire department.

In addition, the KNX/IoT technology developed for AAL House allows the integration of services like medicine notification, monitoring and report of eating habits, etc. Additional services for elders like water tap overflow identification and respective notification triggering or turning off the oven are also supported.

Figures 18.10 and 18.11 presentthe main control panel of the AAL House through which the aforementioned services are being monitored and controlled.

The AAL House supports also a new trend in home automation which is called *Geofencing* allowing the definition and monitoring of virtual fences within the house (as well as close surrounding areas). This is especially important for people suffering

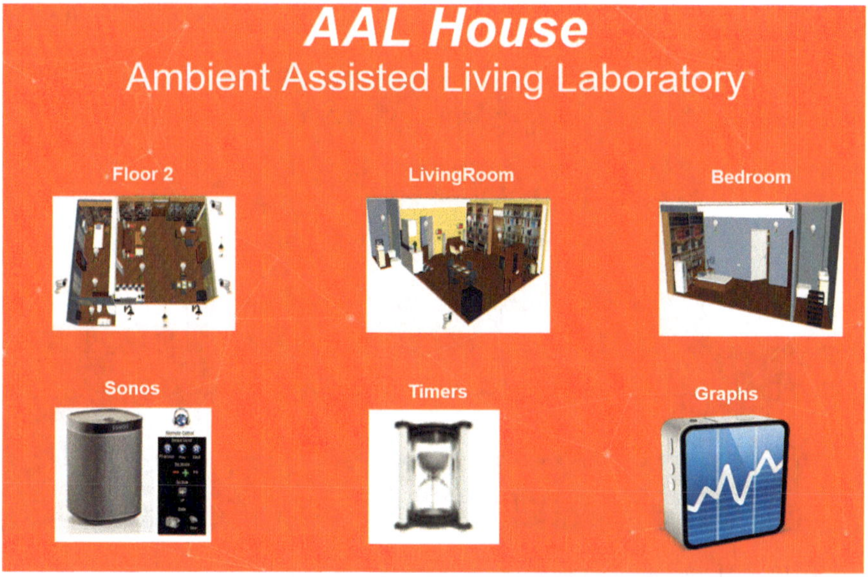


**Fig. 18.11** Ambient Assisted Living House

from dementia related diseases. In addition, the Geofencing service of the AAL House allows KNX related actions such as turning off of lights, electrical appliances, ventilation, central heating, and air-conditioning.

From a technical point of view, all the services offered by the ALL House are strongly dependent on the underlying infrastructure which combines KNX with wireless sensor technologies like the ones presented in Sect. 18.4. The heterogeneous technologies hosted by AAL House are a proof of concept of combining the worlds of KNX of Things and IoT in the context of a BAS.

## 18.10 Conclusion

In this paper, we present the constituent parts of a modern Building Automation Systems (BAS). Nowadays, BAS face a real challenge since their needs and their requirements have shifted from standalone automation services to solutions that will allow BAS to operate seamlessly with IoT devices.

The existence of several protocols from different manufacturers and the fact that during the next years the number of interconnected devices worldwide will be more than 50 billion pose significant challenges for modern BAS. To address those challenges, we present an overview of the existing technologies that could be employed in a BAS and we analyze their pros and cons. Based on this analysis, we propose an abstract architecture for combining KNX of Things (the most popular BAS protocol) with the IoT paradigm. The proposed architecture has been implemented in the context of a real world Ambient Assisted Living environment: a fully operational residence for supporting elders and persons with chronic diseases, where KNX/IoT devices have been employed for developing state-of-the-art services.

The work presented in this chapter is the foundation of future research in the area of BAS. In the near future, we plan to design and implement an extension of the proposed infrastructure, so as to enrich and enhance the offered services with artificial intelligence [12]. For that purpose, a cloud-based platform will be developed that will allow the transition from "smart" to "intelligent" homes and buildings.

## References

1. Echelon Corp, *Neuron chips* (Echelon - Developers, 2009), http://www.echelon.com/developers/lonworks/neuron
2. Wikipedia EN, *LonWorks* (Wikipedia EN) [Online], http://en.wikipedia.org/wiki/LonWorks. Accessed 23 Feb 2011
3. LONWORKS, *FTT-10A Free Topology Transceiver User's Guide, Version 6* (ECHELON Corporation), www.echelon.com
4. Konnex Association, *KNX Handbook for Home and Building Control, Basic Principles*, 6th rev edn. (2013)
5. Konnex Association, *KNX of Things* (2015), www.knx.org
6. Konnex Association, *KNX Tutor Documentation* (2014), www.knx.org
7. W. Kastner, G. Neugschwandtner, S. Soucek, H.M. Newman, *Communication Systems for Building Automation and Control* (IEEE, 2005)
8. Konnex Association, *KNX News KNX Internet of Things, KNX Secure, ETS Inside* (2016), www.knx.org
9. H. Merz, T. Hansemann, C. Hübner, *Building Automation Communication Systems with EIB/KNX, LON and BACnet* (Springer, Berlin and Heidelberg, 2009)
10. S.C. Mukhopadhyay, *Internet of Things Challenges and Opportunities*, vol. 9 (Springer, Cham, 2014)
11. O. Hersent, D. Boswarthick, O. Elloumi, *The Internet of Things Key Applications and Protocols* (Wiley, 2012)
12. J.C. Augusto, C.D. Nugent (eds.), *Designing Smart Homes. The Role of Artificial Intelligence* (Springer, Berlin and Heidelberg, 2006)
13. http://aalhouse.esda-lab.cied.teiwest.gr

# Index





G. Keramidas et al. (eds.), *Components and Services for IoT Platforms*,
DOI 10.1007/978-3-319-42304-3

MIX
Papier aus verantwortungsvollen Quellen
Paper from responsible sources
FSC® C105338

Printed by Books on Demand, Germany